Reagents for Organic Synthesis

Fieser and Fieser's
Reagents for Organic Synthesis

VOLUME FOURTEEN

Mary Fieser
Harvard University

WILEY

A WILEY-INTERSCIENCE PUBLICATION
JOHN WILEY & SONS
NEW YORK ● CHICHESTER ● BRISBANE ●
TORONTO ● SINGAPORE

ISBN 0-471-50400-9
ISSN 0271-616X

Printed in the United States of America

10 9 8 7 6   5 4 3 2 1

PREFACE

This volume reviews synthetic use of reagents reported for the most part from January, 1986 to August, 1988. The manuscript has benefited markedly from the careful scrutiny of John O. Link and Greg Fu, who caught many errors and suggested many improvements in the presentations. My co-workers provided expert proofreading of both galleys and page proofs. They include Philip A. Carpino, Keith DeVries, James R. Gage, Guy V. Lamoureux, Kristin M. Lundy, John K. Lynch, Seiichi P. T. Matsuda, John A. Porco, Jr., John A. Ragan, Greg Reichard, Soroosh Shambayati, Robert F. Standaert, Edward M. Suh, Scott Virgil, Keith Woerpel.

Greg Fu supervised the proofreading and provided the structural formula based on the x-ray data for the new Sharpless catalytic reagent for osmylation of alkenes. Martita F. Barsotti is the photographer for the picture of some present and former co-workers for Reagents.

MARY FIESER

March 15, 1989

CONTENTS

Reagents for Organic Synthesis

A

p-Acetamidobenzenesulfonyl azide, p-AcNHC$_6$H$_4$SO$_2$N$_3$ (**1**, m. p. 108° dec.). The azide is prepared by reaction of the sulfonyl chloride with NaN$_3$ in aqueous CH$_2$Cl$_2$ in the presence of (C$_2$H$_5$)$_4$NCl as phase-transfer catalyst.

Diazo transfer.   This azide is recommended as a relatively safe substitute for tosyl azide for diazo-transfer reactions to reactive methylene groups. Either DBU or N(C$_2$H$_5$)$_3$ is a suitable base. It is also suitable for synthesis of vinyl diazo compounds.[1]

$$C_2H_5O_2CCH_2CH=CHCOOC_2H_5 \xrightarrow[84\%]{\substack{1, N(C_2H_5)_3 \\ CH_3CN, 0°}} C_2H_5O_2C\overset{\overset{\displaystyle N_2}{\|}}{C}CH=CHCOOC_2H_5$$

[1] J. S. Baum, D. A. Shook, H. M. L. Davies, and H. D. Smith, *Syn. Comm.*, **17**, 1709 (1987).

Aceto(carbonyl)cyclopentadienyl(triphenylphosphine)iron (1), **12**, 1–2.

Reactions of this pseudooctahedral complex have been studied in particular detail by the Davies group at Oxford and the Liebeskind group in the United States because of its potential use as a chiral auxiliary for control of the absolute stereochemistry of various reactions of the acyl enolate. Both R-(−)-**1** and S-(+)-**1** are now available commercially (Fluka), but at a prohibitive cost ($125.60 per gram).

α,β-*Unsaturated iron acyls.*[1]   The aldols formed by reaction of an aldehyde with the enolate of **1** can be dehydrated via the acetate to (E)-α,β-unsaturated iron acyls (**2**). These products undergo 1,4-addition with RLi or RNHLi, and the intermediate enolate can be alkylated with high diastereoselectivity. Thus addition

1

of BzlNHLi to **2** followed by alkylation with CH_3I gives **3** (24:1), which on oxidative cleavage provides the *cis*-2,3-disubstituted lactam **4**.

Asymmetric alkylation.[2]   Deprotonation of (−)-**1** provides exclusively an (E)-enolate, which is alkylated to provide a single diastereomeric product. Decomplexation by oxidation [Br_2, I_2, Ce(IV)] in the presence of water provides the corresponding acid with the same configuration. This sequence has been used for synthesis of the drug (−)-captopril (**3**). In this case liberation of the acyl group in the presence of the amine provides the amide **2**.

[1] L. S. Liebeskind, M. E. Welker and R. W. Tengl, *Am. Soc.*, **108**, 6328 (1986).
[2] S. G. Davies, *Pure Appl. Chem.*, **60**, 13 (1988).

Acetyl hypofluorite, AcOF (1), 12, 3–4.

Oxygenation of pyridines.[1]   Reaction of the reagent with pyridine or 4-methylpyridine results in a 2-acetoxypyridine in high yield. A similar reaction with 3-

methylpyridine gives a 1:1 mixture of 2- and 5-acetoxy-3-methylpyridine, which is hydrolyzed to the corresponding pyridones. Substitution of chlorine at the α-position prevents this oxidation.

[1] S. Rozen, D. Hebel, and D. Zamir, *Am. Soc.*, **109**, 3789 (1987).

3-Acylthiazolidin-2-thiones, chiral, 11, 518–519; 12, 4.

Bicyclic alkaloids. Nagao *et al.*[1] have developed a general synthesis of chiral bicyclic alkaloids with a nitrogen atom at the ring juncture, such as pyrrolizidines [5.5], quinolizidines [6.6], and indolizidines [6.5], based on a highly diastereoselective alkylation of 3-ω-chloroacyl-(4S)-isopropyl-1,3-thiazolidine-2-thiones (**1**, $m = 1, 2$) with 5-acetoxy-2-pyrrolidinone (**2**, $n = 1$) or 6-acetoxy-2-piperidinone (**2**, $n = 2$). Thus the tin enolate of **1** ($m = 1$), prepared with $Sn(OTf)_2$ and *N*-

ethylpiperidine, is alkylated by **2** ($n = 1$) to give **3** ($n = m = 1$) in 64% chemical yield and 97% de. Reaction of **3** with lithium aluminum hydride in THF at 0° to reduce the amide linkage and then at reflux to effect reductive annelation provides the [5.5]-bicyclic **4** [(−)-trachelanthamide] in 44% yield and 99% optical purity. The same sequence but with **1** ($m = 2$) and **2** ($n = 2$) provides the quinolizidine (−)-epilupinine. The naturally occurring (+)-epilupinine can be synthesized by using the (4R)-isopropyl-1,3-thiazolidine-2-thione.

This asymmetric alkylation of cyclic acylimines can provide optically active precursors to carbapenems.[2] Thus reaction of the 4-acetoxy-2-azetidinone **5** with the chiral 3-acyl-(4S)-ethyl-1,3-thiazolidine-2-thione **6** provides the substituted azetidinone **7**, an intermediate in a total synthesis of (−)-1-β-methylcarbapenem.

[1] Y. Nagao, W.-M. Dai, M. Ochiai, S. Tsukagoshi, and E. Fujita, *Am. Soc.*, **110**, 289 (1988).
[2] Y. Nagao, T. Kumagai, S. Tamai, T. Abe, Y. Kuramoto, T. Taga, S. Aoyagi, Y. Nagase, M. Ochiai, Y. Inoue, and E. Fujita, *Am. Soc.*, **108**, 4673 (1986).

Alkylaluminum halides.

[4 + 2]-*Dipolar cycloaddition.*[1]   Arylthiomethyl chlorides (**1**) in the presence of a Lewis acid can undergo a [4+2]cycloaddition to a tetrasubstituted alkene. They can be prepared by reaction of thiophenol with $BrCH_2Cl$ in the presence of DBU in CH_3CN. $C_2H_5AlCl_2$ is preferred over $AlCl_3$, $SnCl_4$, or $TiCl_4$ as the Lewis acid. This reaction provides a short synthesis of cuparene (**2**).

Michael reactions of silyl enol ethers.[2]   The silyl enol ether of 1-acetylcy-clohexene (**1**) undergoes two consecutive Michael reactions with an α,β-enone or -enal in the presence of this Lewis acid to form 1-decalones.

β-*Lactams.*   In the presence of $(CH_3)_2AlCl$, lithium ester enolates react with enolizable aldimines to afford β-lactams in 60–95% yield as a mixture of *cis*- and *trans*-isomers.[3]

Example:

cis/trans = 76:24

[1] H. Ishibashi, T. S. So, H. Nakatani, K. Minami, and M. Ikeda, *Chem. Pharm. Bull.*, **33**, 90 (1985); *J. C. S. Chem. Comm.*, 827 (1988).
[2] H. Hagiwara, A. Okano, T. Akama, and H. Uda, *ibid.*, 1333 (1987).
[3] M. Wada, H. Aiura, and K. Akiba, *Tetrahedron Letters*, **28**, 3377 (1987).

Alkyl chloromethyl ethers.

α-*Methylenecyclobutanones.*[1]   The simplest route to norsarkomycin methyl ester (**3**) involves α-methylenation of the cyclobutanone **1**. However, reaction of the enolate of **1** with monomeric formaldehyde under the best conditions proceeds in low yield to give the ketol (**2**). A more useful route to α-methylenecyclobutanones involves alkylation of the silyl enol ether of the cyclobutanone with $ROCH_2Cl$ to give an α-(alkoxymethyl)cyclobutanone (**4**), which undergoes elimination of ROH when heated with $KHSO_4$ (**9**, 415). The R group is chosen so that the boiling point of ROH is significantly different from that of the α-methylenecyclobutanone.

[1] J. Vidal and F. Huet, *J. Org.*, **53**, 611 (1988).

Alkyldimesitylboranes.

Boron–Wittig reaction (**12**, 12–13).[1]   The direct reaction of the anion of an alkyldimesitylborane at −78° with an aromatic aldehyde followed by oxidation results in an (E)-alkene in low yield. The intermediate adduct can be isolated in about 80% yield as the silyl ether of a *syn*-1,2-diol by addition of $ClSi(CH_3)_3$ to the reaction, and this product on desilylation (HF, CH_3CN) affords (E)-alkenes with high selectivity. Somewhat lower (E)-selectivity obtains in a one-pot reaction. In contrast, addition of trifluoroacetic anhydride (slight excess) to the reaction at −78° to −110° results in a (Z)-alkene with almost comparable selectivity (Z/E ∼ 9:1).

[1] A. Pelter, D. Buss, and E. Colclough, *J. C. S. Chem. Comm.*, 297 (1987).

Alkyl isopropyl sulfide–Sulfuryl chloride.

ortho-Alkylation of phenols (**5**, 131–132; **12**, 213).[1]   This combination of re- agents converts phenols into a phenoxysulfonium chloride (**a**), which forms an ylide

$$ + \ RCH_2SCH(CH_3)_2 \xrightarrow{\ SO_2Cl_2\ } $$

1, R = CH_2COOCH_3

(b) when treated with triethylamine. This ylide (b) rearranges at 25° to an *o*-alkylphenol (2), which is a useful precursor to coumarin (3).
Chroman (4) is prepared by a related process.

[1] S. Inoue, H. Ikeda, S. Sato, K. Horie, T. Ota, O. Miyamoto, and K. Sato, *J. Org.*, **52**, 5495 (1987).

Alkyllithium reagents.

Tandem **1,2-** *and* **1,4-***additions to quinones.*[1]   The lithium alkoxide formed by 1,2-addition of an alkyllithium to a *p*-benzoquinone can react as a Michael acceptor with some nucleophiles in the presence of HMPT or DMPU (**13**, 122). The process involves lithium–metal exchange followed by intramolecular delivery

of the nucleophile to the β-carbon atom. The reaction shows high regioselectivity, and only a single conjugate adduct is formed (equation II). In addition to delivery of carbon nucleophiles, this process can transfer hydride ion from DIBAL or LiAlH$_4$ (equation III).

Substituted cyclobutenediones. These useful precursors to quinones (**13**, 209–210) can be prepared from commercially available dialkoxycyclobutenediones (**1**, dialkyl squarates). Thus a wide variety of organolithium reagents add to **1** at −78°, and the adducts (**a**) are hydrolyzed under mild conditions to the cyclobutenediones **2**.[2,3] Protection of **a** as the *t*-butyldimethylsilyl ether (**b**) permits a second addition

of an alkyllithium to provide the diadduct **3**, which is hydrolyzed by acid to dialkylcyclobutenediones (**4**).

Addition to thionolactones; cyclic ethers.[4]   A wide variety of alkyllithium reagents add to the C=S group of thionolactones. The adducts, after reaction with CH_3I, can be isolated in high yield as mixed methyl thioketals. The methylthio group can be removed by reduction with triphenyltin hydride (AIBN) to give cyclic ethers. The reaction is not dependent on the ring size and can be stereoselective, as shown by the synthesis of the ether lauthisan (**2**) from a thionolactone (**1**).

The radical formed on reduction of the methylthio group can be used to effect intramolecular cyclization of an unsaturated ring.

Example:

Addition to thioamides.[5]   Alkyl- or aryllithiums add to the carbon–sulfur bond of aromatic thioamides to give adducts that are hydrolyzed to unsymmetrical ketones. Reduction of the adducts with $LiAlH_4$ before hydrolysis provides α-alkylated amines.

$$C_6H_5\overset{O}{\underset{\|}{C}}{-}R \xleftarrow[\text{55-80\%}]{\substack{1)\ RLi \\ 2)\ H_3O^+}} C_6H_5\overset{S}{\underset{\|}{C}}{-}N\diagdown O \xrightarrow[\text{90\%}]{\substack{1)\ RLi \\ 2)\ LiAlH_4 \\ 3)\ H_2O,\ NaOH}} C_6H_5\diagdown\underset{R}{\overset{}{CH}}{-}N\diagdown O$$

Addition to chiral acylsilanes.[6]   Addition of alkyllithium or Grignard reagents to α-chiral aldehydes shows only modest (about 5:1) *syn*-selectivity. In contrast, the same reagents add to chiral acylsilanes with high *syn*-selectivity to give, after

protiodesilylation, the adducts of the nucleophiles to the corresponding aldehydes.

[1] M. Solomon, W. C. L. Jamison, M. McCormick, D. Liotta, D. A. Cherry, J. E. Mills, R. D. Shah, J. D. Rodgers, and C. A. Maryanoff, *Am. Soc.*, **110**, 3702 (1988).
[2] M. W. Reed, D. J. Pollart, S. T. Perri, L. D. Foland, and H. W. Moore, *J. Org.*, **53**, 2477 (1988).
[3] L. S. Liebeskind, R. W. Fengel, K. R. Wirtz, and T. T. Shawe, *ibid.*, **53**, 2482 (1988).
[4] K. C. Nicolaou, D. G. McGarry, P. K. Somers, C. A. Veale, and G. T. Furst, *Am. Soc.*, **109**, 2504 (1987).
[5] Y. Tominaga, S. Kohra, and A. Hosomi, *Tetrahedron Letters*, **28**, 1529 (1987).
[6] M. Nakada, Y. Urano, S. Kobayashi, and M. Ohno, *Am. Soc.*, **110**, 4826 (1988).

Alkynyliodonium tetrafluoroborates, $RC\equiv C{-}I^+C_6H_5BF_4^-$ (**1**).
   Preparation[1]:

$$RC\equiv C{-}Si(CH_3)_3 \ + \ C_6H_5IO \xrightarrow[\text{55-85\%}]{\substack{1)\ BF_3\cdot O(C_2H_5)_2,\ CH_2Cl_2 \\ 2)\ NaBF_4,\ H_2O}} \textbf{1}$$

Cyclopentene annelation.[2]   The reaction of the anion of a 1,3-dicarbonyl compound with 1-decynyl(phenyl)iodonium tetrafluoroborate results in an annelated 3-pentylcyclopentene in reasonable yield. The product is considered to result from

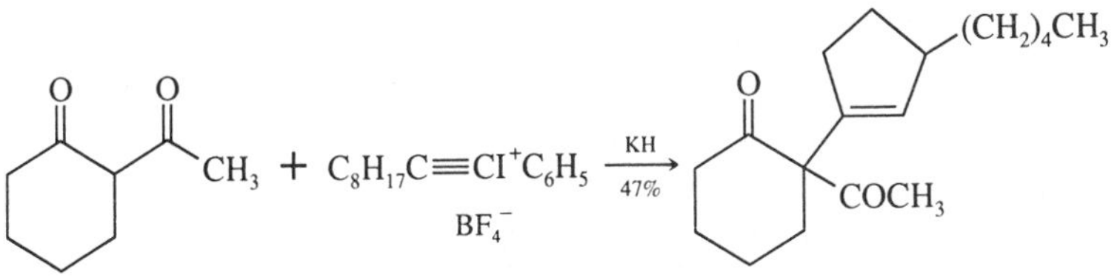

Michael addition to the salt to form an alkylidenecarbene, which undergoes intra-molecular C—H insertion to form a cyclopentene.

The reaction can also be used for synthesis of furans by a [2 + 3]annelation. Example:

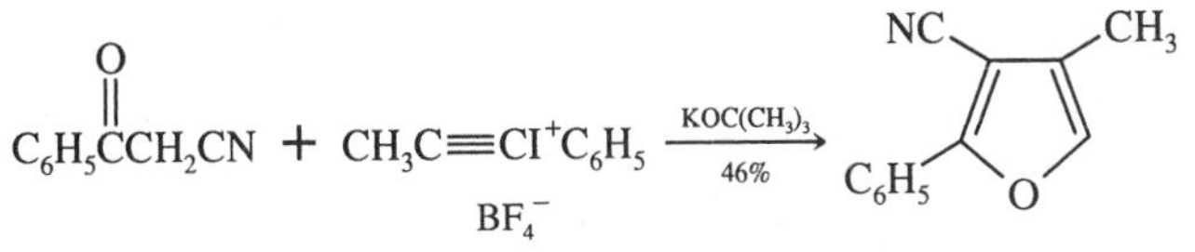

[1] M. Ochiai, M. Kunishima, K. Sumi, Y. Nagao, M. Arimoto, H. Yamaguchi, and E. Fujita, *Tetrahedron Letters*, **26**, 4501 (1985).
[2] M. Ochiai, M. Kunishima, Y. Nagao, K. Fuji, M. Shiro, and E. Fujita, *Am. Soc.*, **108**, 8281 (1986).

B-Allyl-9-borabicyclo[3.3.1]nonane (1).

Amino acid synthesis (*cf.* **12**, 15). The reaction of **1** with chiral α-imino esters provides an enantio- and diastereoselective synthesis of amino acid derivatives.[1] The imine (**2**), prepared from (S)-phenethylamine, reacts with **1** to give **3**, which

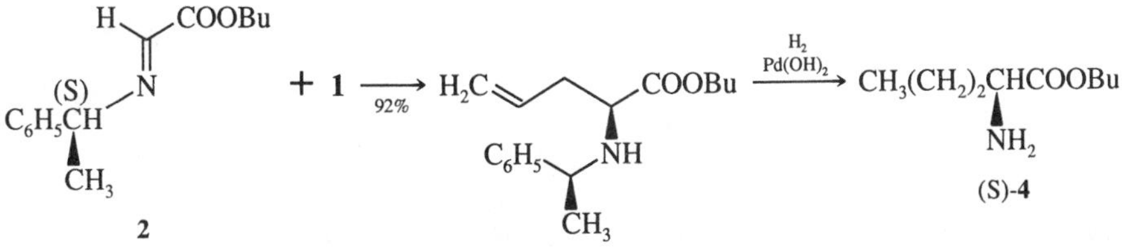

is hydrogenated to the butyl ester of L-norvaline (**4**).

The addition of crotyl-9-BBN to the chiral imino ester **2** provides a *syn*-selective synthesis of optically active amino esters.[1]

Example:

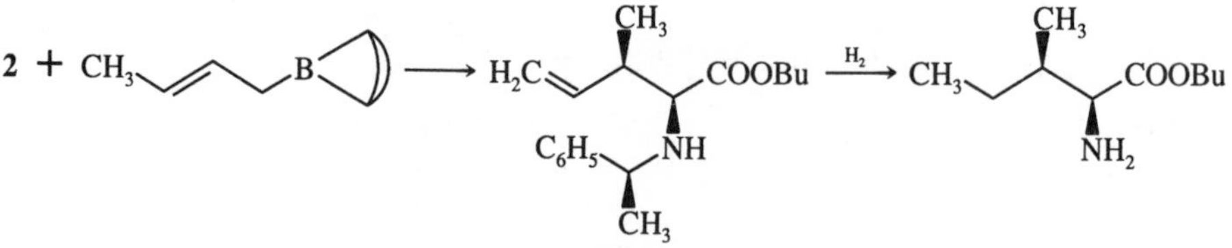

(93:3:3:1)

[1] Y. Yamamoto, S. Nishii, K. Maruyama, T. Komatsu, and W. Ito, *Am. Soc.*, **108**, 7778 (1986).

B-Allylisopinocamphenylborane (1).

Allyboration of chiral aldehydes (**12**, 17).   The borane **1**, prepared from (+)-α-pinene, reacts with a chiral, α-substituted aldehyde such as **2** with 96:4 diastereofacial selectivity. Reaction of **2** with *ent*-**1**, prepared from (−)-α-pinene, shows reversed facial selectivity (5:95).[1]

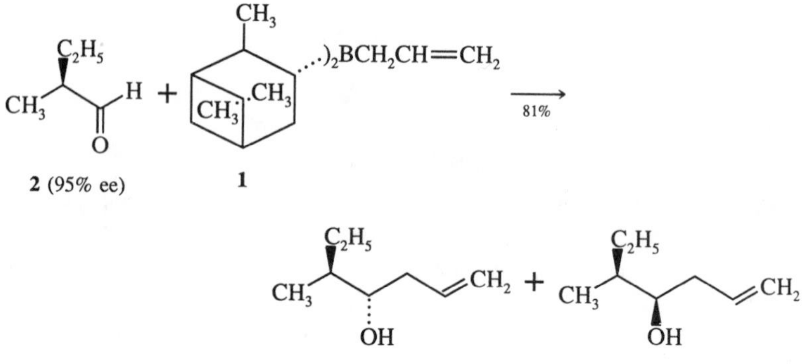

2 (95% ee)   **1**

3   96:4   **4**

2 + *ent*-1 ⟶ 4 + 3
95:5

[1] H. C. Brown, K. S. Bhat, and R. S. Randad, *J. Org.*, **52**, 320 (1987).

Allyl methyl carbonate, $CH_2=CHCH_2OCO_2CH_3$ (1).

The reagent, b.p. 127–130°, is prepared by reaction of allyl alcohol in ether with methyl chloroformate and pyridine (86% yield).

Dehydrogenation of alcohols.[1]   Allylic or secondary alcohols can be oxidized to the ketones by reaction with **1** catalyzed by $RuH_2[P(C_6H_5)_3]_4$ in benzene. The reaction involves decarboxylation of the intermediate allyl alkyl carbonate followed by reductive elimination of propene (equation I).

(I) $R^1R^2CHOH \xrightarrow{\ 1\ } [R^1R^2CHOCO_2CH_2CH=CH_2] \xrightarrow{Ru(II)}$

$$\underset{R^2}{\overset{R^1}{\diagdown}}C=O \ + \ CO_2 \ + \ CH_3OH \ + \ CH_3CH=CH_2$$

Dehydrogenation of primary alcohols gives aldehydes, but in only moderate yield, but primary allylic alcohols are oxidized in high yield (80–95%). Secondary alcohols are also oxidized in high yield. The method, being neutral, is recommended for oxidation of substrates containing acid- or base-sensitive groups.

1,4- and 1,5-Diols are converted by this reaction (with excess **1**) into lactones in 75–95% yield.

Examples:

[1] I. Minami and J. Tsuji, *Tetrahedron*, **43**, 3903 (1987).

2-Allylphospholidine oxides, chiral.

The chiral phospholidines **1** and **2** are prepared by reaction of (−)-ephedrine with allylphosphonyl dichloride and $N(C_2H_5)_3$ and are separated by chromatography.

The anions (BuLi) of **1** and **2** undergo enantioselective 1,4-addition to enones. Ozonolysis of the adducts provides optically active keto aldehydes in 70–75% ee.

Examples:

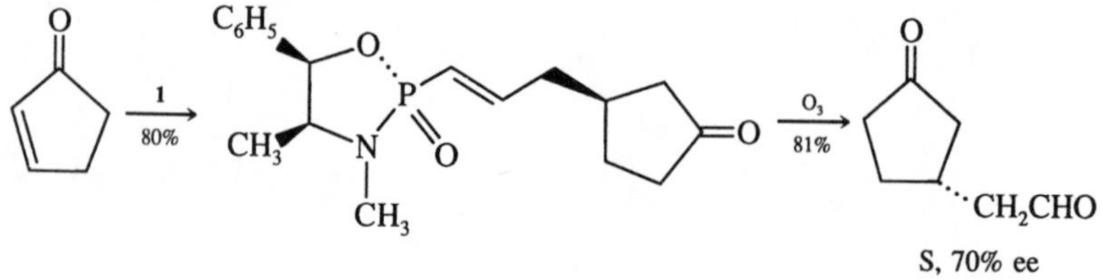

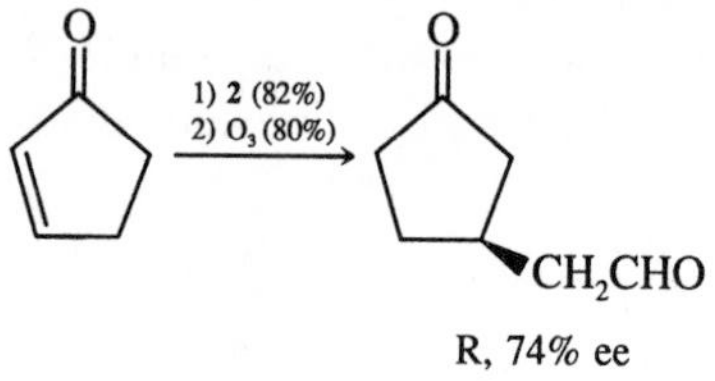

S, 70% ee

R, 74% ee

The enantioselectivity can be increased by changing the methyl group on nitrogen to an isopropyl group. Thus highly diastereoselective additions to cyclic enones are obtained with the phospholidine **3**, prepared in several steps from norpseudoephedrine.[1]

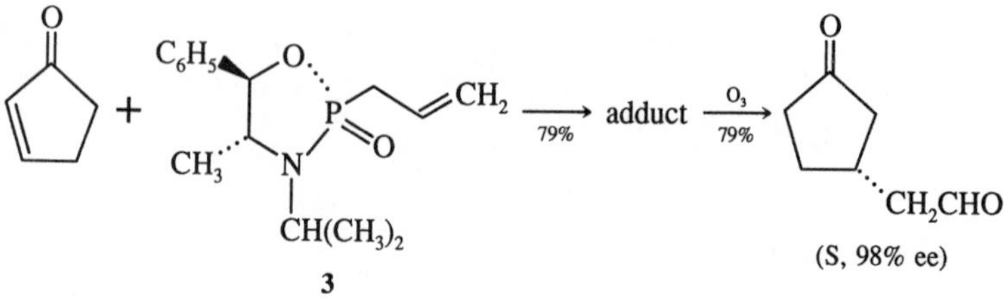

(S, 98% ee)

3

---

[1] D. H. Hua, R. Chan-Yu-King, J. A. McKie, and L. Myer, *Am. Soc.*, **109**, 5026 (1987).

Allyltributyltin.

Isoquinoline alkaloids. The regioselective allylation of N-acyl heterocycles (**13**, 10) can be used for synthesis of isoquinoline alkaloids. Thus simultaneous reaction of the dihydroisoquinoline (**1**) with a diunsaturated acyl chloride (**2**) and allyltributyltin affords the 1,2-adduct (**3**), which undergoes a Diels–Alder cyclization

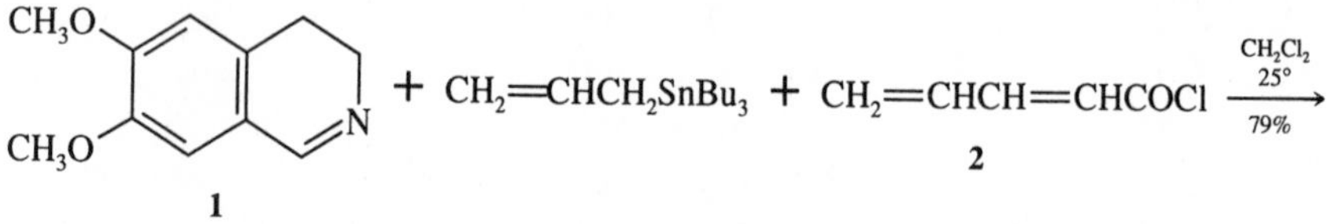

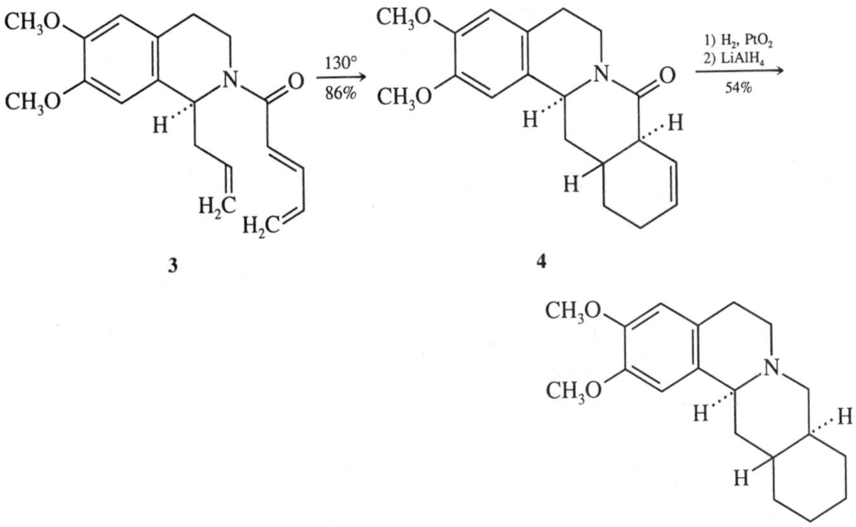

3

4

5

to **4**. Hydrogenation of **4** followed by reduction of the keto group provides the berbane alkaloid **5**.

Isoquinoline also undergoes this simultaneous allylation–acylation to afford 1,2-adducts that can undergo Diels–Alder cyclization (equation I).[1]

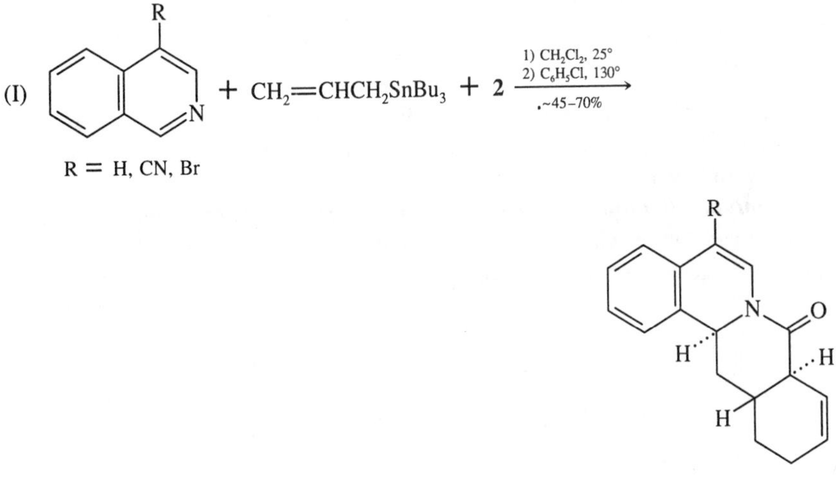

(I)

R = H, CN, Br

(~90:10)

γ-*Alkoxy allyltins; syn,vic-diols.*   Either (Z)- or (E)-γ-methoxyallyltributyltin (**1**) reacts with aldehydes in the presence of a Lewis acid catalyst to form *syn*-adducts selectively (5–25:1), equation (I).[2]

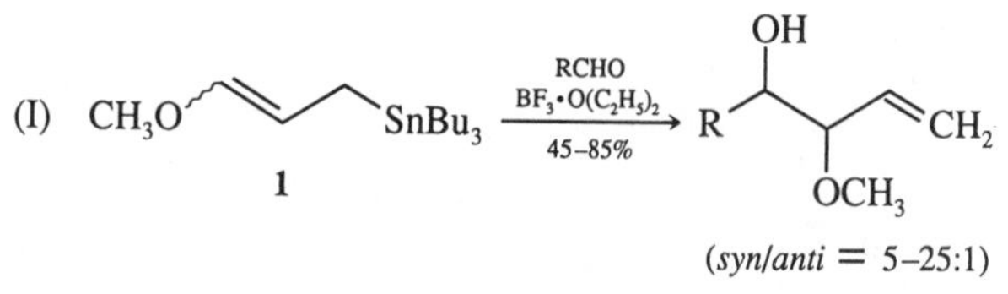

$$(I) \quad CH_3O\diagup\diagdown SnBu_3 \xrightarrow[\substack{BF_3 \cdot O(C_2H_5)_2 \\ 45-85\%}]{RCHO} R\diagup\diagup CH_2$$

1

(syn/anti = 5–25:1)

Keck *et al.*[3] have obtained similar results in the reaction of (Z)-γ-silyloxyallyltributyltin with α- and β-alkoxyaldehydes (equation II). A single adduct is usually formed, that with a *syn*, *vic*-diol unit.

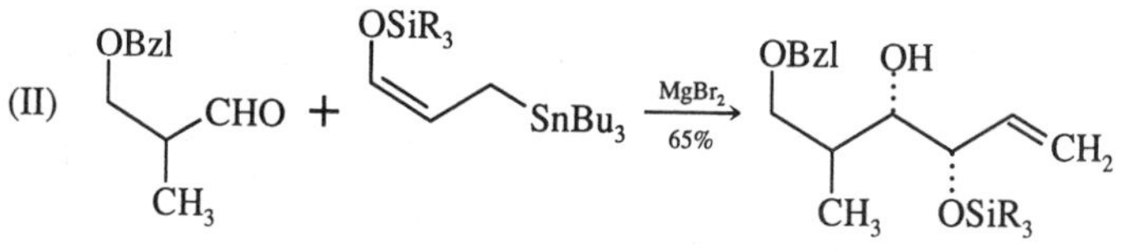

$$(II)$$

Three-component coupling.   Addition of AIBN to a benzene solution of an allylic stannane, an alkyl iodide, and an electron-deficient alkene initiates a radical process that results in 1,2-addition of the alkyl and allyl groups to the alkene.[4]
   Example:

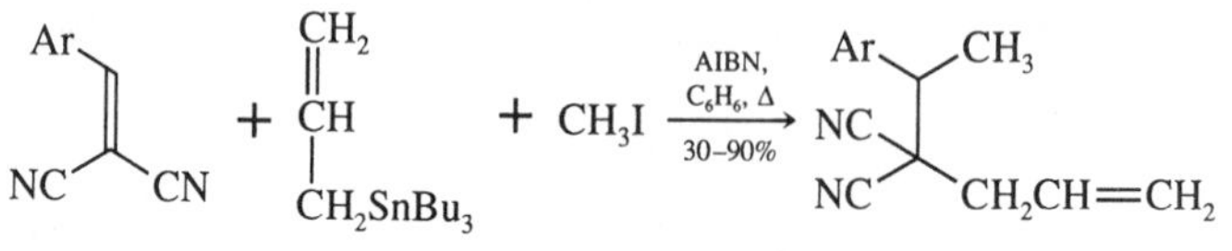

The reactivity of RI is primary > secondary > tertiary. No reaction occurs with RBr or RCl. The allyl group of the stannane is introduced selectively α to the two electron-withdrawing groups.

Photolytic allylation of α-phenylseleno ketones.[5]   α-Phenylselenocycloalkanones on irradiation with allyltributyltin undergo coupling to give α-allylcycloalkanones in generally good yield (equation I).

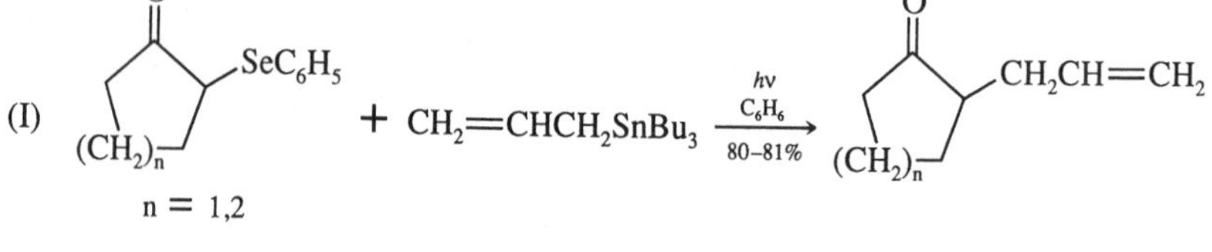

$$(I)$$

n = 1,2

This radical allylation provides a route to prostaglandins (equation II). The resulting 6-methylene-PGE$_1$ derivative (**3**) is converted to a 6-oxo-PGE$_1$ derivative by ozonolysis (50% yield).

(II)

1

2

3

Preparation.[6] The reagent can be prepared in 94% yield by the reaction of allylmagnesium bromide with bis(tributyltin) oxide in ether. Sonication is useful for initiation of the Grignard reaction.

[1] R. Yamaguchi, A. Otsuji, and K. Utimoto, *Am. Soc.*, **110**, 2186 (1988).
[2] M. Koreeda and Y. Tanaka, *Tetrahedron Letters*, **28**, 143 (1987).
[3] G. E. Keck., D. E. Abbott, and M. R. Wiley, *ibid.*, **28**, 139 (1987).
[4] K. Mizuno, M. Ikeda, S. Toda, and Y. Otsuji, *Am. Soc.*, **110**, 1288 (1988).
[5] T. Toru, Y. Mamada, T. Ueno, E. Maekawa, and Y. Ueno, *Am. Soc.*, **110**, 4815 (1988).
[6] N. G. Halligan and L. C. Blaszczak, *Org. Syn.*, submitted (1987).

Allyltrifluorosilanes. These silanes can be obtained in acceptable yield by reaction of allyltrichlorosilanes with $ZnF_2 \cdot 4H_2O$.

Homoallylic alcohols.[1] In the presence of fluoride ion these silanes react with aldehydes to form homoallylic alcohols in high yield (equation I). Crotyltrifluorosilanes undergo a similar reaction with aldehydes, but in this case the stereose-

(I) $(CH_3)_2C{=}CHCH_2SiF_3 + C_6H_5CHO \xrightarrow[96\%]{\substack{CsF, \\ THF}} CH_2{=}CHC{-}CHC_6H_5$

with substituents CH_3 and CH_3 OH on the product.

lectivity is controlled by the geometry of the crotyl group, with the (E)-isomer showing high *anti*-selectivity and the (Z)-isomer high *syn*-selectivity. In contrast, the reaction of both (E)- and (Z)-crotyltrimethylsilane is *syn*-selective (**12**, 146).

$$(II) \quad CH_3CH{=}CHCH_2SiF_3 \; + \; C_6H_5CHO \xrightarrow[93\%]{CsF} CH_2{=}CHCH{-}\overset{CH_3}{\underset{OH}{CHC_6H_5}}$$

(E/Z = 90:10)	*anti/syn* = 88:12
(E/Z = 21:79)	*anti/syn* = 19:81

The report suggests that the actual reagent may be a pentacoordinate allyl-siliconate, such as $CH_3CH{=}CHCH_2Si^-F_4Cs^+$, which reacts with an aldehyde via a six-membered cyclic transition state.

[1] M. Kira, M. Kobayashi, and H. Sakurai, *Tetrahedron Letters*, **28**, 4081 (1987).

Allyltrimethylsilane.

Indolizidine alkaloids.[1]   The key step in a new stereocontrolled synthesis of these alkaloids, such as castanospermine (**5**), depends upon the diastereoselective reaction of an azagluco aldehyde with allylmetal reagents catalyzed by Lewis acids (**12**, 21–22). Thus reaction of allyltrimethylsilane with the aldehyde **1** and $TiCl_4$ (excess) in CH_2Cl_2 at $-85°$ results in the product **2**, formed by selective chelation of the α-amino aldehydo group with $TiCl_4$. The product can be converted into **5**

in four steps as shown. The same sequence provides 6-epicastanospermine (**7**) from the aldehyde **6** derived from D-mannose.

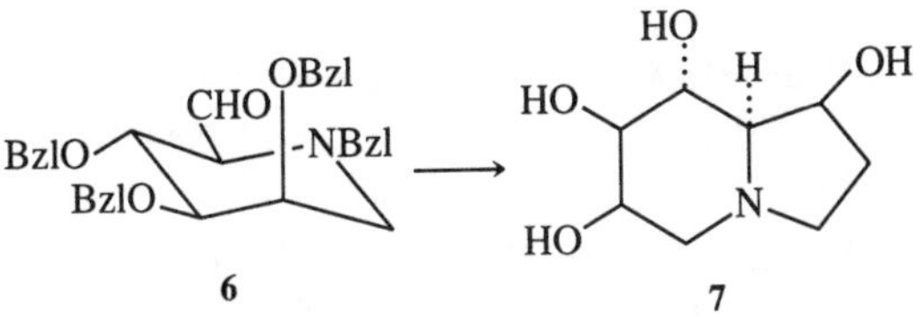

6 → 7

Cycloheptane annelation. [2] A new route to cycloheptanes is based on a Lewis acid-catalyzed, intramolecular addition of an allylsilane group to a 3-vinylcycloalkenone (equation I). This annelation has been applied to a synthesis of the ses-

(I)

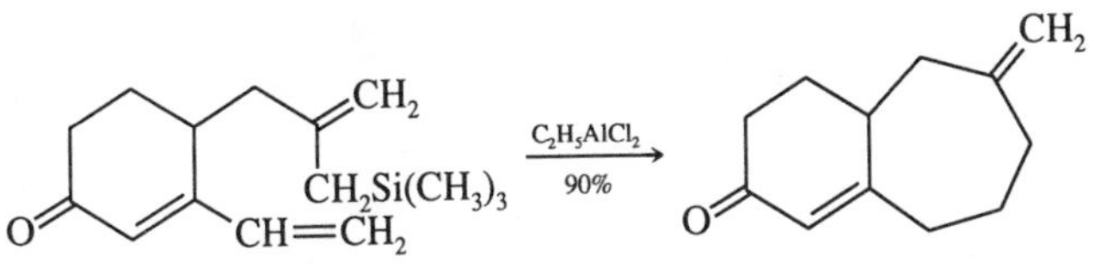

quiterpene 8-epiwiddrol (**3**) from the allylsilane **1**. Cyclization of (Z)-**1** results in a single product (**2**), which is a suitable precursor to 8-epiwiddrol. Cyclization of (E)-**1** results in an inseparable 1:1 mixture of **2** and its 8-epimer. Attempts to

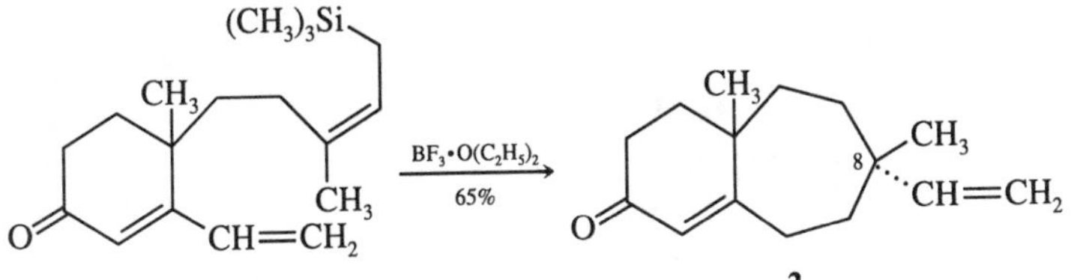

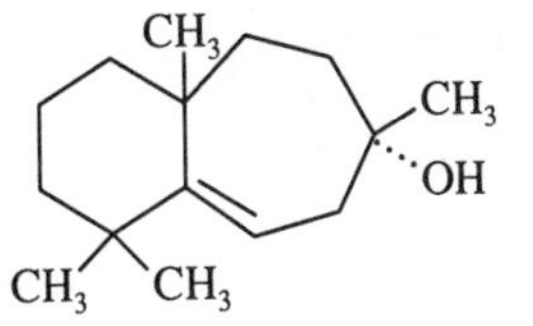

improve the diastereoselectivity by use of other Lewis acids failed to provide the product required for synthesis of widdrol itself.

[1] H. Hamana, N. Ikota, and B. Ganem, *J. Org.*, **52**, 5492 (1987).
[2] G. Majetich and K. Hull, *Tetrahedron*, **43**, 5621 (1987).

Allyltriphenyltin.

Addition to β-alkoxy aldehydes.[1]   The stereochemistry of this reaction is dependent on the Lewis acid catalyst as well as the β-oxygen substituent. Thus the reaction of **1** in the presence of $TiCl_4$ affords **2** and **3** in the ratio 96:1. The stereoselectivity decreases when the benzyl group is replaced by a methyl group, and is lost when R^1 = $SiMe_2$-t-Bu. These chemical results are consistent with NMR studies of the chelates of β-alkoxy aldehydes with $TiCl_4$, $SnCl_4$, and $MgBr_2$, which indicate that the preferred conformations of **1a** and **1b** with $TiCl_4$ are clearly different. The low level of stereoselectivity of **1c** follows from the failure to form a bidentate complex with $TiCl_4$.

Example:

$$n\text{-}C_6H_{14} \quad \overset{\quad}{\underset{OR^1 \quad O}{\diagup\diagdown\diagup}} H \quad + \quad CH_2{=}CHCH_2Sn(C_6H_5)_3 \quad \xrightarrow{TiCl_4}$$

1a, R^1 = $CH_2C_6H_5$
b, R^1 = CH_3
c, R^1 = t-$BuMe_2Si$

$$R \diagdown\diagup\diagdown\diagup\diagdown\diagup{=}CH_2 \quad + \quad R\diagdown\diagup\diagdown\diagup\diagdown\diagup{=}CH_2$$
$$\overset{}{\underset{R^1O \quad\quad OH}{}} \quad\quad\quad\quad \overset{}{\underset{R^1O \quad\quad OH}{}}$$

$$\mathbf{2} \quad\quad\quad 96{:}1 \quad\quad \mathbf{3}$$
$$3.8{:}1$$
$$1.1{:}1$$

[1] G. E. Keck and S. Castellino, *Am. Soc.*, **108**, 3847 (1986); G. E. Keck, S. Castellino, and M. R. Wiley, *J. Org.*, **51**, 5478 (1986).

Alumina.

Conjugate addition of RNO_2 to enones.[1]   Primary nitroalkanes and α,β-enones when activated by alumina form conjugate addition products that are oxidized *in situ* by alkaline hydrogen peroxide to 1,4-diketones. A similar reaction of nitromethane with a vinyl ketone provides 1,4,7-triketones.

$$CH_3CH_2NO_2 \; + \; CH_2{=}CHCCH_3 \;\; \xrightarrow[60\%]{\overset{1)\,Al_2O_3}{2)\,H_2O_2,\,K_2CO_3}} \;\; CH_3\diagup\diagdown\diagup\diagdown CH_3$$

$$CH_3NO_2 \; + \; C_2H_5CCH{=}CH_2 \;\; \xrightarrow[50\%]{} \;\; C_2H_5\diagup\diagdown\diagup\diagdown\diagup C_2H_5$$

[1] R. Ballini, M. Petrini, E. Marcantoni, G. Rosini, *Synthesis*, 231 (1988).

Aluminum chloride

Cyclopentenones. Some time ago, Martin *et al.*[1] reported that α,β-unsaturated acid chlorides react with acetylene in the presence of 1 equiv. of AlCl$_3$ to give 5-chlorocyclopentenones. More recent research[2] shows that this reaction is a convenient route to 4- and 5-substituted cyclopentenones after zinc reduction (equation

I). Substituted acetylenes can also be used as in a short synthesis of methylenomycin B (**1**).

1 (47%)   (40%)

Condensation of unsaturated alcohols with aldehydes. Homoallylic alcohols couple with aldehydes in the presence of AlCl$_3$ or AlBr$_3$ to form 4-halotetrahydropyrans. The products have the all-*cis* (2,4,6-equatorial) conformation. This condensation can also afford cyclic seven-membered ethers (oxepanes).[3] A similar reaction occurs with allylsilanes.

[1] G. J. Martin, Cl. Rabiller, and G. Mabon, *Tetrahedron Letters*, 3131 (1970).
[2] C. J. Rizzo, N. K. Dunlap, and A. B. Smith, III, *J. Org.*, **52**, 5280 (1987).
[3] L. Coppi, A. Ricci, and M. Taddei, *ibid.*, **53**, 911 (1988).

(S)-1-Amino-2-methoxymethyl-1-pyrrolidine (SAMP).

Enantioselective α-hydroxylation of carbonyl compounds.[1] The lithium enolates of the SAMP-hydrazones of ketones undergo facile and diastereoselective oxidation with 2-phenylsulfonyl-3-phenyloxaziridine (**13**, 23–24) to provide, after ozonolysis, (R)-α-hydroxy ketones in about 95% ee. High enantioselectivity in hydroxylation of aldehydes requires a more demanding side chain on the pyrrolidine ring such as —$C(C_2H_5)_2OCH_3$, which also results in reversal of the configuration.

Enantioselective synthesis of $R^1R^2CHNH_2$.[2] Alkyllithiums add stereoselectively to the C=N bond of SAMP hydrazones (**2**) of aldehydes. Reductive cleavage of the N—N bond of the products (**3**) affords either (R)- or (S)-**4** with recovery of

SAMP. The enantiomer of **4** can be obtained by use of RAMP or by exchange of the R^1 and R^2 groups.

[1] D. Enders and V. Bhushan, *Tetrahedron Letters*, **29**, 2437 (1988).
[2] D. Enders, H. Schubert, and C. Nübling, *Angew. Chem. Int. Ed.*, **25**, 1109 (1986).

2-Arenesulfonyloxaziridines.

Oxidation of silyl enol ethers.[1] Oxidation of silyl enol ethers to α-hydroxy aldehydes or ketones is usually effected with *m*-chloroperbenzoic acid (**6**, 112). This oxidation can also be effected by epoxidation with 2-(phenylsulfonyl)-3-(*p*-nitrophenyl) oxaziridine in $CHCl_3$ at 25–60° followed by rearrangement to α-silyloxy carbonyl compounds, which are hydrolyzed to the α-hydroxy carbonyl compound (Bu_4NF or H_3O^+). Yields are moderate to high. Oxidation with a chiral 2-arenesulfonyloxaziridine shows only modest enantioselectivity.

[1] F. A. Davis and A. C. Sheppard, *J. Org.*, **52**, 954 (1987).

Arene(tricarbonyl)chromium complexes.

Biphenylbis(tricarbonyl)chromium (1).[1] Lithium anthracenide converts biphenyl complexed by two $Cr(CO)_3$ groups into a fairly stable dianion of structure 2, in which two η^5-cyclohexadienyl anions are coordinated with $Cr(CO)_3$. This dianion is alkylated by an alkyl halide to form 3, in which one benzene ring is

neutral and coordinated with $Cr(CO)_3$ and the other ring is a C_1-substituted η^5-cyclohexadienyl anion coordinated with $Cr(CO)_3$. Iodine converts 3 in quantitative yield into a 2-alkylbiphenyl (4). Protonation (CF_3COOH) of 3 results in loss of one $Cr(CO)_3$ group to provide 5, which on exposure to sunlight and O_2 loses the second $Cr(CO)_3$ group to provide 5-alkyl-5-phenyl-1,3-cyclohexadienes (6).

Diastereoselective Cr(CO)$_3$ complexes.[2] Anisoles substituted in the *ortho*-position by a hydroxyl-substituted side chain can undergo diastereoselective complexation with $Cr(CO)_3$. Thus **1** and **3** undergo ligand exchange with (naphthalene)$Cr(CO)_3$ to form mainly the complexes **2** and **4**, respectively, with

two chiral centers. These chromium compounds can be used for stereoselective reactions in the side chain. Thus the complex **5**, obtained by benzylation and hydrolysis of the ketal group of **4**, can be converted into **6**, which contains three asymmetric centers. In addition, the complexed anisole ring can also undergo stereoselective reactions.

Isomerization of **1,3-dienes** (**12**, 36).[3]   The 1,5-hydrogen shift in isomerization of 1,3-dienes catalyzed by (naphthalene)Cr(CO)₃ (**1**) can be used for synthesis of aryl-substituted exocyclic alkenes, which are not readily available by coupling of aryl halides with exocyclic vinyl halides.

Example:

$$R_3SiO \qquad \xrightarrow[\substack{CH_3COCH_3,\ 20° \\ MnO_2 \\ 73\%}]{H_2,\ 1} \qquad R_3SiO$$

(E/Z = 1:1)

[1] L. D. Schulte and R. D. Rieke, *J. Org.*, **52**, 4827 (1987).
[2] M. Uemura, T. Minami, and Y. Hayashi, *Am. Soc.*, **109**, 5277 (1987).
[3] M. Sodeoka, S. Satoh, and M. Shibasaki, *ibid.*, **110**, 4823 (1988).

Azidotrimethylsilane.

Addition to carbonyl compounds.[1]   In the presence of ZnCl₂ or SnCl₂, N₃Si(CH₃)₃ adds to aldehydes or ketones to form *gem*-di azides. Reactions catalyzed by NaN₃ and 15-crown-5 provide α-silyloxy azides exclusively. The adducts of aldehydes in both reactions are obtained in higher yield than the adducts of ketones.

[1] K. Nishiyama and T. Yamaguchi, *Synthesis*, 106 (1988).

Azobiscyclohexanecarbonitrile (ACN), (**1**), m.p. 114–115° dec. Preparation.[1] This azo compound undergoes particularly rapid decomposition to radicals.

β-*Stannyl enones as radical traps.*[2]   This radical initiator is more efficient than AIBN for generation of a radical from the iodo acetal **2, which reacts with the β-stannyl enone **3** with β-scission to provide **4** in 72% isolated yield. This product has been converted to PGF₂α (**5**) in 54% yield.

[1] C. G. Overberger, M. T. O'Shaughnessy, and H. Shalit, *Am. Soc.*, **71**, 2661 (1949); C. Walling, *Tetrahedron Symp.*, **41**, 3887 (1985).
[2] G. E. Keck and D. A. Burnett, *J. Org.*, **52**, 2958 (1987).

B

Benzeneselenenyl chloride.

2-Chloro-1-alkenes.[1] A regioselective route to these chloroalkenes involves thermodynamically controlled addition of C_6H_5SeCl to a 1-alkene followed by chlorination to provide a (2-chloroalkyl)phenylselenium dichloride (**2**). These products undergo elimination when treated with $NaHCO_3$ in a two-phase system to provide 2-chloro-1-alkenes (**3**).

$$RCH{=}CH_2 \xrightarrow[\substack{75-85\%}]{\substack{1)\ C_6H_5SeCl,\ CHCl_3,\ >20° \\ 2)\ SO_2Cl_2}} R \overset{Cl}{\diagdown} \underset{\underset{Cl}{|}}{\overset{\overset{Cl}{|}}{SeC_6H_5}} \xrightarrow[\substack{85-95\%}]{\substack{NaHCO_3, \\ C_6H_6/H_2O,\ 100°}} R\overset{\overset{Cl}{|}}{C}{=}CH_2 + (C_6H_5Se)_2$$

2   **3**

[1] L. Engman, *Tetrahedron Letters*, **28**, 1463 (1987).

Benzeneselenenyl triflate.

Glycosylation of thioglycosides.[1] 1-Thioglycosides when activated with 1.5 equiv. of C_6H_5SeOTf, prepared *in situ* from C_6H_5SeCl and AgOTf, react at 0° to −40° with alcohols to form glycosides in 60–95% yield. The reaction is generally β-selective.

[1] Y. Ito and T. Ogawa, *Tetrahedron Letters*, **29**, 1061 (1988).

Benzenesulfenyl chloride.

Rearrangement of dienynols to vinylallene sulfoxides. A few years ago, Okamura *et al.* (**11**, 39)[1] reported the rearrangement of a dienynol to an allenyldiene with transfer of chirality of the propargylic alcohol. This rearrangement has now been used for an enantioselective synthesis of a sesquiterpene, (+)-sterpurene (**3**).[2] Thus reaction of the optically active propargylic alcohol **1** with C_6H_5SCl at 25° results in a vinylallene (**a**) that cyclizes to the optically active sulfoxide **2**. Nickel-

1, $\alpha_D − 13°$ **a**

2, α_D − 110° (61:39)   **3**, α_D + 65°

catalyzed coupling of **2** with the Grignard reagent (**9**, 147) provides a diene that is reduced by sodium-ammonia to (+)-**3**.

[1] W. H. Okamura, R. Peter, and W. Reischli, *Am. Soc.*, **107**, 1034 (1985).
[2] R. A. Gibbs and W. H. Okamura, *ibid.*, **110**, 4062 (1988).

2-Benzoyloxynitroethylene, $O_2NCH=CHOCOC_6H_5$ (**1**).

Preparation:

$$CH_3NO_2 + (CH_3)_2NCH(OCH_3)_2 \longrightarrow [O_2NCH=CHN(CH_3)_2] \xrightarrow[34\%]{\substack{1) KOH, 0° \\ 2) C_6H_5COCl}} 1$$

Diels-Alder reactions.[1] 1,3-Dienes react with **1** to form adducts that undergo ready loss of the benzoyloxy group to provide nitrobenzenes.

Examples:

(1:1)

The adducts can also be reduced to provide *cis*-amino alcohols (equation I).

[1] G. A. Kraus, J. Thurston, P. J. Thomas, R. A. Jacobson, and Y. Su, *Tetrahedron Letters*, **29**, 1879 (1988).

O-Benzylhydroxylamine.

Radical reaction of oxime ethers. A report[1] that an oxime ether, unlike the parent ketone, can serve as a radical trap has led to a more detailed study in two laboratories.[2,3] One study[2] focused on intermolecular addition of radicals to O-benzylformaldoxime (**1**). Reduction of radicals rather than the desired addition proved to be a problem when tributyltin hydride or hexamethylditin in combination with AIBN was used for generation of the radical source. This side reaction can be largely avoided by radical initiation with bis(trimethylstannyl)benzopinacolate (**2**), which liberates trimethyltin radicals when heated above 60°. This reagent can form primary, secondary, tertiary, and aryl radicals from iodides, bromides, and selenides. Using this initiator, a variety of radicals add to O-benzylformaldoxime in >50% yield (equation I).

$$(CH_3)_3SnO \quad OSn(CH_3)_3$$
$$| \qquad |$$
$$(C_6H_5)_2C-C(C_6H_5)_2 \quad \textbf{(2)}$$

(I)

X = I	76%
= Br	56%
= SeC₆H₅	78%

$$X = I \quad\quad 76\%$$
$$= Br \quad\quad 56\%$$
$$= SeC_6H_5 \quad 78\%$$

This homologation of an oxime ether may have wider application for coupling of alkyl halides to other substrates such as acrylic esters (equation II).

(II)

The other study[3] of oxime ethers focused on radical cyclizations to alkoxyaminocyclopentanes and -cyclohexanes (equations III and IV). Yields decrease as chain length of the precursors is increased, mainly because of reduction. The cyclizations

(III)

cis   52:48   trans

(IV)

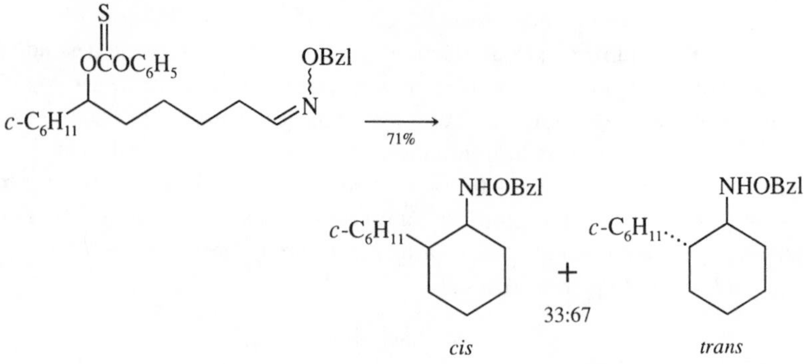

show low stereoselectivity; in general, aldoximes cyclize more readily than ketoximes. The cyclization is more useful in conversion of carbohydrate precursors to carbocycles (equation V). Thus **3**, obtained from glucose cyclizes to only two of the four possible products in high yield.

(V)

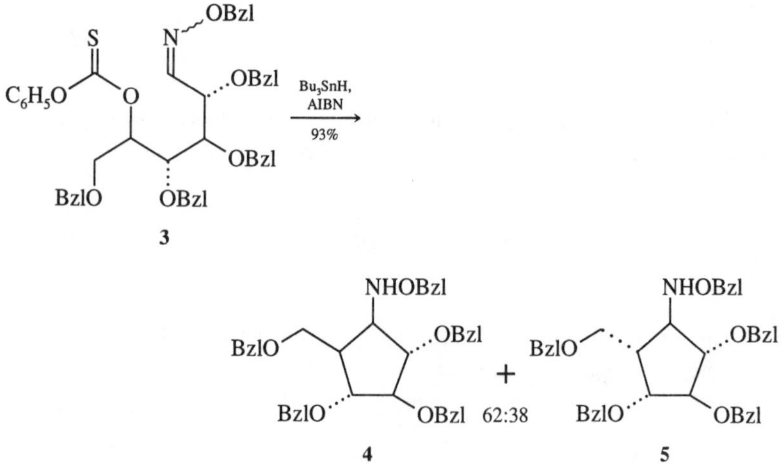

[1] E. J. Corey and S. G. Pyne, *Tetrahedron Letters*, **24**, 2821 (1983).
[2] D. J. Hart and F. L. Seely, *Am. Soc.*, **110**, 1631 (1988).
[3] P. A. Bartlett, K. L. McLaren, and P. C. Ting, *ibid.*, **110**, 1633 (1988).

N-Benzyl-N-(methoxymethyl)-N-trimethylsilylmethylamine,

$$(CH_3)_3SiCH_2N—CH_2C_6H_5 \ (\mathbf{1}).$$
$$\underset{CH_2OCH_3}{|}$$

The amine is prepared by reaction of N-benzyl-N-trimethylsilylmethylamine with HCHO in CH_3OH.

[3 + 2]Cycloadditions.[1]   Desilylation of **1** in CH_3CN with LiF results in an azomethine ylide (**a**), which undergoes cycloaddition reactions with dipolarophiles and activated alkenes to give pyrrolidines.

[1] A. Padwa and W. Dent, *J. Org.*, **52**, 235 (1987).

Benzyloxymethoxymethyllithium, $C_6H_5CH_2OCH_2OCH_2Li$ (**1**).
    Generation:

Hydroxymethylation.[1]   The reagent undergoes asymmetric conjugate addition to tolylsulfinylcycloalkenones such as **2**. The product after reductive desulfinylation (**3**) and hydrogenolysis furnishes (S)-(−)-3-hydroxymethylcycloalkanones (**4**) in high enantiomeric purity.

[1] G. H. Posner, M. Weitzberg, and S. Jew, *Syn. Comm.*, **17**, 611 (1987).

1-Benzyl-(3S)-*p*-tolylsulfinyl-1,4-dihydropyridine.
This chiral NADH model is prepared by reaction of 3-pyridylmagnesium bromide with menthyl (S)-*p*-tolylsulfinate (**11**, 312) to give the sulfoxide **2**, which is a precursor to **1**.

1) C₆H₅CH₂Br (84%)
2) Na₂S₂O₄ (42%)

2

CH₂C₆H₅

1, α_D + 80°

Asymmetric reduction of ketones.   Pioneering work by Ohno *et al.* (**6**, 36; **7**, 15) has established that 1-benzyl-1,4-dihydronicotinamide is a useful NADH model for reduction of carbonyl groups, but only low enantioselectivity obtains with chiral derivatives of this NADH model. In contrast, this chiral 1,4-dihydropyridine derivative (**1**) reduces α-keto esters in the presence of Mg(II) or Zn(II) salts in >90% ee (equation I).[1] This high stereoselectivity of **1** results from the beneficial effect

$$(I)\quad C_6H_5\overset{O}{\underset{\|}{C}}COOCH_3 \xrightarrow{1,\ CH_3CN} C_6H_5\overset{H}{\underset{R}{}}\overset{OH}{}COOCH_3$$

+ Mg(ClO₄)₂   77%   93.9% ee
+ Zn(ClO₄)₂   43%   96% ee

of a chiral center adjacent to the C₄ reaction center and from the similarity of a sulfinyl group to a carboxamide group as an electron-withdrawing group.

[1] T. Imanishi, Y. Hamano, H. Yoshikawa, and C. Iwata, *J. C. S. Chem. Comm.*, 473 (1988).

(R)-(+)- and (S)-(−)-1,1'-Binaphthyl-2,2'-diamine (1).
Racemic **1** is prepared by reaction of 2-naphthol with hydrazine at 160° and is

NH₂NH₂, 180°
60–65%

(±)-**1**

(R)-(+)-1   (S)-(−)-1

resolved with d-10-camphorsulfonic acid, which forms a 1:1 salt with (R)-(+)-**1**.[1]

Asymmetric lactonization.[2]   The diamide **3**, prepared from (R)-**1** and **2**, on treatment with TFA is converted into the δ-lactone **4** in 98% de by a highly diastereoselective reaction with the pro-S-carbonyl group of **3**.

However, similar cyclization of the related diamide **6** results in the γ-lactone **7**, formed with moderate diastereoselectivity with the pro-R carbonyl group. The unexpected asymmetric bias is ascribed to the highly strained diamide ring of **6**.

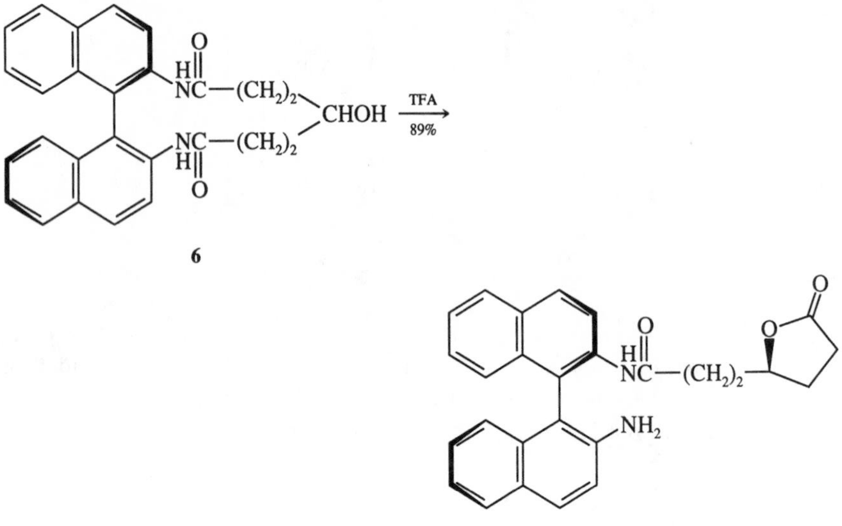

6

7 (71% de)

[1] K. J. Brown, M. S. Berry, and J. R. Murdoch, *J. Org.*, **50**, 4345 (1985).
[2] A. Sakamoto, Y. Yamamoto, and J. Oda, *Am. Soc.*, **109**, 7188 (1987).

1,1'-Binaphthyl-2,2'-dicarboxylic acid (**1**). (RS)-**1** can be resolved with brucine, which selectively forms a salt with (R)-**1** that is insoluble in acetone.

Resolution of α-alkylbenzyl alcohols, C$_6$H$_5$CH(OH)R.   (S)- or (R)-**1** can be used to resolve alcohols such as **2** by selective formation of complexes with one enantiomer. Thus (S)-**1** selectively forms a 2:1 complex with the (R)-enantiomer

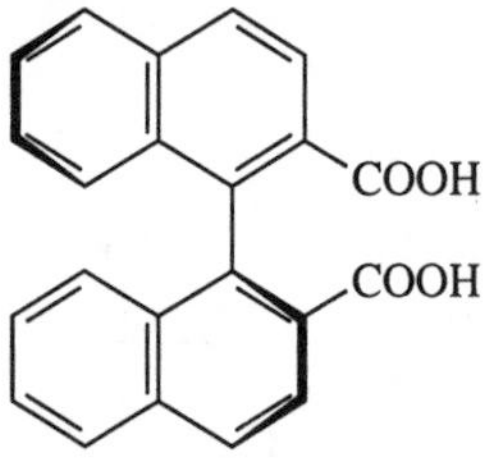

(S)-**1**

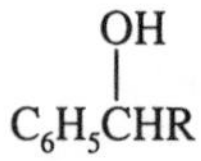

OH
|
C$_6$H$_5$CHR

2a = C$_2$H$_5$, 70% ee (R)
  b = Pr,   90% ee (R)

of **2**, which on decomposition with aqueous NaOH gives free (R)-**2** in 70–90% ee. However, clathrates are not formed when R is a bulky group.

[1] S. Kanoh, Y. Hongoh, S. Katoh, M. Motoi, and H. Suda, *J. C. S. Chem. Comm.*, 405 (1988).

Birch reduction.

Regioselective control.[1]   The site of reduction of a polynuclear arene by Li/ NH_3 can be controlled by the presence of an electron-withdrawing or electron-donating group. A trimethylsilyl group is useful for this purpose, since it is readily removable and is a strong activating group. Thus 2-methoxynaphthalene is reduced under Birch conditions to 2-methoxy-3,4-dihydronapthalene as the major product. In contrast, the silylated derivative **1** is reduced in the silylated ring to provide **2**. Desilylation of **2** gives 2-methoxy-5,8-dihydronaphthalene (**3**).

[1] P. W. Rabideau and G. L. Karrick, *Tetrahedron Letters*, **28**, 2481 (1987).

Bis(acetonitrile)dichloropalladium(II).

Coupling of vinyl halides with vinyltin reagents.[1]   This cross-coupling can be effected at room temperature in DMF in the presence of $(CH_3CN)_2PdCl_2$ or $Cl_2Pd[P(C_6H_5)_3]_2$. Pure (E,E)-, (E,Z)-, (Z,E)-, or (Z,Z)-1,3 dienes can be prepared by this reaction, since the geometry of each partner is retained. The reaction tolerates many functional groups.

1,3-Enynes.   In the presence of this Pd(II) catalyst, vinyl iodides couple with alkynyltrimethylstannanes with retention of the geometry of the vinyl iodide to provide 1,3-enynes in high yields.[2] Since the conjugated triple bond can be reduced to a (Z)-double bond stereospecifically, the coupling also provides a route to (Z,E)- or (Z,Z)-1,3-dienes. An example is the synthesis of bombykol (**1**), the sex attractant of the silkworm moth.

Coupling of vinyl epoxides with $RSn(CH_3)_3$.[3]   This palladium complex catalyzes 1,2- and 1,4-addition of organostannanes to vinyl epoxides to provide homoallylic or allylic alcohols, with the latter usually predominating. The presence of water increases the selectivity and the yields.

Examples:

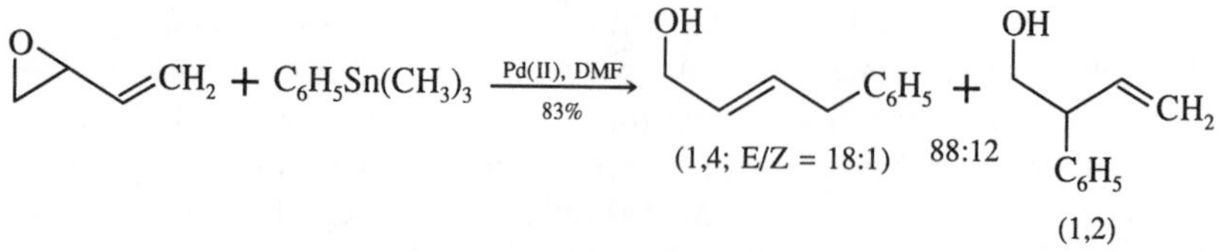

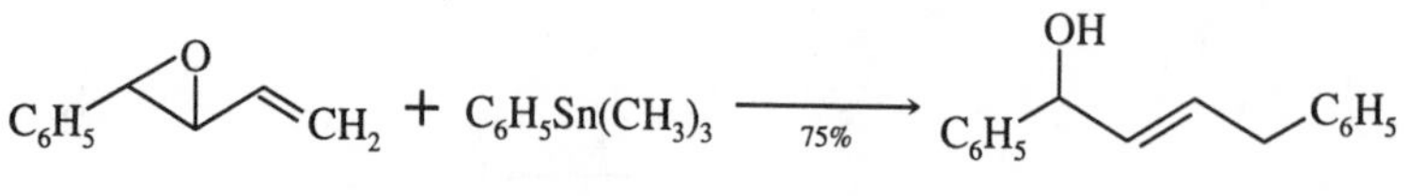

(1,4; E/Z = 18:1)   88:12   (1,2)

Cross double carbonylation of amines and alcohols.[4] Oxamates can be prepared by double carbonylation of amines and alcohols in the presence of $(CH_3CN)_2PdCl_2$ as catalyst with O_2 and CuI as oxidant and co-catalyst. This reaction is particularly efficient when applied to β-amino alcohols.

Example:

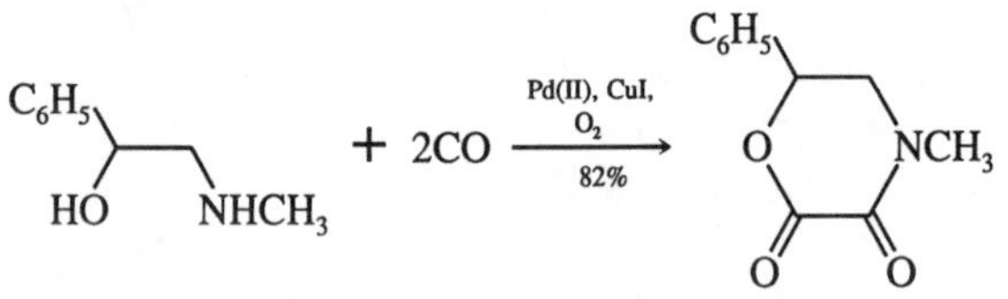

[1] J. K. Stille and B. L. Groh. *Am. Soc.*, **109**, 813 (1987).
[2] J. K. Stille and J. H. Simpson, *ibid.*, **109**, 2138 (1987).
[3] A. M. Echavarren, D. R. Tueting, and J. K. Stille, *ibid.*, **110**, 4039 (1988).
[4] S.-I. Murahashi, Y. Mitsue, and K. Ike, *J. C. S. Chem. Comm.*, 125 (1987).

Bis(cyclooctadiene)nickel, $Ni(COD)_2$.

[4 + 4]*Cycloadditions of bisdienes.*[1] This reaction, catalyzed efficiently by $Ni(COD)_2$–$P(C_6H_5)_3$ (usually 1:2), provides a useful route to the eight-membered ring system found in taxane diterpenes.

Examples:

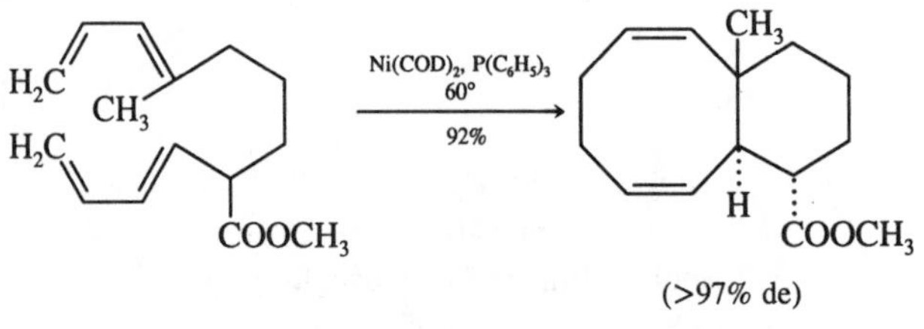

(>97% de)

These examples (and others) indicate that *trans*-fused products are favored when the two dienes are joined by a four-carbon chain, whereas *cis*-fused products are favored by connection with a three-carbon chain. The first example also shows that high stereoinduction is possible in this intramolecular reaction.

Cyclization of enynes with isocyanides.[2] In the presence of Ni(COD)$_2$ (1 equiv.) complexed with PBu$_3$, 1,6-enynes and an isocyanide cyclize to 1-imino-2-cyclopentenes, which can be hydrolyzed to cyclopentenones.

Example:

| R = C$_6$H$_5$ | 92% | 2:1 |
| R = Bu | 47% | 3:1 |

Yields are lower in cyclizations with *t*-BuNC and in the absence of an oxygen function between the two unsaturated bonds.

[1] P. A. Wender and N. C. Ihle, *Am. Soc.*, **108**, 4678 (1986); P. A. Wender and M. L. Snapper, *Tetrahedron Letters*, **28**, 2221 (1987).
[2] K. Tamao, K. Kobayashi, and Y. Ito, *Am. Soc.*, **110**, 1286 (1988).

Bis(cyclopentadienyl)dihydridozirconium, Cp$_2$ZrH$_2$ (1). Preparation.[1]

Selective oxidation of allylic alcohols.[2] This zirconocene complex when used in catalytic amount can effect an Oppenauer-type oxidation of alcohols, including allylic ones, in the presence of a hydrogen acceptor, usually benzaldehyde or cyclohexanone. This system oxidizes primary alcohols selectively in the presence of secondary ones. Thus primary allylic alcohols are oxidized to the enals with retention of the configuration of the double bond in 75–95% yield. The method is not useful for oxidation of propargylic alcohols.

[1] P. C. Wailes and H. Weigold, *J. Organometal. Chem.*, **24**, 405 (1970).
[2] Y. Ishii, T. Nakano, A. Inada, Y. Kishigami, K. Sakurai, and M. Ogawa, *J. Org.*, **51**, 240 (1986); T. Nakano, Y. Ishii, and M. Ogawa, *ibid.*, **52**, 4855 (1987).

Bis(dibenzylideneacetone)palladium(0), Pd(dba)$_2$.

Stille coupling (12, 56).[1]   The key step in a synthesis of (E)-neomanoalide (4) involves palladium-catalyzed coupling of an allyl halide with an α-stannylfuran. Thus 1 and 2 couple in the presence of Pd(dba)$_2$ and P(C$_6$H$_5$)$_3$ to form 3 as a 1:1 mixture of (E)- and (Z)-isomers in 66% yield. Conversion of 3 to 4 involves reduction (DIBAH) and selective singlet oxygen oxidation of a 2-silylfuran to a butenolide.

[1] S. Katsumura, S. Fujiwara, and S. Isoe, *Tetrahedron Letters*, **28**, 1191 (1987).

(R)- and (S)-2,2'-Bis(diphenylphosphino)-1,1'-binaphthyl (BINAP). Supplier: Fluka.

Enantioselective catalytic hydrogenation.   The ruthenium(II) complexes of (R)- and (S)-1, bearing a chiral BINAP ligand, catalyze asymmetric hydrogenation of N-acyl-1-alkylidenetetrahydroisoquinolines to give (1R)- or (1S)-tetrahydroisoquinolines in 95–100% ee.[1] Thus the (Z)-enamide (2), prepared by acylation of 3,4-dihydropapaverine, is hydrogenated in the presence of (R)-1 to (1R)-tetrahydroisoquinolines (3). The enantiomeric (1S)-3 is obtained on use of (S)-1 as catalyst.

A wide range of α,β-unsaturated carboxylic acids, including substituted acrylic acids, undergo enantioselective homogeneous hydrogenation when catalyzed by

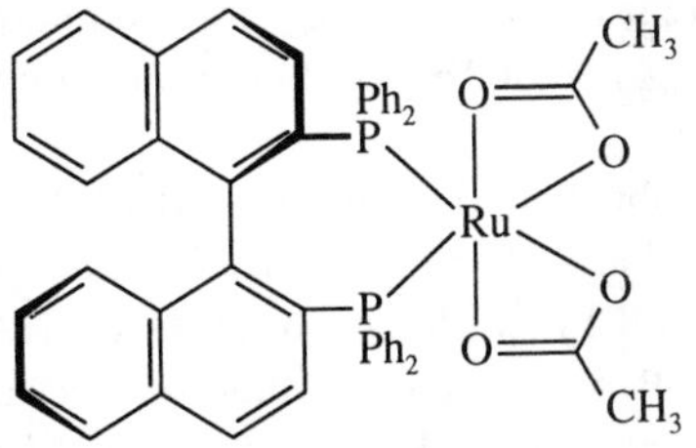

(R)-1

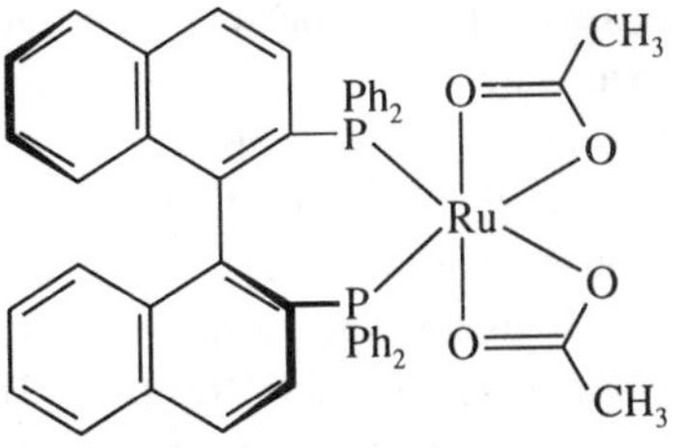

(S)-1

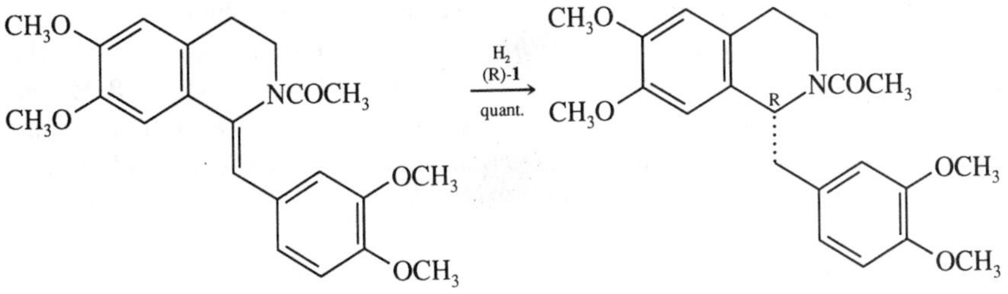

(Z)-2

(1R)-3 (>99.5% ee)

ruthenium(II) carboxylates possessing the BINAP ligand, such as (R)- and (S)-**1**.[2] Chemical yields are nearly quantitative; optical yields depend upon the double-bond geometry and the substitution pattern, but optical yields of greater than 90% can be obtained by suitable change of hydrogen pressure. This system can also effect selective hydrogenation of only the α,β-double bond. The reaction also provides a route to optically active hydroxy acids or the corresponding lactones. A few β,γ-unsaturated acids can be hydrogenated enantioselectively.

Ru(II)-BINAP complexes (**1**) can effect enantioselective hydrogenation of prochiral allylic and homoallylic alcohols, without hydrogenation of other double bonds in the same substrate.[3] The alcohols geraniol (**2**) and nerol (**3**) can be reduced to either (R)- or (S)-citronellol (**4**) by choice of either (R)- or (S)-**1**. Thus the stereochemical outcome depends on the geometry of the double bond and the chirality

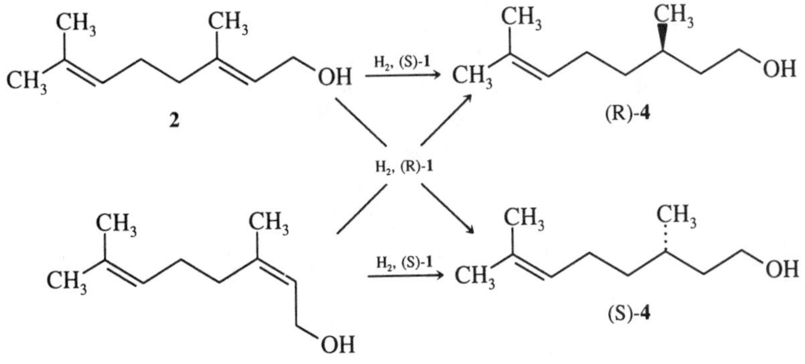

2

(R)-4

(S)-4

3

of the catalyst. In either case the optical yields are 96–99% ee. Homogeraniol is hydrogenated with (S)-**1** in the same sense as **2** and in 92% ee, but in this case the isolated double bond is also reduced.

Somewhat more effective catalysts are obtained by replacing BINAP with TolBINAP, which is 2,2′-bis(di-*p*-tolylphosphino)-1,1′-binaphthyl.[4] The presently preferred catalysts are complexes of $Ru(OCOCF_3)_2$ with (R)- or (S)-TolBINAP, obtained by treatment of $Ru(OAc)_2$·TolBINAP with 2 equiv. of trifluoroacetic acid. Such catalysts promote hydrogenation of typical enamides in 98% ee and 98% yield. This reaction can be used to provide asymmetric synthesis of isoquinoline alkaloids as well as of morphinans used as substitutes for morphine.

Asymmetric hydrogenation of prochiral ketones.[5] Ketones substituted in the α- or β-position by diverse polar groups, particularly OH,OR,NR₂,COOR, can undergo highly enantioselective hydrogenation catalyzed by BINAP–Ru complexes. A key factor of asymmetric induction is undoubtedly chelation of the carbonyl group and the hetero atom to the Ru atom.

Examples:

(96% ee)

(98% ee)

Enantioselectivity is generally reduced by the presence of two functional groups, probably because of competitive ligation; generally, the effect of an ester group overrides that of an alkoxyl or siloxyl group.

Two-step asymmetric hydrogenation of β-diketones shows that the overall stereochemistry is determined by the catalyst and by the chirality of the intermediate hydroxy ketone. Thus partial hydrogenation of acetylacetone (**2**) catalyzed by Ru–

(R)-**1** results in the (R)-hydroxy ketone (**3**) as expected. Further hydrogenation with the same catalyst gives R,R-**4** and *meso*-**4** in the ratio 99:1. In contrast, hydrogenation of **3** catalyzed by Ru–(S)-**1** gives the same diols, but in the ratio

$$
\begin{array}{c}
\underset{\underset{\textstyle \mathbf{5}}{\underset{\textstyle CH_3 \quad CH_3}{}}}{\overset{O \quad\quad O}{\bigvee\!\!\bigvee}} \xrightarrow[100\%]{\underset{Ru\text{-}(S)\text{-}1}{H_2}} \underset{\underset{\textstyle meso\text{-}\mathbf{6}}{CH_3 \quad CH_3}}{\overset{HO \quad\quad HO}{\bigwedge}}_{74:26} + \underset{\underset{\textstyle S,S\text{-}\mathbf{6}\ (100\%\ ee)}{CH_3 \quad CH_3}}{\overset{OH \quad\quad OH}{\bigwedge}}
\end{array}
$$

15:85. Hydrogenation of diacetyl (**5**) catalyzed by (S)-**1**–Ru gives a 74:26 mixture of *meso*-**6** and S,S-**6**. Evidently in this reduction catalyst control favoring formation of *meso*-diols dominates over substrate control favoring formation of *l* or *d*-diols.

Stereoselective hydrogenation of 1,3-diketones.[6] Hydrogenation of 1,3-alkane-diones catalyzed by $Ru_2Cl_4[(R)\text{-}\mathbf{1}][N(C_2H_5)_3]$ results in *anti*-1,3-diols with high diastereoisomeric and enantiomeric excesses (equation I). Under the same conditions 1-phenyl-1,3-butanedione (**2**) is reduced mainly to the β-hydroxy ketone **3** in 98%

$$
\text{(I)} \quad CH_3 \underset{O \quad\ O}{\bigvee\!\!\bigvee} R \xrightarrow[\ ]{\overset{Ru\text{-}(R)\text{-}1}{50°}} CH_3 \underset{OH \quad OH}{\bigwedge} R \ + \ CH_3 \underset{OH \quad OH}{\bigwedge} R
$$

R = CH$_3$	98%	>99% ee	99:1
R = *t*-Bu	84%	98% ee	91:9

$$
\text{(II)} \quad CH_3 \underset{\underset{\textstyle \mathbf{2}}{O \quad\ O}}{\bigvee\!\!\bigvee} C_6H_5 \xrightarrow[\ ]{\overset{Ru\text{-}(R)\text{-}1}{50°}} \underset{\underset{\textstyle 3\ (98\%\ ee)}{OH \quad O}}{CH_3 \bigwedge C_6H_5} \ +\ \underset{\underset{\textstyle anti\text{-}\mathbf{4}}{92:2}}{CH_3 \underset{OH \quad OH}{\bigwedge} C_6H_5}
$$

$$
\underset{58\%}{\underset{\longrightarrow}{\text{Ru-(R)-1, 100°}}}
$$

ee, but under more drastic conditions it also is reduced to the *anti*-1,3 diol (**4**) in 99% ee. Alkanones are not reduced under the mild conditions used for the alkanediones, and hydrogenation of 1,2- and 1,4-diketones give complex mixtures. Evidently hydrogenation is facilitated by chelation of 1,3-diones with the ruthenium catalyst.

Asymmetric hydrogenation of β-keto esters.[7] The $Ru(OAc)_2(BINAP)$ complexes are ineffective catalysts for asymmetric hydrogenation of β-keto esters, but on treatment with HX (2 equiv.) are converted into complexes with the empirical formula $RuX_2(BINAP)$, which are effective catalysts for this enantioselective hydrogenation. Complexes of (R)-BINAP catalyze hydrogenation to (R)-β-hydroxy esters in >99% ee, whereas the enantiomeric (S)-β-hydroxy esters are obtained by use of complexes of (S)-BINAP in equally high enantioselectivity. The enantioselectivity is independent of the nature of the ester group or of the chain length. The $Ru_2Cl_4(BINAP)_2[N(C_2H_5)_3]$ complex (**13**, 36–37) is equally effective for this enantioselective hydrogenation.

Catalytic hydrogenation of ethyl α-methyl-β-oxobutanoate (**1**) catalyzed by $RuBr_2[(R)\text{-BINAP}]$ gives a 1:1 mixture to two β-hydroxy esters, both of which have the (3R)-configuration (equation I).

(I)

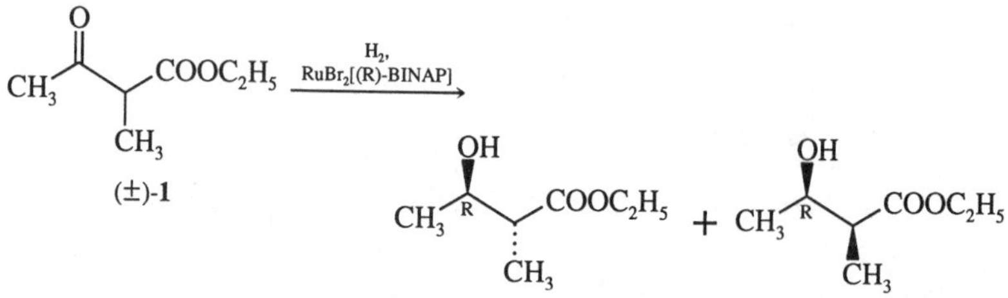

2 (49%, 97% ee)   3 (51%, 96% ee)

Asymmetric isomerization.[8]   Cationic Rh-phosphine complexes such as **1** can effect isomerization of allylic substrates by a 1,3-hydrogen migration (**12**, 56–57).

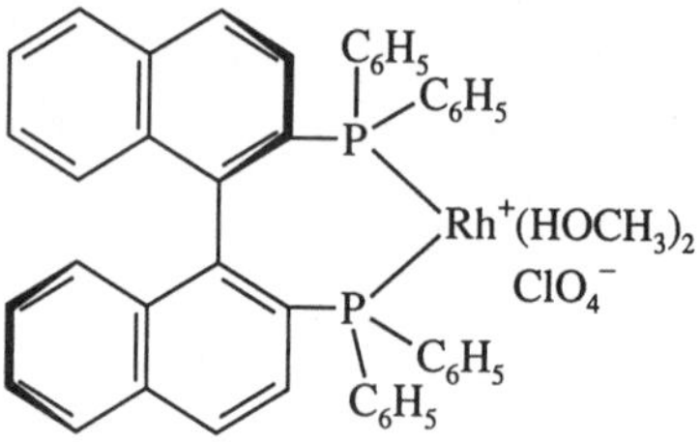

(R)-**1**

Such an isomerization of 4-hydroxy-2-cyclopentenone (**2**) results in 1,3-cyclopentanedione (**3**) via the keto enol. On exposure of racemic **2** to the optically active Rh-BINAP complex (R)-**1**, the (S)-enantiomer isomerizes more rapidly than (R)-**2** to give, after 14 days at 0°, a mixture of **3** and (R)-**2** in 91% ee.

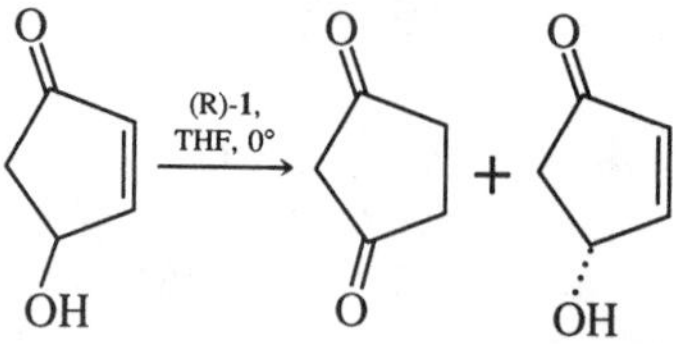

2   3   (R)-**2**, 27%
                              (91% ee)

Kinetic resolution of allylic alcohols.[9]   The (R)- and (S)-BINAP-Ru diacetate complexes can resolve racemic allylic alcohols, both acyclic and cyclic, with high enantiomeric selectivity. Thus hydrogenation of (±)-**2** catalyzed by (S)-**1** at 76% conversion provides (S)-**2** (>99% ee) and *anti*-**3** (49:1, 39% ee). Hydrogenation of (S)-**2** catalyzed by either (R)- or (S)-**1** provides *anti*-**3** (>23:1). Similar results obtain with (±)-**4**.

(±)-**2**   →^{H_2, (S)-1}   (S)-**2**   +   (2R,3R)-**3** (49:1)

↓ H_2, (R)- or (S)-1

(2S,3S)-**3** (>23:1)

(±)-**4**   →^{H_2, (R)-1}   (S)-**4**   +   (1R,3R)-**5** (95% ee)

↓ H_2, (S)-1

[1] R. Noyori, M. Ohta, Y. Hsiao, M. Kitamura, T. Ohta, and H. Takaya, *Am. Soc.*, **108**, 7117 (1986).

[2] T. Ohta, H. Takaya, M. Kitamura, K. Nagai, and R. Noyori, *J. Org.*, **52**, 3174 (1987).

[3] H. Takaya, T. Ohta, N. Sayo, H. Kumobayashi, S. Akutagawa, S. Inoue, I. Kasahara, and R. Noyori, *Am. Soc.*, **109**, 1596 (1987).

[4] M. Kitamura, Y. Hsiao, R. Noyori, and H. Takaya, *Tetrahedron Letters*, **28**, 4829 (1987).

[5] M. Kitamura, T. Ohkuma, S. Inoue, N. Sayo, H. Kumobaysahi, S. Akutagawa, T. Ohta, H. Takaya, and R. Noyori, *Am. Soc.*, **110**, 629 (1988).

[6] H. Kawano, Y. Ishii, M. Saburi, and Y. Uchida, *J.C.S. Chem. Comm.*, 87 (1988).

[7] R. Noyori, T. Ohkuma, M. Kitamura, H. Takaya, N. Sayo, H. Kumobayashi, and S. Akutagawa, *Am. Soc.*, **109**, 5856 (1987).
[8] M. Kitamura, K. Manabe, R. Noyori, and H. Takaya, *Tetrahedron Letters*, **28**, 4719 (1987).
[9] M. Kitamura, I. Kasahara, K. Manabe, R. Noyori, and H. Takaya, *J. Org.*, **53**, 708 (1988).

[1,4-Bis(diphenylphosphine)butane]norbornadienerhodium trifluoromethanesulfonate (**1**), **12**, 426. The triflate salt is easier to prepare than the tetrafluoroborate used earlier.

1

Stereoselective hydrogenation of β-hydroxy acrylates.[1] These esters can be prepared by condensation of methyl acrylate with aldehydes catalyzed by DABCO. On hydrogenation catalyzed by **1**, these esters are converted with high selectivity into *anti*-α-methyl-β-hydroxy esters (equation I). Similar directed hydrogenation

obtains when the hydroxyl group is replaced by another polar group: $COOCH_3$, $CONR_2$, NHCOR.

The related chiral rhodium catalyst **4** has been used to effect kinetic resolution of these substrates.[2] In this catalyst the achiral phosphine ligand of **1** is replaced by (R,R)-1,2-bis(o-anisylphenylphosphino)ethane (DIPAMP). Hydrogenation cat-

o-CH$_3$OC$_6$H$_4$ ⋯ P ⟍ C$_6$H$_5$

Rh$^+$⋯Norbornadiene

BF$_4^-$

C$_6$H$_5$ P C$_6$H$_4$OCH$_3$-o

4

alyzed by **4** of the β-hydroxy acrylate **2** to 65% completion results in (S)-(−)-**2** and (2R,3R)-(−)-**3** in 97% ee (equation II).

(II)   2 $\xrightarrow{\text{H}_2,\ 4}$

$$\underset{\text{(S)-2}}{\overset{\text{CH}_2}{\text{CH}_3\text{O}_2\text{C}}\diagup\diagdown\text{C}_6\text{H}_5,\ \text{OH}} + \underset{\text{(2R,3R)-3 (97\% ee)}}{\overset{\text{CH}_3}{\text{CH}_3\text{O}_2\text{C}}\diagup\diagdown\text{C}_6\text{H}_5,\ \text{OH}}$$

Stereoselective hydrogenation of 3-substituted itaconate esters.   Hydrogenation of the dimethyl itaconates (**5**) catalyzed by **1** results in *syn*-dialkylsuccinates

$$\underset{\textbf{5, R} = \text{CH}_3,\ \text{C}_2\text{H}_5,\ \text{C}_6\text{H}_5}{\overset{\text{CH}_2}{\text{CH}_3\text{OOC}}\diagup\diagdown\text{COOCH}_3,\ \text{R}} \xrightarrow{\text{H}_2,\ 1} \underset{\textit{syn}\text{-6 (250:1)}}{\overset{\text{CH}_3}{\text{CH}_3\text{OOC}}\diagup\diagdown\text{COOCH}_3,\ \text{R}}$$

$$\underset{\text{(R,R)-6}}{\overset{\text{CH}_3}{\text{CH}_3\text{OOC}}\diagup\diagdown\text{COOCH}_3,\ \text{R}} \qquad \underset{\text{(S,S)-6}}{\overset{\text{CH}_3}{\text{CH}_3\text{OOC}}\diagup\diagdown\text{COOCH}_3,\ \text{R}}$$

(**6**). The high diastereoselectivity is attributed to the directing influence of the COOCH$_3$ group. Incomplete hydrogenation of **5** with the chiral catalyst **4** results in formation of (R,R)-**6**; further hydrogenation of the unreacted precursor with **1** gives (S,S)-**6**. Thus the two *syn*-2,3-dialkylsuccinates can be prepared in ⩾96% ee.[3]

[1] J. M. Brown, P. L. Evans, and A. P. James, *Org. Syn.*, submitted (1987).
[2] J. M. Brown, I. Cutting, P. L. Evans, and P. J. Maddox, *Tetrahedron Letters*, **27**, 3307 (1986).
[3] J. M. Brown and A. P. James, *J. C. S. Chem. Comm.*, 181 (1987).

Bis(hexafluoroacetylacetonate)copper(II), $Cu(hfa)_2$ **(1).**[1]

Ylides from R_2CN_2.   This reagent is more effective than bis(acetylacetonate)-copper(II) (**5**, 244) for generation of carbenes from diazo compounds.[2] The decomposition proceeds at a lower temperature, even at room temperature. The mild conditions are particularly useful in the preparation of heat-sensitive ylides, such as those of antimony, bismuth, and tellurium.

Example:

$$(C_6H_5SO_2)_2CN_2 \ + \ (C_6H_5)_3Sb \ \xrightarrow[78\%]{\overset{1,\ C_6H_6}{\Delta}} \ (C_6H_5SO_2)_2C{=}Sb(C_6H_5)_3$$

[1] R. L. Belford, A. E. Martell, and M. Calvin, *J. Org. Nucl. Chem.*, **2**, 11 (1956).
[2] C. Glidewell, D. Lloyd, and S. Metcalfe, *Synthesis*, 319 (1988).

(R)- and (S)-3,3'-Bis(triphenylsilyl)binaphthol–Trimethylaluminum.

These two reagents react to form a chiral organoaluminum reagent formulated as (R)- and (S)-**1**.

(R)-1   (S)-1

Asymmetric Diels–Alder reactions.[1]   Diels–Alder reactions of aldehydes with siloxydienes catalyzed by this chiral organoaluminum compound can proceed with high enantioselectivity. Thus reaction of benzaldehyde with the silyloxydiene **2** at

3 (95% ee)   4

$-20°$ in the presence of 10 mole% of (R)-**1** results in the *cis*-dihydropyrone **3** as the major product in 95% ee. The enantioselectivity is dependent on the bulk of the triarylsilyl group of **1**. Thus replacement of the triphenylsilyl group of **1** by tris(3,5-xylyl)silyl groups increases the *enantio*- and *cis*-selectivity, but replacement by trimethylsilyl groups reduces the enantioselectivity.

C-Glycosides.[2]   Reaction of an aldehyde with the diene **2** catalyzed by (R)-**1** results in dihydropyrones (**3**) in 80–97% ee, which can be converted to the glycals

4. These glycals react with trialkylaluminums to give C-glycosides (**5**), in which the *trans*-isomer predominates.

Example:

4			
	Al(CH$_3$)$_3$	96%	96:4
	Al(*i*-Bu)$_3$	74%	94:6
	(C$_2$H$_5$)$_2$AlC≡CBu	85%	100:0

---

[1] K. Maruoka, T. Itoh, T. Shirasaka, and H. Yamamoto, *Am. Soc.*, **110**, 310 (1988).
[2] K. Maruoka, K. Nonoshita, T. Itoh, and H. Yamamoto, *Chem. Letters*, 2215 (1987).

1,2-Bis(*trans*-diphenylpyrrolidino)ethane (1). The chiral diamine is prepared[1] by reaction of (3R,4R)- or (3S,4S)-diphenylpyrrolidine with oxalyl chloride and tri- ethylamine followed by reduction of the resulting dione with lithium aluminum hydride.

(3R,4R)-(−)-**1**, α_D − 143°

Asymmetric addition of Grignard reagents to C_6H_5CHO.[2]   Grignard reagents pretreated with (−)-1 at −78° add to benzaldehyde with only slight enantiose-lectivity (~20% ee). But if the addition is effected in the presence of 2,4,6-tri-methylphenoxyaluminum dichloride (2), the carbinol is obtained in 40–70% ee.

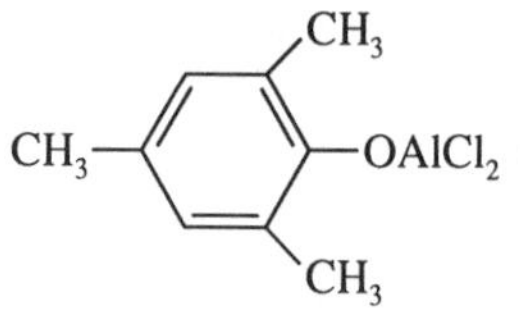

$$C_6H_5CHO + RMgBr \xrightarrow[60-90\%]{\substack{1, 2, \\ -100°}} C_6H_5 \overset{H\quad OH}{\underset{R\quad R}{\bigwedge}}$$

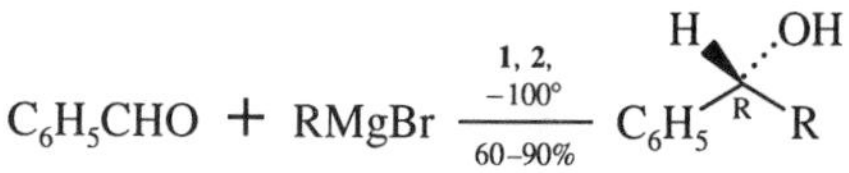

(40–70% ee)

Asymmetric osmylation of alkenes.[3]   In the presence of 1 equiv. each of 1 and OsO_4, alkenes undergo highly enantioselective *cis*-dihydroxylation. Highest enantiofacial selectivity (90–99%) is shown in osmylation of *trans*-di- and trisub-

$$OsO_4, (+)-1$$

$$OsO_4, (-)-1$$

A

stituted alkenes, and the enantioface differentiation is completely controlled by the chiral diamine (A). Optical yields are only moderate in the case of *cis*-alkenes.
   Examples:

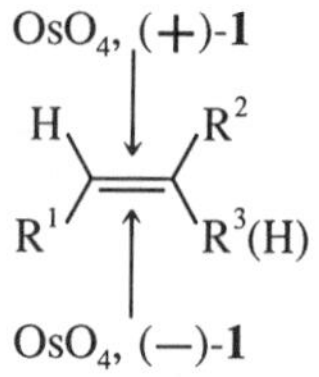

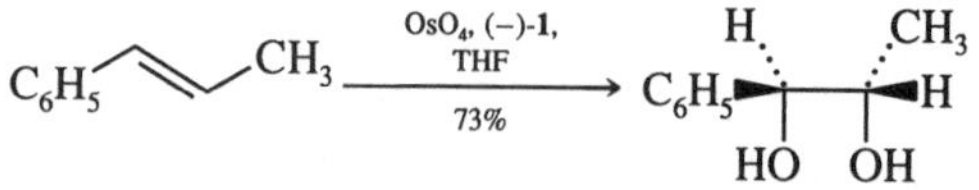

(S,S) 99% ee

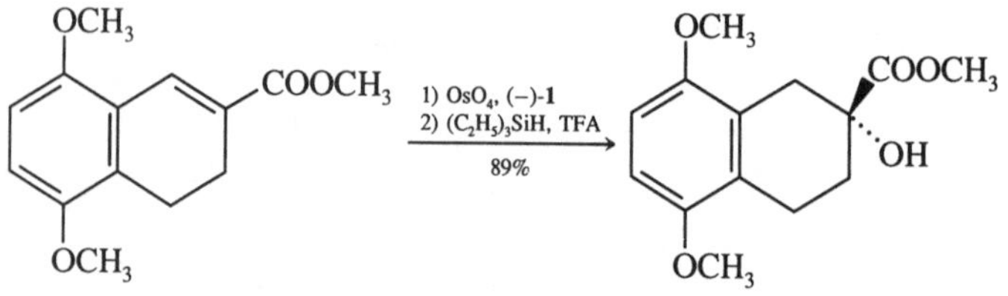

(85% ee)

[1] K. Tomioka, M. Nakajima, and K. Koga, *Chem. Lett.*, 65 (1987).
[2] *Idem, Tetrahedron Lett.*, **28**, 1291 (1987).
[3] *Idem, Am. Soc.*, **109**, 6213 (1987).

trans-4,5-Bis(hydroxydiphenylmethyl)-2,2-dimethyl-1,3-dioxacyclopentane (1).

Preparation from (+)-diethyl tartrate:

(−)-1, m.p. 196°, α_D − 60°

Resolution of bicyclic enones.[1]   This optically active diol selectively forms a hydrogen-bonded insoluble complex with one enantiomer of the bicyclic enone **2**, which when heated liberates (−)-**2** of 100% ee. This resolution is useful for a few

other bicyclic enones related to **2**, but is not generally applicable, even though the diol (**1**) forms complexes with a wide variety of substrates such as ethers, epoxides, and lactones.

[1] F. Toda and K. Tanaka, *Tetrahedron Letters*, **29**, 551 (1988).

Bis(pentamethylcyclopentadienyl)dimethyltitanium(IV), $(C_5Me_5)_2Ti(CH_3)_2$.

This compound when heated (80°) eliminates methane to give the Fischer carbene $(C_5Me_5)_2Ti{=}CH_2$ (**1**).[1]

Titanium enolates.[2]   This Fischer carbene converts epoxides into titanium enolates. In the case of cyclohexene oxide, the product is a titanium enolate of cyclohexanone. But the enolates formed by reaction with 1,2-epoxybutane (equation I) or 2,3-epoxybutane differ from those formed from 2-butanone (Equation II). Apparently the reaction with epoxides does not involve rearrangement to the ketone but complexation of the epoxide oxygen to the metal and transfer of hydrogen from the substrate to the methylene group.

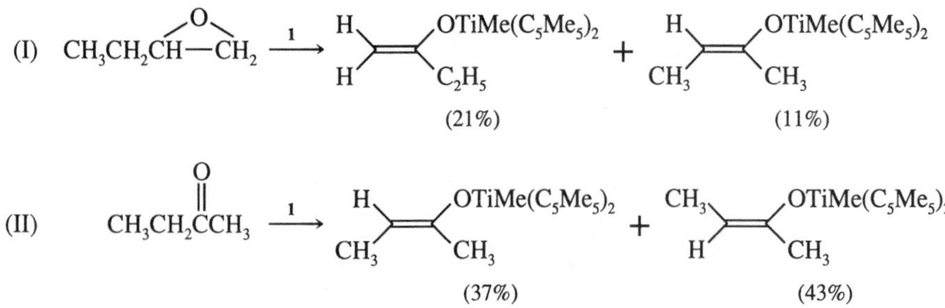

(I) CH$_3$CH$_2$CH—CH$_2$ $\xrightarrow{1}$ +

(21%)   (11%)

(II) CH$_3$CH$_2$CCH$_3$ $\xrightarrow{1}$ +

(37%)   (43%)

[1] C. McDade, J. C. Green, and J. E. Bercaw, *Organometallics*, **1**, 1629 (1982).
[2] C. P. Gibson, G. Dabbagh, and S. H. Bertz, *J. C. S. Chem. Comm.*, 603 (1988).

Bis(phenylthio)methyllithium, LiCH(SC$_6$H$_5$)$_2$ (**1**). The reagent is obtained in high yield by reaction of bis(phenylthio)methane in THF with BuLi at 0°.[1]

Ring expansion.[2]   The adducts (**2**) of cycloalkanones with **1** on reaction with methyllithium or *sec*-butyllithium at 0° rearrange to ring-expanded α-phenylthio ketones.

Examples:

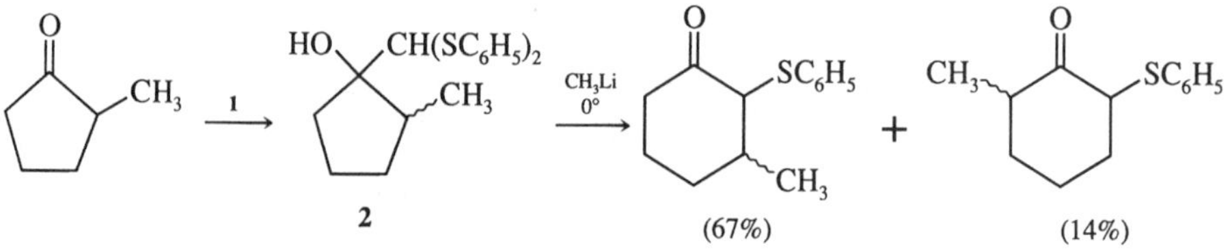

2   (67%)   (14%)

[1] E. J. Corey and D. Seebach, *J. Org.*, **31**, 4097 (1966).
[2] W. D. Abraham, M. Bhupathy, and T. Cohen, *Tetrahedron Letters*, **28**, 2203 (1987).

N,N'-Bis(salicylidene)ethylenediaminonickel(II), Ni(salen) (**1**). Supplier: Aldrich.

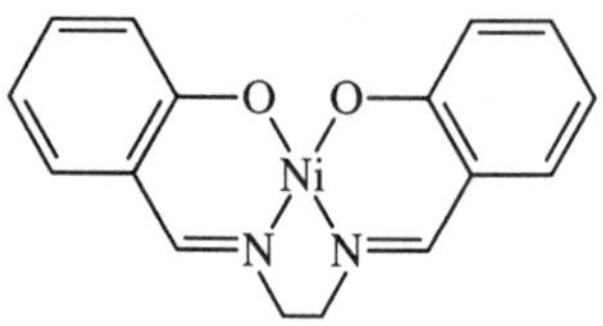

1

Epoxidation with sodium hypochlorite.[1]   Ni(salen)$_2$ is an effective catalyst for oxidation of some alkenes with NaOCl under phase-transfer conditions. Styrenes are converted mainly to epoxides but purely aliphatic alkenes are converted to a mixture of products.

[1] H. Yoon and C. J. Burrows, *Am. Soc.*, **110**, 4087 (1988).

Bis(trimethylsilyl) selenide, $[(CH_3)_3Si]_2Se$ **(1).**
  Preparation:[1]

$$Se \xrightarrow{Li(C_2H_5)_3BH} [Li_2Se] \xrightarrow[95\%]{2(CH_3)_3SiCl} 1, \text{ b.p. } 46° \text{ (5 mm.)}$$

Selenoaldehydes.[2] In the presence of 5–10 mole% of butyllithium, this disilyl selenide converts aldehydes into selenoaldehydes with formation of the disiloxane $[(CH_3)_3SiOSi(CH_3)_3]$. The active reagent is presumed to be lithium trimethylsilyl-selenide, $(CH_3)_3SiSeLi$, which can be regenerated by a Peterson-type elimination (equation I).

$$(\text{I}) \quad 1 \xrightarrow{BuLi} (CH_3)_3 SiSeLi \xrightarrow{RCHO} \overset{\overset{Se}{\|}}{RCH} + (CH_3)_3SiOLi$$

$$\downarrow 1$$

$$(CH_3)_3SiSeLi + [(CH_3)_3Si]_2O$$

The selenoaldehyde can be obtained in yields of 50–85%, as shown by trapping with cyclopentadiene (*cf.*, **13**, 286).

The reagent also converts sulfoxides and selenoxides into sulfides and selenides, respectively, but does not reduce amine oxides.

Aldehydes can be converted into thioaldehydes by a similar reaction with bis(trimethylsilyl) sulfide catalyzed by BuLi (equation II). This disilyl sulfide has been used indirectly for conversion of aldehydes into thioaldehydes via boron trisulfide (**11**, 63).

$$(\text{II}) \quad RCHO + [(CH_3)_3Si]_2S \xrightarrow{BuLi} \left[\overset{\overset{S}{\|}}{\underset{R \quad H}{\diagup}} \right] \xrightarrow{80-97\%}$$

(endo/exo = 4-20:1)

[1] M. R. Detty and M. D. Seidler, *J. Org.*, **47**, 1354 (1982).
[2] M. Segi, T. Nakajima, S. Suga, S. Murai, I. Ryu, A. Ogawa, and N. Sonoda, *Am. Soc.*, **110**, 1976 (1988).

Bis(trimethylstannyl)benzopinacolate
$$\underset{(C_6H_5)_2C-C(C_6H_5)_2}{\overset{(CH_3)_3SnO \quad OSn(CH_3)_3}{|\qquad\quad|}} \quad (1).$$

Radical homologation. This tin pinacolate is known to generate trimethyltin radicals at 60° and appears to be superior to tributyltin hydride as a source of stannyl radicals for addition of alkyl halides to O-benzylformaldoxime (equation I).[1] Iodides, bromides, and selenides can be used as radical precursors. The same

(I) ⬡—I + CH$_2$=NOBzl ⟶ ⬡—CH$_2$NHOBzl

+ **1**, C$_6$H$_6$, Δ   76%
Bu$_3$SnH, AIBN   25%

paper reports that the reaction of cyclohexyl iodide, **1**, and ethyl acrylate, in 1:1:1 ratio, produces the adduct **2** in 83% yield after aqueous potassium fluoride workup.

⬡—I + CH$_2$=CHCOOCH$_2$CH$_3$ + **1** $\xrightarrow[83\%]{C_6H_6,\ 60°}$ ⬡—CH$_2$CH$_2$COOCH$_2$CH$_3$

2

[1] D. J. Hart and F. L. Seely, *Am. Soc.*, **110**, 1631 (1988).

9-Borabicyclo[3.3.1]nonane (9-BBN).

Hydroboration of allylsilanes; 1,2- or 1,3-diols.[1] A dimethylphenylsilyl group can markedly affect the regiochemistry of hydroboration of an adjacent double bond by $BH_3 \cdot THF$ or 9-BBN. Since a dimethylphenylsilyl group can be converted to a hydroxyl group (**12**, 210), hydroboration can provide a route to *syn-* or *anti*-diols. The conversion of a β-silyl alcohol to a 1,2-diol (last example) is effected with $Hg(OAc)_2$ or Br_2 followed by peracid oxidation. The older method, fluorination followed by peracid oxidation, can result in β-elimination.

$RCOOH \rightarrow RCHO$. This reduction can be effected by reaction of the acid with **1** to form an acyloxy-9-BBN, which is not isolated but reduced by excess Li 9-BBNH (obtained by reaction of 9-BBN with LiH) in THF to give aldehydes in 85–99% yield. Reduction of aromatic acids is slow and yields are about 80%.[2]

[1] I. Fleming, *Pure Appl. Chem.*, **60**, 71 (1988).
[2] J. S. Cha, J. E. Kim, S. Y. Oh, and J. D. Kim, *Tetrahedron Letters*, **28**, 4575 (1987).

Borane–Dimethyl sulfide.

Hydroboration of allylic amines.[1] Hydroboration of primary and secondary allylic amines presents problems because amino groups interact with boron reagents. Hydroboration proceeds normally when the amino group is protected by trimethylsilyl groups, and deprotection can be effected by protonolysis in CH_3OH.

Examples:

[1] A. Dicko, M. Montury, and M. Baboulene, *Tetrahedron Letters*, **28**, 6041 (1987).

Boron(III) bromide.

Trisubstituted alkenes.[1] The (Z)-2-bromo-1-alkenylboranes (**1**), obtained by bromoboration of 1-alkynes with BBr_3 (**13**, 43), undergo coupling with organozinc chlorides in the presence of $Cl_2Pd[P(C_6H_5)_3]_2$ to provide, after protonolysis, disubstituted alkenes (**3**). However, the intermediate alkenylborane (**2**) can undergo a second coupling with an alkyl halide in the presence of a base, usually lithium methoxide (**13**, 290), to provide a trisubstituted alkene (**4**).

Example:

[1] Y. Satoh, H. Serizawa, N. Miyaura, S Hara, and A. Suzuki, *Tetrahedron Letters*, **29**, 1811 (1988).

Boron(III) chloride.

Chloromethyl ethers.[1] BCl_3 cleaves methoxymethyl (MOM) ethers to give chloromethyl ethers.

Glycosyl chlorides.[2] Methyl glycosides can be converted directly into the corresponding glycosyl chlorides by reaction with BCl_3 in CH_2Cl_2 at $-78°$ without effect on benzyl or acetyl protecting groups. The glycosyl chlorides can be used, without isolation, for glycosylation.

[1] D. A. Goff, R. N. Harris, III, J. C. Bottaro, and C. D. Bedford, *J. Org.*, **51**, 4711 (1986).
[2] G. R. Perdomo and J. J. Krepinsky, *Tetrahedron Letters*, **28**, 5595 (1987).

Boron trifluoride etherate.

Substitution of lactols.[1] In the presence of a Lewis acid, particularly BF_3 etherate, some organometallic reagents in which the metal is Al, Zn, or Sn can

$(trans/cis = 50-100:1)$

react with γ- or δ-lactols by formal substitution of the hydroxyl group to provide 2,5- and 2,6-*trans*-disubstituted tetrahydrofurans or tetrahydropyrans.

Epoxy alcohol to aldol rearrangement.[2]   Lewis acids, particularly BF_3 or $TiCl_4$, effect a 1,2-rearrangement of α,β-epoxy alcohols to β-hydroxy ketones. The migratory aptitude of the α-substituent is α-trimethylsilylalkenyl > aryl > alkyl. The rearrangement is stereospecific with respect to the epoxide and generally involves *anti*-migration. Use of chiral epoxides affords chiral aldols with high stereoselectivity.

Example:

(100% de)

This rearrangement can also be used to obtain chiral α,α-disubstituted aldols.

(99% ee)

Macrocyclization.[3]   A new route to cembranolides (**3**) involves intramolecular coupling of an alkoxyallyltin derivative (**1**) with an acetylenic aldehyde catalyzed by $BF_3 \cdot O(C_2H_5)_2$ (*cf.* **12**, 513–514). Thus in the presence of BF_3 etherate **1** cyclizes to **2** with *syn*-selectivity. The product is converted to the cembranolide **3** by hydrolysis of the enol ether and oxidation.

1

2 (*syn/anti* > 7:1)

3

[1] K. Tomooka, K. Matsuzawa, K. Suzuki, and G. Tsuchihashi, *Tetrahedron Letters*, **28**, 6339 (1987).

[2] K. Maruoka, M. Haregawa, H. Yamamoto, and K. Suzuki, M. Shimazaki, and G. Tsuchihashi, *Am. Soc.*, **108**, 3827 (1986); K. Suzuki, M. Miyazawa, and G. Tsuchihashi, *Tetrahedron Letters*, **28**, 3515, 5891 (1987).

[3] J. A. Marshall, B. S. DeHoff and S. L. Crooks, *Tetrahedron Letters*, **29**, 527 (1987).

Boron tris(trifluoromethanesulfonate). This triflate is obtained by reaction of BCl_3 with triflic acid in SO_2ClF at $-78°$. Distillation at reduced pressure provides a solid, m. p. 45°, b. p. 68–73°/0.5 mm. It is extremely hygroscopic, and is soluble in CH_2Cl_2, CH_3NO_2, CH_3CN. Aluminum and gallium triflate are poorly soluble in the common solvents. All three triflates can function as Friedel–Crafts catalysts, but the boron triflate is the most effective as a soluble catalyst.[1]

[1] G. A. Olah, O. Farooq, S. M. F. Farnia, J. A. Olah, *Am. Soc.*, **110**, 2560 (1988).

Bromine.

Bromolactamization (**11**, 76).   This reaction can be used for stereoselective preparation of 3,4-disubstituted β-lactams.[1] Thus the β,γ-unsaturated hydroxamic acid **1**, prepared in several steps from tiglic acid, on reaction with bromine and K_2CO_3 in aqueous CH_3CN cyclizes mainly to a *trans*-β-lactam (**2**). In contrast, the protected α-amino-β,γ-unsaturated hydroxamic acid **3**, prepared in several steps from L-methionine, cyclizes on reaction with bromine mainly to a *cis*-β-lactam (**4**).

1

2 (*trans*/*cis* = 7:3)

3

4 (*cis*/*trans* = 7:3)

[1] G. Rajendra and M. J. Miller *J. Org.*, **52**, 4471 (1987).

1-Bromo-2-phenylthioethylene, $BrCH=CHSCH_2H_5$ (**1**). (E)- and (Z)-**1** are obtained in a 4:1 ratio by addition of C_6H_5SH to propiolic acid followed by addition of bromine and decarboxylative dehalogenation.[1]

Olefin synthesis.[2] This reagent is useful for preparation of (E)- and (Z)-nonsymmetrical alkenes. Thus in the presence of a Pd(II) catalyst, (E)- or (Z)-**1** couples with a Grignard reagent selectively with substitution of the bromine. Substitution of the phenylthio group by a second Grignard reagent is less facile and is effected with a nickel catalyst, usually $NiCl_2[(C_6H_5)_2PCH_2CH_2P(C_6H_5)_2]$. Stereospecificity is about 99% for the (E)-isomer and overall yields are in the range 85–90%. Overall yields and isomeric purity are somewhat less in the case of (Z)-isomers.

Sequential cross-coupling can be used also for synthesis of 1,1-disubstituted alkenes from 2,3-dibromopropene (Aldrich). The allylic halogen couples directly with a Grignard reagent; vinylic coupling is slower and requires a catalyst (equation I).

[1] F. Montanari, *Gazz, Chim. Ital.*, **87**, 1086 (1957).
[2] F. Naso, *Pure Appl. Chem.*, **60**, 79 (1988).

N-Bromosuccinimide.

Dihydroxylation. The key step in the synthesis of a natural mycotoxin from the dehydropentacyclic precursor **1** requires dihydroxylation of the nuclear double bond. Direct osmylation with catalytic OsO_4 and N-methylmorpholine N-oxide

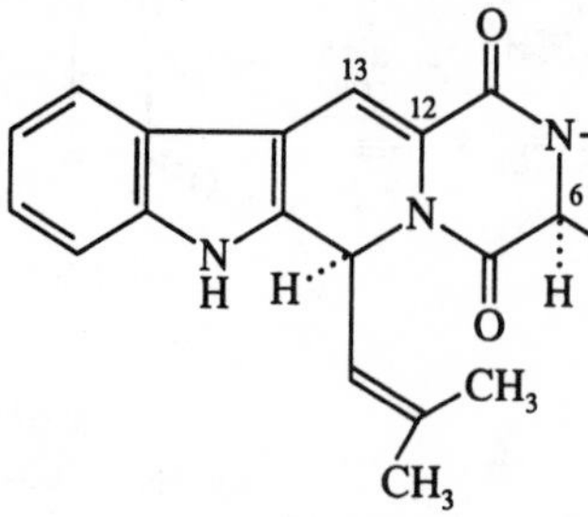

1

proceeds in low yield (32%) but does provide the expected α-*cis*-diol by attack from the less-hindered side. Oxidation with $KMnO_4$ fails completely, but reaction with NBS (1.1 equiv.) in THF/H_2O (5:1) unexpectedly gives two *trans*-diols. The major diol (12α,13β, 65% yield) is assumed to arise via an α-bromonium ion, which is attacked by H_2O at C_{13} to form a *trans*-bromohydrin (12α-Br,13β-OH) followed by solvolysis. The minor diol (12β,13α, 30% yield) is assumed to arise from a β-bromonium ion.

[1] S. Kodato, M. Nakagawa, M. Hongu, T. Kawate, and T. Hino, *Tetrahedron*, **44**, 359 (1988).

(5S,6R) - and (5R,6S) - 4 - *t* - Butoxycarbonyl - 5,6 - diphenyl - 2,3,5,6 - tetrahydro - 4*H* - 1,4-oxazin-2-one, 1a, 1b. Supplier: Aldrich.
Preparation:

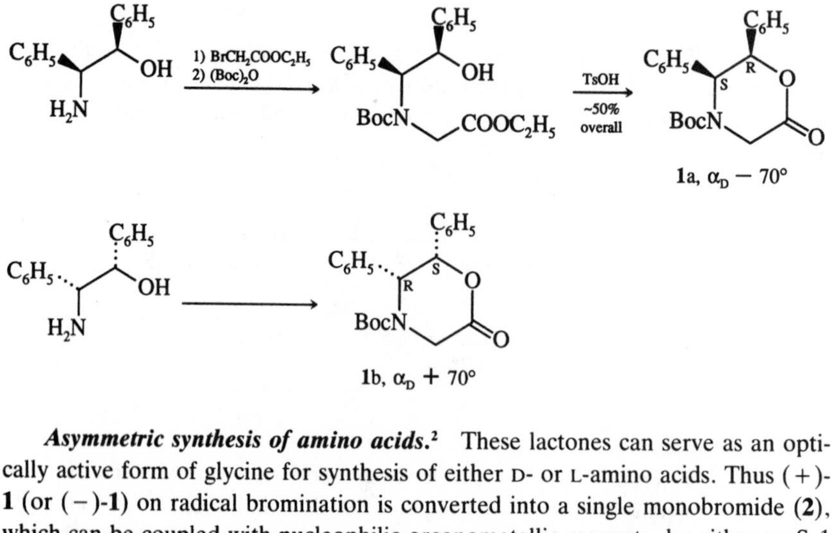

1a, α_D − 70°

1b, α_D + 70°

Asymmetric synthesis of amino acids.[2]   These lactones can serve as an optically active form of glycine for synthesis of either D- or L-amino acids. Thus (+)-**1** (or (−)-**1**) on radical bromination is converted into a single monobromide (**2**), which can be coupled with nucleophilic organometallic reagents, by either an S_N1

(retention) or an S_N2 (inversion) reaction. The stereoselectivity of coupling can be controlled by the particular nucleophile employed, the Lewis acid, and the solvent. After coupling, the amino acid is released by catalytic hydrogenation or dissolving-metal reduction (Scheme I).

Scheme (I)

[1] R. M. Williams, P. J. Sinclair, D. Chen, and D. Zhai, *Org. Syn.*, submitted (1988).
[2] R. M. Williams, P. J. Sinclair, D. Zhai, and D. Chen, *Am. Soc.*, **110**, 1547 (1988).

t-**Butoxydiphenylsilyl chloride, (1)** $(C_6H_5)_2Si{\overset{OC(CH_3)_3}{\underset{Cl}{\big\langle}}}$ This silyl chloride is prepared (81% yield) by reaction of $(C_6H_5)_2SiCl_2$ with *t*-butyl alcohol and $N(C_2H_5)_3$ in CH_2Cl_2 (reflux).

Protection of alcohols.[1] *t*-Butyldiphenylsilyl ethers (**6**, 51) are useful for protecton of alcohols, but are more resistant to acid hydrolysis and fluorolysis than *t*-butoxydiphenylsilyl ethers. In contrast, *t*-butoxydiphenylsilyl ethers are relatively acid-stable but are readily cleaved by fluoride ion in CH_2Cl_2. The ethers are stable to most alkyllithiums and to Swern or PCC oxidation.

Primary alcohols are selectively silylated by **1** and $N(C_2H_5)_3$ in CH_2Cl_2 at 25°. Silylation of secondary alcohols is effected by catalysis with DMAP, and even tertiary alcohols can be silylated by **1** in DMF catalyzed by DMAP.

[1] J. W. Gillard, R. Fortin, H. E. Morton, C. Yoakim, C. A. Quesnelle, S. Daignault, and Y. Guindon, *J. Org.*, **53**, 2602 (1988).

t-Butyldimethylsilyl triflate–Triphenylphosphine.

Phosphoniosilylation.[1] This combination reacts with acyclic or cyclic enones to give phosphonium salts, formed by addition of $P(C_6H_5)_3$ to the β-position of the enone and silylation of the carbonyl group. The products can be converted into β-substituted enones by deprotonation (BuLi), a Wittig reaction, and hydrolysis.
Example:

[1] A. P. Kozikowski and S. H. Jung, *J. Org.*, **51**, 3400 (1986).

t-Butyldiphenylchlorosilane.

Protection of primary amines.[1] Primary amines are selectively converted into mono-*t*-BDPSi derivatives by reaction with *t*-butyldiphenylchlorosilane and $N(C_2H_5)_3$ in CH_3CN. These derivatives are stable to basic and hydrolytic reagents as well as alkylating and acylating reagents such as CH_3I and a base. They are cleaved by 80% HOAc at 25° or by Py–HF at 25°.

[1] L. E. Overman, M. E. Okazaki, and P. Mishra, *Tetrahedron Letters*, **27**, 4391 (1986).

t-Butyl hydroperoxide–Chromium(VI) oxide.

Ynones.[1] Sarett's reagent (CrO_3-pyridine) is the classical reagent for oxidation of a methylene group adjacent to a triple bond but a large excess is required and yields are generally low. Oxidation of alkynes by *t*-butyl hydroperoxide and catalytic amounts of SeO_2 (**9**, 79–80) effects oxidation at both centers adjacent to a triple bond. Selective oxidation of alkynes to ynones can be effected in about 50% yield with *t*-butyl hydroperoxide (70%) and a catalytic amount of CrO_3 in benzene. Addition of *p*-toluenesulfonic acid can be beneficial. The oxidation shows some regioselectivity. A methylene group is oxidized in preference to a methyl group, and diketones are generally not obtained.
Example:

[1] J. Muzart and O. Piva, *Tetrahedron Letters*, **29**, 2321 (1988).

t-Butyl hydroperoxide-Dialkyl tartrate-Titanium(IV) isopropoxide.

Catalytic asymmetric epoxidation (**13**, 51–53).[1]   Complete experimental details are available for this reaction, carried out in the presence of heat-activated crushed 3Å or powdered 4Å molecular sieves. A further improvement, both in the rate and enantioselectivity, is use of anhydrous oxidant in isoctane rather than in CH_2Cl_2. The titanium–tartrate catalyst is not stable at 25°, and should be prepared prior to use at −20°. Either the oxidant or the substrate is then added and the mixture of three components should be allowed to stand at this temperature for 20–30 min. before addition of the fourth component. This aging period is essential for high enantioselectivity. Epoxidations with 5–10 mole % of Ti(O-*i*-Pr)$_4$ and 6–12% of the tartrate generally proceed in high conversion and high enantioselectivity (90–95% ee). Some increase in the amount of catalyst can increase the enantioselectivity by 1–5%, but can complicate workup and lower the yield. Increase of Ti(O-*i*-Pr)$_4$ to 50–100 mole % can even lower the enantioselectivity.

Asymmetric trishomoallylic epoxidation.[2]   The 6,7-double bond of **1** is epoxidized stereoselectively to afford **2** with trityl hydroperoxide catalyzed by (−)-diethyl tartrate and Ti(O-*i*-Pr)$_4$ in the presence of molecular sieves. *t*-Butyl hydroperoxide is not useful in this case. The product was used for an enantioselective

route to venustatriol, a tetraoxacyclic squalenoid with antiviral activity. This epoxidation is the first report of a stereoselective epoxidation of a trishomoallylic alcohol. Kishi[3] has reported successful asymmetric epoxidation of a bishomoallylic alcohol.

Diepoxidation of a diene.[4]   Diepoxidation of the diene **1** with *m*-chloroperbenzoic acid gives a mixture of the *d,l*- and *meso*-diepoxides, whereas Sharpless epoxidation results in *d*- or *l*-**2** by a double asymmetric epoxidation. On treatment with base, **2** rearranges to the diepoxide **a** and then cyclizes to the *meso*-tetrahydrofuran **3**, a unit of teurilene, a cytotoxic C_{30}-cyclic ether of red algae. This *meso*-3   a

sequence is an impressive example of the advantage of use of an asymmetric reaction for synthesis of a *meso*-compound.

Chiral sulfoxides (**12**, 92). Chiral sulfoxides are obtained with improved enantioselectivity by substitution of cumene hydroperoxide (CHP, Aldrich) for *t*-butyl hydroperoxide in the Sharpless reagent.[5]

Example:

(S), 89% ee

[1] Y. Gao, R. M. Hanson, J. M. Klunder, S. Y. Ko, H. Masamune, and K. B. Sharpless, *Am. Soc.*, **109**, 5765 (1987).
[2] E. J. Corey and D.-C. Ha, *Tetrahedron Letters*, **29**, 3171 (1988).
[3] T. Fukuyama, B. Vranesic, D. P. Negri, and Y. Kishi, *Tetrahedron Letters*, 2741 (1978).
[4] T. R. Hoye and S. A. Jenkins, *Am. Soc.*, **109**, 6196 (1987).
[5] S. H. Zhao, O. Samuel, and H. B. Kagan, *Compl. Rend.*, **304**, 273 (1987); *idem, Org. Syn.* submitted (1987).

t-Butyl hydroperoxide–Dioxobis(2,4-pentanedionate)molybdenum [$MoO_2(acac)_2$], **11**, 91; **12**, 89.

Epoxidation.[1] This combination is known to oxidatively cleave double bonds but to effect epoxidation when catalyzed by a metalloporphyrin. Epoxidation of alkenes can also be effected by catalysis with a simple amine. The choice of the amine depends on the olefin. N,N-Dimethylethylenediamine is the most efficient ligand for epoxidation of a 1-alkene (68% yield). Pyridine is the best ligand for epoxidation of stilbene (93%), and imidazole is preferred for epoxidation of $C_6H_5CH=CHCH_3$ (71% yield).

[1] J. Kato, H. Ota, K. Matsukawa, and T. Endo, *Tetrahedron Letters*, **29**, 2843 (1988).

t-Butyl isocyanide.

Intramolecular nitrile oxide cycloaddition.[1] The conjugate addition of *t*-butyl isocyanide to a nitroalkene can generate a nitrile oxide, which can be trapped intramolecularly by a double bond to form an isoxazoline.

Example:

t-BuNC
CH$_3$CN, 80°
48%

O=CNH-t-Bu

(2:1)

[1] J. Knight and P. J. Parsons, *J.C.S. Chem. Comm.*, 189 (1987).

Butyllithium.

Generation in situ.[1] Butyllithium (primary, secondary, or tertiary) can be generated by sonication of a mixture of lithium wire and a butyl chloride at 15° in dry THF. The corresponding butane is evolved under these conditions and LiCl precipitates; the reaction is generally complete within 15 min. The highly useful lithium diisopropylamide can be prepared by sonication of a mixture of diisopropylamine, lithium, and butyl chloride in dry THF or ether. The yield is 91% and the solution can be used directly for deprotonation. Other lithium amides, even LiTMP, can be prepared in the same way.

Payne rearrangement. The Payne rearrangement[2] of a primary *cis*-2,3-epoxy alcohol to a secondary 1,2-epoxy alcohol usually requires a basic aqueous medium, but it can be effected with BuLi in THF, particularly when catalyzed by lithium salts. As a consequence, the rearrangement becomes a useful extension of the Sharpless epoxidation, with both epoxides available for nucleophilic substitutions. Thus the more reactive rearranged epoxide can be trapped *in situ* by various organometallic nucleophiles. Cuprates of the type RCu(CN)Li are particularly effective for this purpose, and provide *syn*-diols (**3**).[3]

This Payne rearrangement of an optically active epoxide (**4**) was used for a synthesis of pure (+)-*exo*-brevicomin (**6**) in 31% overall yield.

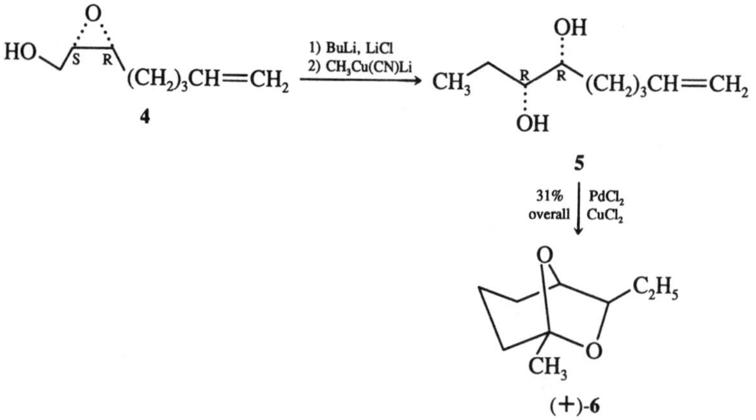
HO
S
R
O
$(CH_2)_3CH{=}CH_2$
4
1) BuLi, LiCl
2) $CH_3Cu(CN)Li$
CH_3
OH
R
R
$(CH_2)_3CH{=}CH_2$
OH
5
31% overall
$PdCl_2$
$CuCl_2$
O
C_2H_5
O
CH_3
(+)-6

2-Azaallyl anion cycloadditions (**13**, 163).[4] Nonstabilized 2-azaallyl anions (**1**) are readily generated by transmetallation of N-(trialkylstannyl)methylimines, prepared as shown in equation I, with a base such as butyl- or methyllithium. The

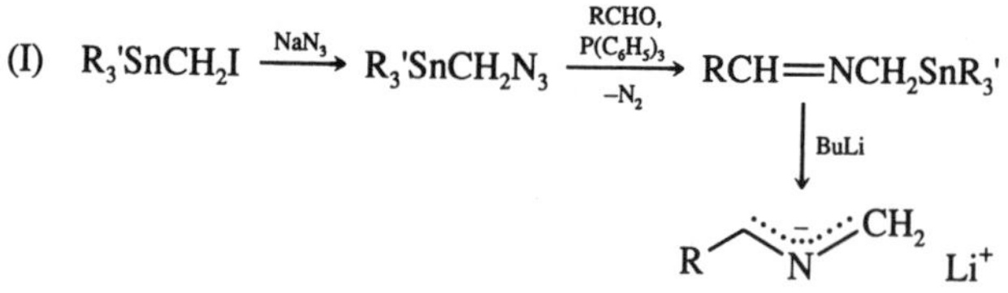
(I) $R_3'SnCH_2I$
NaN_3
$R_3'SnCH_2N_3$
RCHO, $P(C_6H_5)_3$
$-N_2$
$RCH{=}NCH_2SnR_3'$
BuLi
R
N
CH_2
Li^+

1

anions undergo [3 + 2]cycloaddition with alkenes to form pyrrolidines.

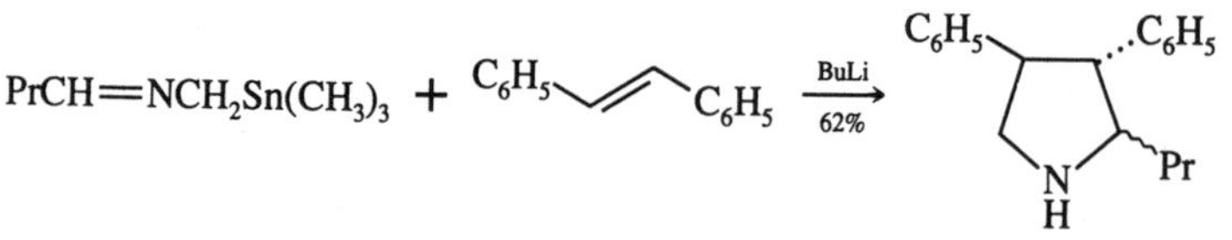
$PrCH{=}NCH_2Sn(CH_3)_3$ +
C_6H_5
C_6H_5
BuLi
62%
C_6H_5
C_6H_5
N
H
Pr

(1:1)

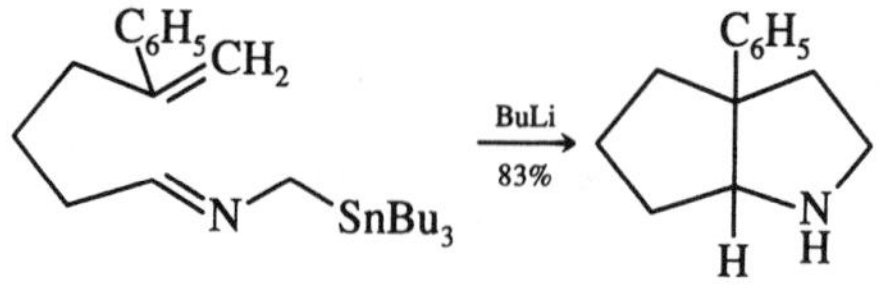
C_6H_5
CH_2
N
$SnBu_3$
BuLi
83%
C_6H_5
N
H
H

Tetrahydrofurans.[5] The (tributylstannyl)methyl ethers (**1**) of homoallylic alcohols (**9**, 475) on treatment with butyllithium undergo tin–lithium exchange to α-alkoxy lithium products. In the presence of excess BuLi, these products undergo cyclization to *cis*-2,4-disubstituted tetrahydrofurans in moderate yield.

Example:

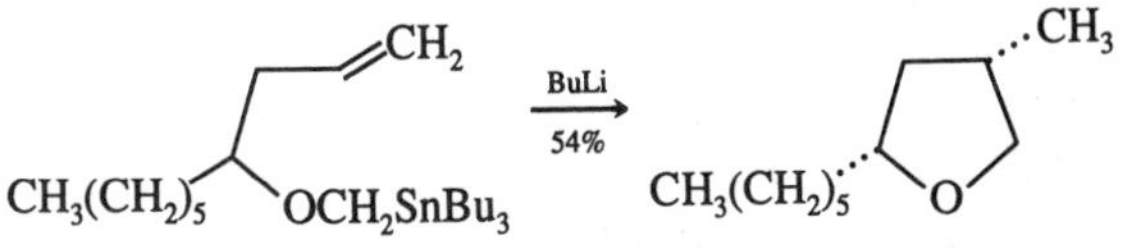

$$CH_3(CH_2)_5 \overset{\text{BuLi}}{\underset{54\%}{\longrightarrow}} CH_3(CH_2)_5$$

1   **2 (11:1)**

This cyclization occurs more readily and in higher yield when an allylic ether is used as a leaving group, with formation of vinyltetrahydrofurans.

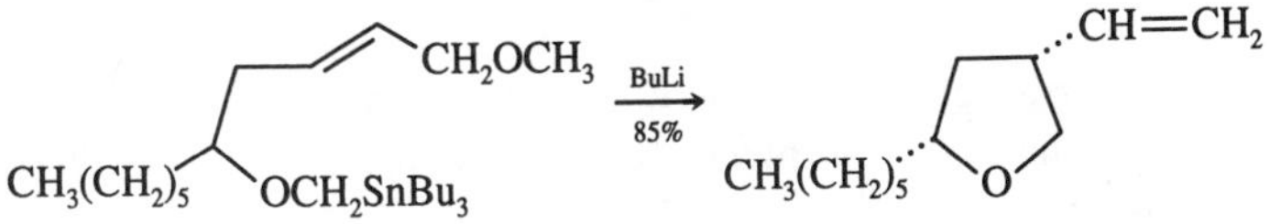

(13:1)

[2,3]*Wittig rearrangements.* On reaction with BuLi, the 17-membered cyclic allylic propargylic ether **1** undergoes Wittig rearrangement with contraction of the ring to give the propargylic alcohol **2** and the epimer as a 4.5:1 mixture.[6] The alcohol (**2**) is a useful precursor to the cembranoid epimulol (**3**). The method is a

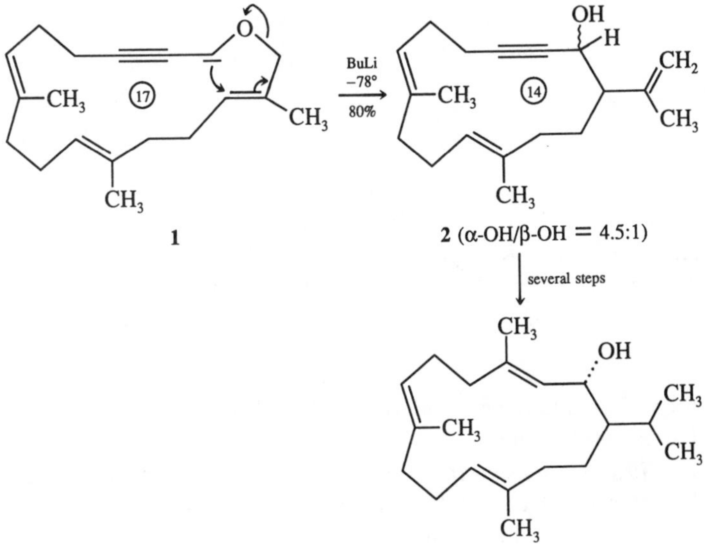

1   **2 (α-OH/β-OH = 4.5:1)**

several steps

3

particularly useful route to medium-size rings, which are not easily obtained by direct cyclization.

This ring contraction has been used to provide the ten-membered ring of the germacranolide aristolactone (**4**) from a 13-membered allylic propargylic ether.

4

The efficient rearrangement of these cyclic ethers may stem from the favorable juxtaposition of the reactive centers. Rearrangement of related acyclic substrates is notably less efficient.

Asymmetric Wittig rearrangements.[7]   High 1,2-asymmetric induction obtains in the [2,3]Wittig rearrangement of the chiral dioxolanes **1** and **3**. In each case the

vinyl group of the rearranged homoallylic alcohol is *syn* to the C–O bond of the original asymmetric center.

[1,2]-Wittig rearrangement.   The [1,2]Wittig rearrangement of β-alkoxyalkyl allyl ethers shows a high *syn*-selectivity, although the chemical yields are low because of concomitant 1,4-rearrangement.[8]

Examples:

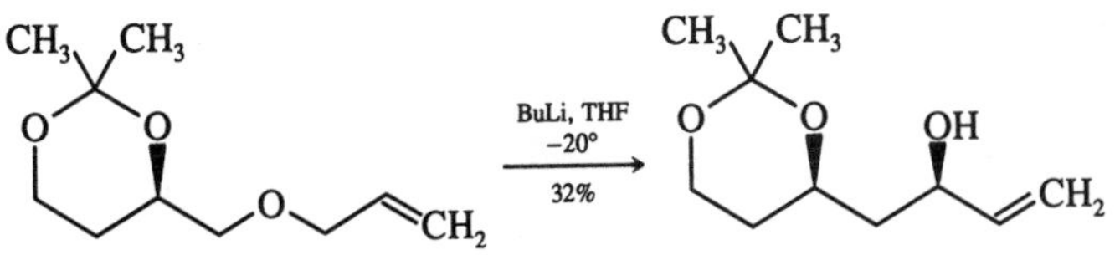

(12:1)

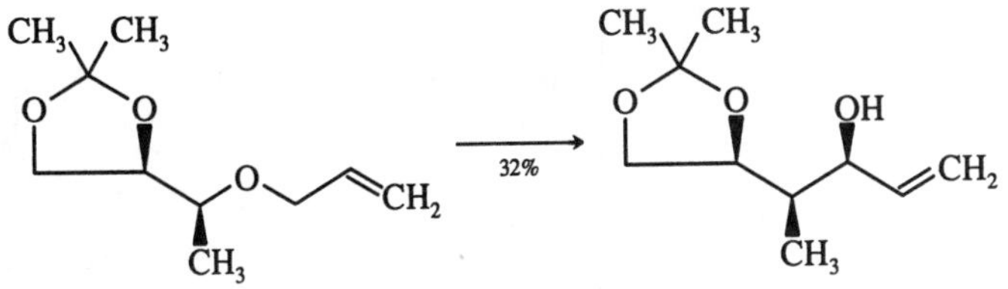

(15:1)

(α-Hydroxyalkyl)silanes.[9]   These α-alkoxysilanes can be prepared by addition of Bu_3SnLi to an aldehyde followed by silylation with cyanotrimethylsilane (or trimethylsilyl triflate). Transmetallation of the adduct (BuLi) results in rearrange-

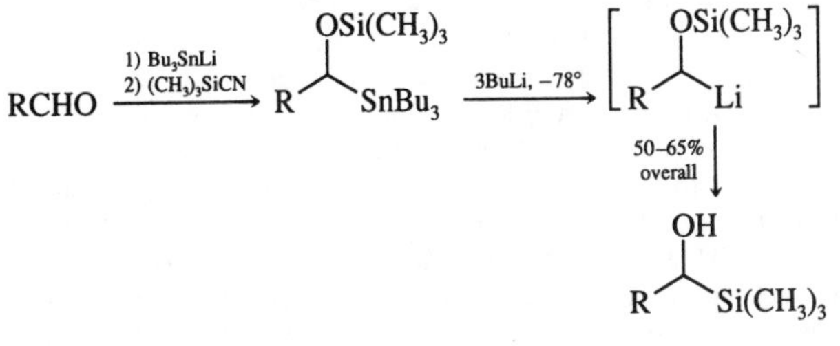

ment of the silyl group from oxygen to carbon to give α-alkoxysilanes in 50–65% overall yield. This rearrangement is the reverse of the Brook rearrangement of a C—Si bond to an O—Si bond, and is possible because of the facile transmetallation of Sn to Li. This rearrangement has been applied to a ketone; in this case, use of trimethylstannyllithium is required for successful rearrangement (equation I).

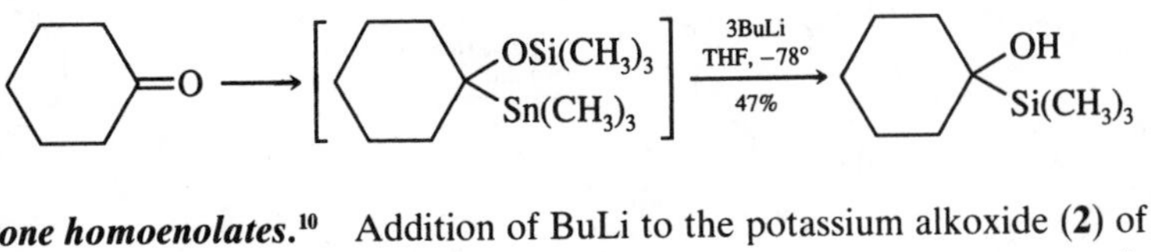

Ketone homoenolates.[10]   Addition of BuLi to the potassium alkoxide (**2**) of a 3-hydroxy-1-alkene (**1**) results in partial isomerization via **a** to the corresponding

3-alkanone (**3**). Under usual conditions the homoenolate (**b**) is also formed, as shown by alkylation to provide **4**.
Example:

[1] J. Einhorn and J. L. Luche, *J. Org.*, **52**, 4124 (1987).
[2] G. B. Payne, *ibid.*, **27**, 3819 (1962).
[3] P. C. Bulman Page, C. M. Rayner, and I. O. Sutherland, *J.C.S. Chem. Comm.*, 356 (1988).
[4] W. H. Pearson, D. P. Szura, and W. G. Harter, *Tetrahedron Letters*, **29**, 761 (1988).
[5] C. A. Broka, W. J. Lee, and T. Shen, *J. Org.*, **53**, 1336 (1988).
[6] J. A. Marshall, T. M. Jensen and B. S. DeHoff, *ibid.*, **51**, 4316 (1986); **52**, 3860 (1987); J. A. Marshall, J. Lebreton, B. S. DeHoff, and T. M. Jensen, *ibid.*, **52**, 3883 (1987).
[7] R. Brückner and H. Priepke, *Angew. Chem. Int. Ed.*, **27**, 278 (1988).
[8] S. L. Schreiber and M. T. Goulet, *Tetrahedron Letters*, **28**, 1043 (1987).
[9] R. J. Linderman and A. Ghannam, *J. Org.*, **53**, 2878 (1988).
[10] T. Cuvigny, M. Julia, L. Jullien, and C. Rolando, *Tetrahedron Letters*, **28**, 2587 (1987).

Butyllithium–Tetramethylethylenediamine.

Angular anthracyclinones.[1] The *ortho*-metalation of a benzyl alcohol has been used to provide a synthesis of angular benzanthraquinone. Thus the *cis*-tetralol

4

(1) after dilithiation condenses with the aldehydo amide **2** to give, after acid-catalyzed cyclization, the phthalide **3**, a precursor to the natural anthracyclinone **4**.

[1] K. Katsuura and V. Snieckus, *Can. J. Chem.*, **65**, 124 (1987).

sec-Butyllithium.

   Fries rearrangement.[1]   Rearrangement of phenyl esters with Lewis acids results in a mixture of *ortho-* and *para*-phenolic ketones. In contrast, reaction of an *o*-bromophenyl ester with *sec*-butyllithium results in exclusive formation of the *ortho*-phenolic ketone by an intramolecular acyl rearrangement.[2]

   Example:

[1] A. H. Blatt, *Org. React.*, **1**, 342 (1942).
[2] J. A. Miller, *J. Org.*, **52**, 322 (1987).

(2R)-2-*t*-Butyl-6-methyl-2*H*,4*H*-1,3-dioxin-4-one,   **(1)**,m.p.48.5°,

α_D-217°; Supplier: Fluka. The dioxinone is prepared by acetalization of pivaldehyde with (R)-3-hydroxybutanoic acid followed by bromination and hydrodebromination (60% yield).[1]

   Stereoselective hydrogenation.[2]   Catalytic hydrogenation of **1** or of the bromo derivatives (**2**) occurs with complete diastereoselectivity (>98% de) with H transfer to the face opposite to that of the *t*-Bu group to regenerate the stereogenic center of (R)-hydroxybutanoic acid (**4**). The result is unexpected because the two faces of the double bond are similar in respect to steric hindrance.

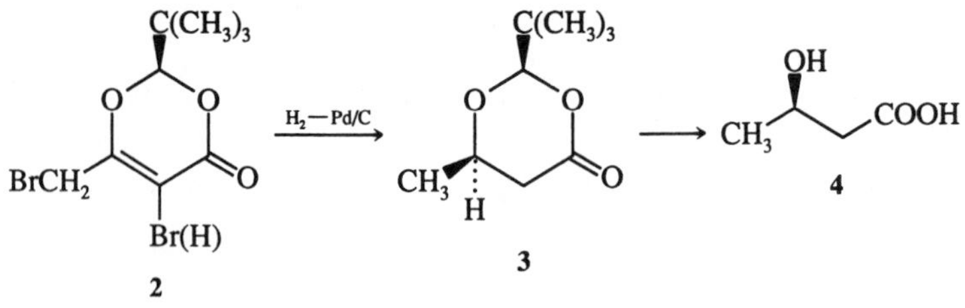

Stereoselective conjugate addition.   The dioxinone **1** undergoes conjugate addition with lithium dialkylcuprates to give a single adduct in which the alkyl group is *cis* to the *t*-butyl group, with no reaction occurring at the acetal carbon. Cuprate addition to the less hindered face of **1** is explained by pyramidalization of the three trigonal centers.[3]

Example:

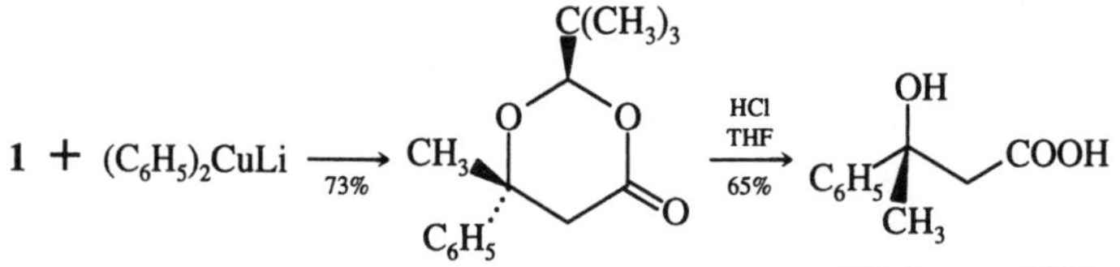

[1] J. Zimmerman and D. Seebach, *Helv.*, **70**, 1104 (1987).
[2] Y. Noda and D. Seebach, *ibid.*, **70**, 2137 (1987).
[3] D. Seebach, J. Zimmermann, U. Gysel, R. Ziegler, and T.-K. Ha, *Am. Soc.*, **110**, 4763 (1988).

t-**Butyl trichloroacetimidate,** (**1**), b.p. 65–69°/12 mm. The reagent is prepared by reaction of trichloroacetonitrile in ether with potassium *t*-butoxide in *t*-butyl alcohol; yield 70%.

t-*Butyl esters and ethers.*[1]   Acids or alcohols are converted to the *t*-butyl esters or ethers by reaction with **1** catalyzed by BF_3 etherate. Cyclohexane or CH_2Cl_2 is the preferred solvent; the reagent can decompose in more polar solvents. Yields are generally greater than 70%.

[1] A. Armstrong, I. Brackenridge, R. F. W. Jackson, and J. M. Kirk, *Tetrahedron Letters*, **29**, 2483 (1988).

C

Calcium–Ammonia.

Reduction of benzyl ethers.[1]   Benzyl ethers are cleaved in high yield by calcium (2 equiv.) in liquid ammonia. By proper control of the amount of metal, selective reduction of benzyl ethers is possible in the presence of a triple bond or a *t*-butyldimethylsilyl ether group. However, there is little selectivity between benzyl ethers and thiophenyl, epoxide, or keto groups in this reduction.

[1] J. R. Hwu and co-workers, *J. Org.*, **51**, 4731 (1986).

Camphor-10-sulfonic acid, 13, 62.

Asymmetric conjugate addition of Grignard reagents.[1]   The sultam **1**, derived from (+)-camphor-10-sulfonic acid, is a useful chiral auxiliary for asymmetric induction in conjugate addition of RMgCl to enones. Thus C_2H_5MgCl reacts with **2** to give **3**, after protonation, with high diastereoselectivity. A second sterogenic

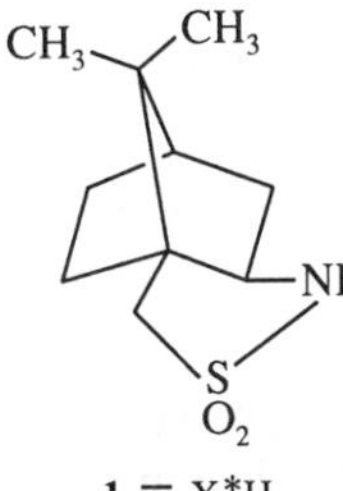

$1 = X_N^*H$

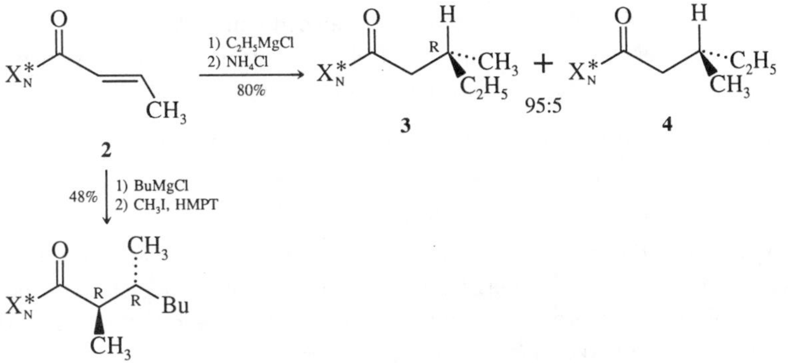

5 (95% de)

71

center can be generated by conjugate addition followed by enolate trapping to give 5, with the (2R,3R)-configuration.

[1] W. Oppolzer, G. Poli, A. J. Kingma, C. Starkemann, and G. Bernardinelli, *Helv.*, **70**, 2201 (1987).

Camphoryloxaziridines, (+)-(2R,8aS)-**1** and (−)-(2S,8aR)-**1**, **13**, 64–65.

Diastereoselective hydroxylation of enolates of chiral amides. Davis and co-workers[1] have examined the asymmetric hydroxylation of the tetrasubstituted enolates of a chiral amide (**2**) with these chiral camphoryloxaziridines. Oxidation of the lithium enolate of **2** with (+)-**1** proceeds with only moderate diastereoselectivity (48.4% de), which is somewhat less than that observed on hydroxylation with the achiral 2-(phenylsulfonyl)-3-phenyloxaziridine (**4**). Oxidation of the enolate of **2**

(+)-**1**	60%	48.4% de
(−)-**1**	55%	88.3% de
(+)-**1** + HMPT	65%	89% de

$$C_6H_5SO_2N—CHC_6H_5 \text{ (4)} \quad 54\% \quad 55\% \text{ de}$$

with (−)-**1**, however, proceeds with high diastereoselectivity, evidently because of matched chirality. Of more interest, addition of HMPT to the reaction with (+)-**1** (mismatched pair) improves the diastereoselectivity from 48% to 89%, but has no affect on the diastereoselectivity of the matched pair. HMPT is effective even when added after enolate formation. It probably does not affect the geometry of the enolate, but may affect the structure of the aggregate in solution.

[1] F. A. Davis, T. G. Ulatowski, and M. S. Haque, *J. Org.*, **52**, 5288 (1987).

Carbon tetrachloride.

α-Chloro carboxylic acids. α-Chloro unsaturated carboxylic acids can be obtained by reaction of the dianion of the acids with CCl₄ without attack on the double bond.[1] This reaction provides a route to unsaturated chloroketenes (**a**), which undergo facile intramolecular [2 + 2]cycloadditions.[2]

Example:

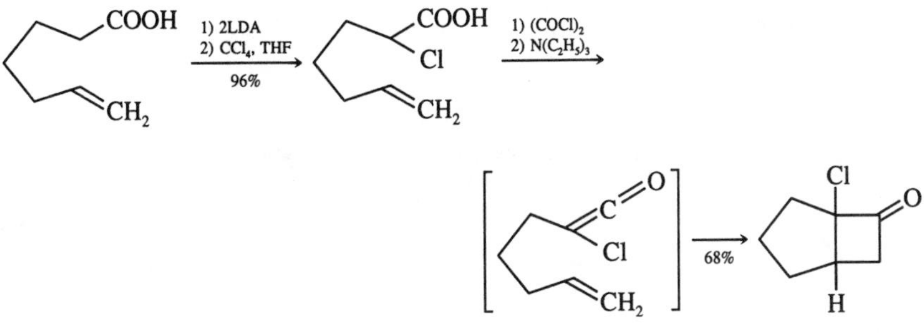

a

[1] R. T. Arnold and S. T. Kulenovic, *J. Org.*, **43**, 3687 (1978).
[2] B. B. Snider and Y. S. Kulkarni, *ibid.*, **52**, 307 (1987).

2,2'-Carbonylbis(3,5-dioxo-4-methyl-1,2,4-dioxazolidine), 1.

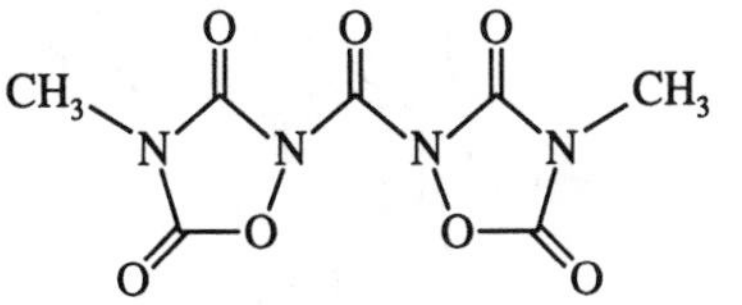

1, m.p. 204°

The reagent is prepared[1] by reaction of phosgene with 3,5-dioxo-4-methyl-1,2,4-oxadiazolidine in refluxing toluene (82% yield). It serves as a coupling reagent for esterification and preparation of amides and carbamates.

β-Keto esters.[2] In the presence of a trace of tertiary amine, an acid reacts with **1** in THF at 0° to form a 2-acyl-3,5,dioxo-1,2,4-oxadiazolidine (**2**). This activated form of an acid reacts with the lithium enolate of an ester in THF at −75°

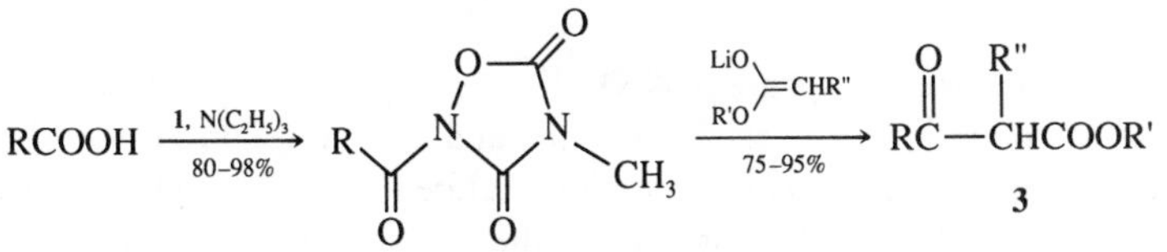

2

to form a β-keto ester in good yield. This procedure is useful for synthesis of unusual γ-amino-β-hydroxy esters such as the protected derivative of (3S,4R,5S)-statine (**5**) from D-alloisoleucine(**4**).

5 (96% de)

[1] M. Denarié, D. Grenouillat, T. Malfroot, J.-P. Senet, G. Sennyey, and P. Wolf, *Tetrahedron Letters*, **28**, 5823, 5827 (1987).

[2] P. Jouin, J. Poncet, M.-N. Dufour, I. Maugras, A. Pantaloni, and B. Castro, *ibid.*, **29**, 2661 (1988).

3-Carboxypyridinium dichromate, $Cr_2O_7^=$ (**1**), m. p. 215–217°. The reagent is prepared in 85% yield in a reaction of nicotinic acid with 2 equiv. of CrO_3 in water.

Oxidation.[1] The rate of oxidation of alcohols with **1** can be markedly increased by addition of pyridine. In fact **1** in combination with pyridine is an excellent reagent for oxidation of 1,4-hydroquinones to the quinones. When used in the absence of pyridine, **1** can effect selective oxidation of benzylic alcohols in the presence of a secondary alcohol.

[1] F. P. Cossio, M. C. Lopez, and C. Palomo, *Tetrahedron*, **43**, 3963 (1987).

Cerium(IV) ammonium nitrate (CAN).

1,4-*Diketones*. Some years ago Heiba and Dessau reported that $Mn(OAc)_3$ promotes oxidative addition of isopropenyl acetate to ketones to give 1,4-diketones in 20–35% yield (**6**, 356). Use of CAN as the oxidant results in higher yields (65–80%) and a regioselective reaction at the more substituted α-position of the ketone.[1] Use of vinyl acetate results in the dimethyl acetal of 4-oxo aldehydes.

$$CH_3COCH_2CH_3 \;+\; CH_2=C\underset{OAc}{\overset{CH_3}{\diagdown}} \quad \xrightarrow[78\%]{\overset{CAN}{CH_3OH}} \quad CH_3\diagdown\diagup\diagup\diagdown CH_3$$

(with CH₃ and two C=O groups as drawn)

Cyclohexanone + $CH_2=CHOAc$ $\xrightarrow{69\%}$ product with OCH_3 groups

Oxidative bisdecarboxylation.[2]   A new route to lactones is based on the ability of CAN to effect oxidative decarboxylation of α-hydroxymalonic acids to carboxylic acids (**11**, 143–144) and of α-alkoxymalonic acids to lactones.
Example:

$$\text{Cyclohexyl-}CH=CH_2 \;+\; O=C(COOC_2H_5)_2 \xrightarrow[80\%]{\overset{SnCl_4}{C_6H_6}} \text{spiro product} \begin{array}{l} COOC_2H_5 \\ COOC_2H_5 \end{array}$$

$$91\% \downarrow \begin{array}{l} 1)\ KOH \\ 2)\ CAN,\ CH_3CN,\ H_2O \end{array}$$

[1] E. Baciocchi, G. Civitarese, and R. Ruzziconi, *Tetrahedron Letters*, **28**, 5357 (1987).
[2] R. G. Salomon, S. Roy, and M. F. Salomon, *ibid.*, **29**, 769 (1988).

Cerium(III) chloride.

Allylmetallic reagents.[1]   The allyl anions obtained by reductive metallation of allyl phenyl sulfides with lithium 1-(dimethylamino)naphthalenide (LDMAN, **10**, 244) react with α,β-enals to give mixtures of 1,2-adducts. The regioselectivity can be controlled by the metal counterion. Thus the allyllithium or the allyltitanium compound obtained from either **1** or **2** reacts with crotonaldehyde at the secondary terminus of the allylic system to give mainly the adduct **3**. In contrast the allylcerium compound reacts at the primary terminus to form **4** as the major adduct.

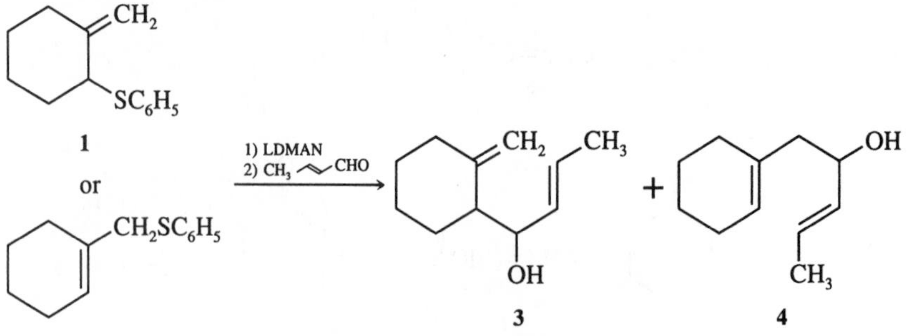

1

or

2

1) LDMAN
2) $CH_3 \diagup CHO$

3 + **4**

+ Ti(O-i-Pr)$_4$, −60°   (92%)   (8%)
+ CeCl$_3$, −78°   (18%)   (82%)

The reaction of allylcerium compounds with acrolein can be used for preparation of either (Z,Z)- or (Z,E)-1,4-dienes, equations (I) and (II). Thus the allylcerium compound generated at −78° from **5** reacts with acrolein at the same temperature to form (Z)-**6**. However, if the allylcerium compound is allowed to warm to −40° before addition to acrolein, (E)-**6** is obtained in 60% yield. Similarly, the reaction of the homoallylic cerium compound from **7** with acrolein can be controlled to give either (Z,Z)- or (Z,E)-**8** selectively.

(I)

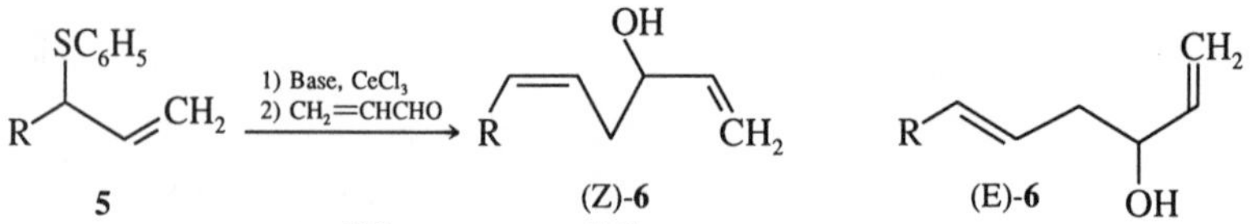

5

1) Base, CeCl$_3$
2) CH_2=CHCHO

(Z)-**6**   (E)-**6**

−78°   72%   —
−40°   —   60%

(II)

7

(Z,Z)-**8** (65%)   or

(Z,E)-**8** (48%)

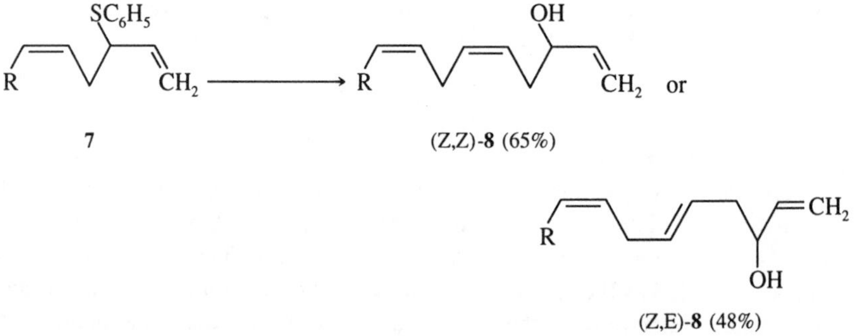

Allylsilanes from esters.[2]   A reagent prepared from CeCl$_3$ and (CH$_3$)$_3$-SiCH$_2$MgCl is effective for conversion of esters to allylsilanes (equation I). The reagent obtained from $CeCl_3$ and $(CH_3)_3SiCH_2Li$ is considerably less efficient for this reaction,[3] but is effective for conversion of acyl chlorides to allylsilanes. Although the nature of the actual reagent is uncertain, $CeCl_3$ is essential in both reactions.

Peterson methylenation (**10**, 433; **11**, 581).   Methylenation with trimethylsilylmethyllithium, $(CH_3)_3SiCH_2Li$, is not widely used in synthesis because of lack of selectivity and moderate yields. However, a modified reagent prepared from $(CH_3)_3SiCH_2Li$ and $CeCl_3$ adds to aldehydes or ketones (even enolizable ones) to form adducts in generally high yield, particularly in the presence of TMEDA. The 2-hydroxysilanes are converted into methylene compounds by aqueous HF (with or without pyridine).[4]

Example:

[1] T. Cohen and B.-S. Guo, *Tetrahedron*, **42**, 2803 (1986); B.-S. Guo, W. Doubleday, and T. Cohen, *Am. Soc.*, **109**, 4710 (1987).

[2] B. A. Narayanan and W. H. Bunnelle, *Tetrahedron Letters*, **28**, 6261 (1987).

[3] M. B. Anderson and P. L. Fuchs, *Syn. Comm.*, **17**, 621 (1987).

[4] C. R. Johnson and B. D. Tait, *J. Org.*, **52**, 281 (1987).

Cesium carbonate.

Intramolecular Michael addition.[1]   Cesium carbonate catalyzes the intramolecular Michael addition of a cyclic β-keto ester to an α,β-ynone to form a cyclic enone after protonation. This reaction proceeds readily when a five- or six-membered ring is formed; higher rings can be formed, but in low yield.[1]

Example:

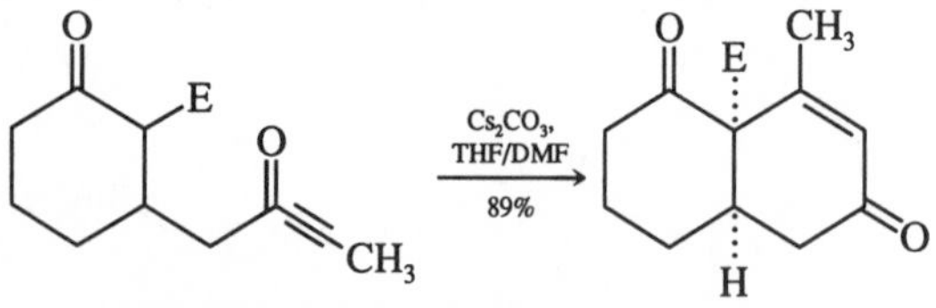

E = COOCH₃

The intermediate in this addition is considered to be an enone-anion, and indeed a sequential addition and alkylation resulting in tricyclic products is possible.[2]

Cleavage of 2-oxazolidinones to amino alcohols.   2-Oxazolidinones and related heterocycles can be cleaved to Boc-amino alcohols by N-*t*-butoxycarbonylation followed by treatment with a catalytic amount of Cs_2CO_3 in CH_3OH at 25°.[3]

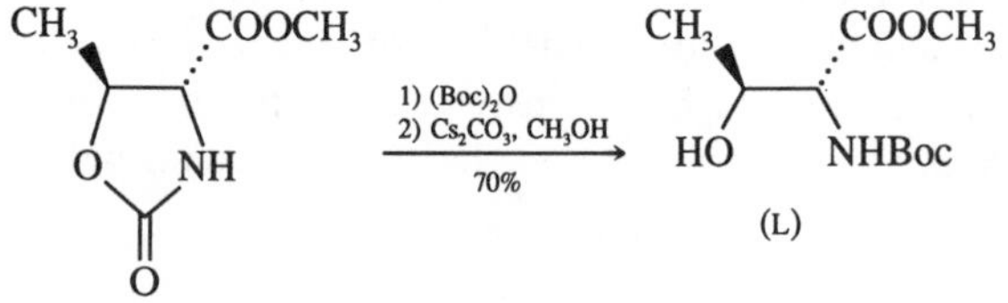

(4S,5R)

---

[1] J.-F. Lavallée, G. Berthiaume, P. Deslongchamps, and F. Grein, *Tetrahedron Letters*, **27**, 5455 (1986).
[2] J.-F. Lavallée and P. Deslongchamps, *ibid.*, **28**, 3457 (1987).
[3] T. Ishizuka and T. Kunieda, *ibid.*, **28**, 4185 (1987).

Cesium *p*-nitrobenzoate, p-$NO_2C_6H_4COOCs$ (1).   The salt is prepared by reaction of the acid with Cs_2CO_3 in aqueous methanol.

D-α-*Hydroxy carboxylic acids*.[1]   These optically active acids can be prepared by a S_N2 reaction between the *t*-butyl esters of L-2-halo carboxylic acids and cesium *p*-nitrobenzoate, which proceeds with complete inversion.

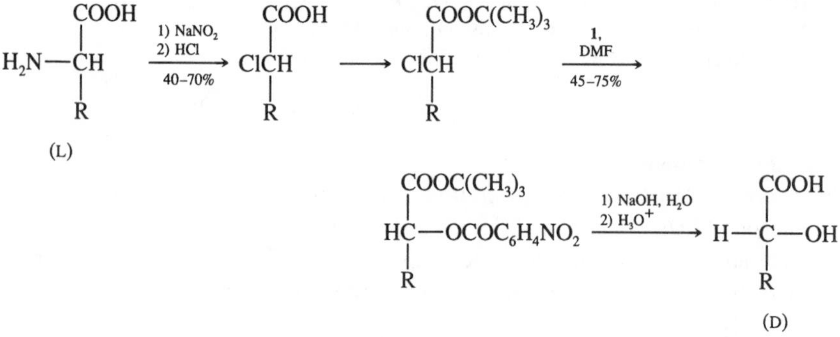

[1] H. Kunz and H.-G. Lerchen, *Tetrahedron Letters*, **28**, 1873 (1987).

Cesium fluoride.

Monoalkylation of a vic-glycol.[1] Selective monoalkylation or monoacylation of the *vic*-glycol group of dimethyl L-tartrate is possible by conversion to the O-stannylene acetal (**1**) by reaction with dibutyltin oxide. The acetal is converted selectively to a mono derivative (**3**) by reaction with an alkyl halide or acyl chloride (excess) and CsF (about 2 equiv.). KF or Bu$_4$NF are less effective than CsF.

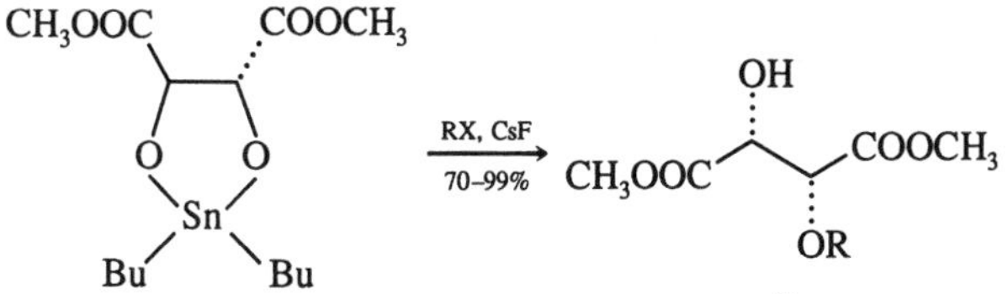

[1] N. Nagashima and M. Ohno, *Chem. Letters*, 141 (1987).

Chlorine oxide, ClO$_2$. The gas is commercially available in aqueous or certain organic solvents, but has received little use by organic chemists.

Oxidation of tertiary amines. Early investigations have shown that tertiary amines are oxidized by ClO$_2$ to iminium ions, and this oxidation can be used to effect cyclization of 2-aminoethanols or 3-aminopropanols to bicyclic oxazolidines or tetrahydro-1,3-oxazines.

Example:

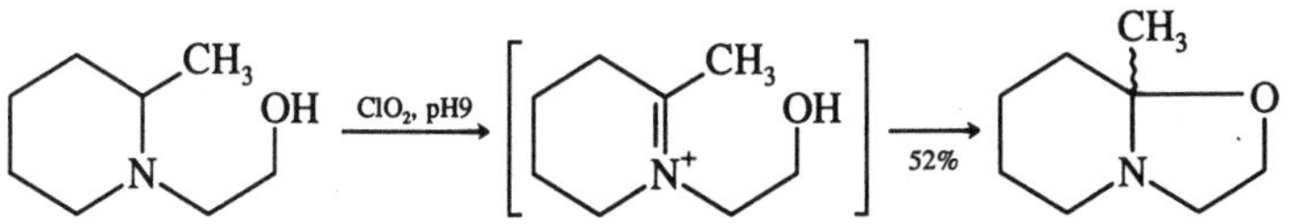

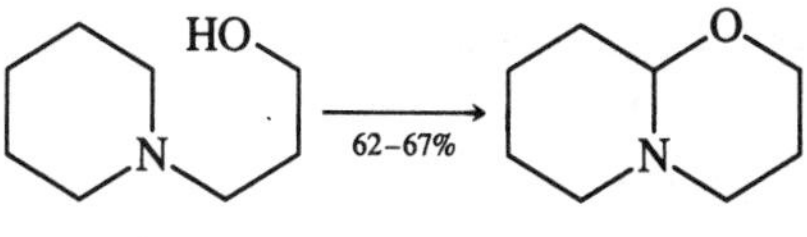

Oxidation of *t*-amines in the presence of NaCN affords α-cyanoamines in 53–83% yield. Tertiary α-cyanoamines are known to afford iminum salts by loss of cyanide ion and this oxidative cyanation is used in a synthesis of the indolizidine

$$(C_2H_5)_3N \xrightarrow[\text{69\%}]{\text{ClO}_2, \text{ NaCN}} (C_2H_5)_2\text{NCHCH}_3$$
$$\overset{|}{\text{CN}}$$

68%   CN 65:35

alkaloid elaeocarpidine (**2**). Thus reaction of the α-cyanoamine **1** with AgOTs and TsOH affords the ion **a**, which is trapped by tryptamine to give **2** in 38% yield.[1]

[1] C.-K. Chen, A. G. Hortmann, and M. R. Marzabadi, *Am. Soc.*, **110**, 4829 (1988).

Chlorocyanoketene, $\overset{\text{Cl}}{\underset{\text{NC}}{\diagdown}}\text{C}{=}\text{C}{=}\text{O}$ **(1), 8, 88.** This ketene is exceptionally prone to polymerization and must be generated *in situ*. Generation from chlorocyano-acetyl chloride is possible, but yields of [2 + 2] adducts are low. It is best generated slowly by thermolysis of the pseudoisopropyl ester of the azidofuranone **3**, prepared from mucochloric acid (**2**), which is less prone to detonation than the pseudomethyl ester used originally (**8, 88**).[1]

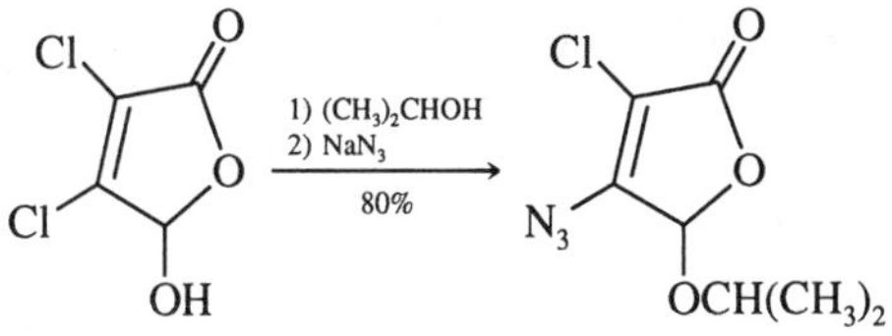

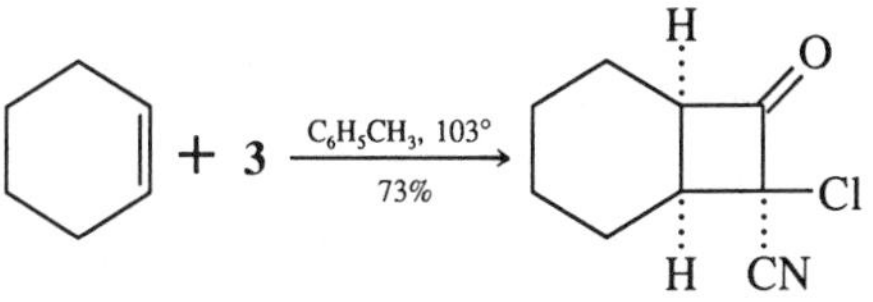

[1] P. L. Fishbein and H. W. Moore, *Org. Syn.*, submitted (1988).

Chlorobis(cyclopentadienyl)hydridozirconium, $Cp_2Zr(H)Cl$ (**1**).
Schwartz's reagent is available commercially but is expensive. It has been prepared by reduction of Cp_2ZrCl_2 with $LiAl(O-t-Bu)_3H$ or sodium bis(2-methoxyethoxy)aluminum hydride (**6**, 176). A new, simplified preparation uses $LiAlH_4$. Although this reductant leads to a mixture of $Cp_2Zr(H)Cl$ and Cp_2ZrH_2, the latter product is converted rapidly to $Cp_2Zr(H)Cl$ when washed with CH_2Cl_2. This procedure can provide large amounts of reagent of about 95% purity in 77–92% yield.[1]

Regiospecific hydrocyanation of alkenes.   Reaction of *t*-butyl isocyanide with the adducts of **1** with alkenes results in products (**2**) that are converted by iodine (excess) into hydrocyanides (**3**) and *t*-butyl iodide with release of $ClCp_2ZrI$. $(CH_3)_3SiN{=}C$ can be used in place of $(CH_3)_3CN{=}C$, but yields are generally lower.[2]

Example:

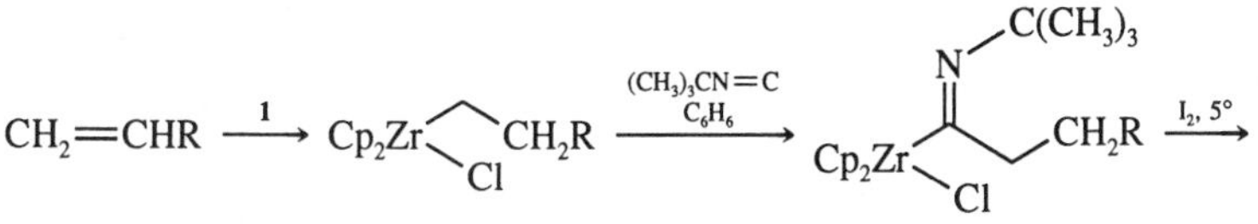

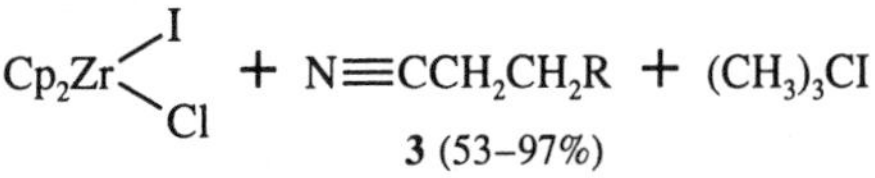

[1] S. L. Buchwald, S. J. LaMaire, R. B. Nielsen, B. T. Watson, and S. M. King, *Tetrahedron Letters*, **28**, 3895 (1987).
[1] S. L. Buchwald and S. J. LaMaire, *ibid.*, **28**, 295 (1987).

Chlorodiisopinocampheylborane, (Ipc$_2$BCl, **1**), **13**, 72.

Reduction of alkyl aryl ketones.[1] Unlike the sluggish reductions using Ipc$_2$BH, the chloro derivative reduces ketones at a reasonable rate at $-25°$, and is stable under nitrogen at $0°$ for at least a year. Both (+)- and (−)-**1** are commercially available (Aldrich). The chiral reagent reduces aryl ketones rapidly and with almost complete induction. n-Alkyl phenyl ketones are reduced by (−)-**1** to (S)-alcohols in 97–98% ee, regardless of the length of the alkyl chain. The reagent reduces straight-chain aliphatic ketones in at best only 32% ee, but alkyl α-tertiary ketones (e.g., pinacolone) and hindered alicyclic ketones are reduced rapidly to (S)-alcohols in 90–95% ee.

[1] H. C. Brown, J. Chandrasekharan, and P. V. Ramachandran, *Am. Soc.*, **110**, 1539 (1988).

Chlorodiisopropylsilane, Cl(i-Pr)$_2$SiH (**1**). The silane is prepared in 76% yield by reaction of isopropylmagnesium chloride with trichlorosilane.

Hydrosilylation of β-hydroxy ketones.[1] β-Silyloxy ketones (**2**), prepared by silylation of β-hydroxy ketones with **1** under the usual conditions (DMAP or Py), on treatment with a Lewis acid form a mixture of siladioxanes, **3a** and **3b**, which on desilylation with HF is converted into a mixture of *anti-* and *syn*-diols (**4**). The

highest *anti*-selectivity for this intramolecular hydrosilylation is obtained with SnCl$_4$, with BF$_3$ etherate a close second. Surprisingly, ZnCl$_2$ is mildly *syn*-selective.

[1] S. Anwar and A. P. Davis, *Tetrahedron*, **44**, 3761 (1988).

Chlorodimethoxyborane, ClB(OCH$_3$)$_2$. Preparation from BCl$_3$ and B(OCH$_3$)$_3$.[1]

Butenolides.[2] 2-Alkylfurans can be converted into 5-alkyl-2(3H)-butenolides by lithiation followed by alkylation with ClB(OCH$_3$)$_2$. The products are unstable, but are oxidized *in situ* by m-chloroperbenzoic acid exclusively to the butenolides **2**.

[1] E. Wiberg and H. Smedsrud, *Zeit. Anorg. Alg. Chem.*, **225**, 204 (1935).
[2] A. Pelter and M. Rowlands, *Tetrahedron Letters*, **28**, 1203 (1987).

Chloromethyllithium, $ClCH_2Li$ **(1).**

Chlorohydrins; epoxides. Chloromethyllithium is thermally unstable, but can be obtained by reaction of lithium wire with bromochloromethane. The halide–lithium exchange is markedly accelerated by sonication [(((], which also promotes a subsequent reaction of the chloromethyllithium with an aldehyde or a ketone to form epoxides via chlorohydrins.[1] Overall yields are 70–90%.

Methylenation.[2] Aldehydes or ketones react with $ClCH_2Li$, prepared *in situ* from $ClCH_2I$ and CH_3Li, to form an adduct that on lithiation decomposes to the alkene. Yields are 50–95%.

[1] C. Einhorn, C. Allavena, and J.-L. Luche, *J.C.S. Chem. Comm.*, 333 (1988).
[2] J. Barluenga, J. L. Fernandez–Simon, J. M. Concellon, and M. Yus, *ibid.*, 1665 (1986).

Chloromethyl trimethylsilylmethyl sulfide, $(CH_3)_3SiCH_2SCH_2Cl$ **(1).** The sulfide is obtained in 69% yield by reaction of hydrogen chloride with a mixture of trimethylsilylmethanethiol and S-trioxane at 0°.

[3 + 2] *Cycloadditions; tetrahydrothiophenes.*[1] The reaction of **1** with CsF in CH_3CN generates a 1,3-dipole (**a**), which undergoes [3 + 2]cycloadditions with alkenes to give tetrahydrothiophenes in 65–85% yield.

Example:

[1] A. Hosomi, Y. Matsuyama, and H. Sakurai, *J.C.S. Chem. Comm.*, 1073 (1986).

Chloromethyl *p*-tolyl sulfone, $ClCH_2SO_2C_6H_4CH_3$-*p* (**1**). The sulfone is prepared by reaction of $CH_3C_6H_4SO_2Na$ and $BrCH_2Cl$ in DMSO.

Vicarious nucleophilic substitution of nitroarenes.[1] The carbanion (**2**, $NaOCH_3$) of **1** reacts with *p*-chloronitrobenzene to give, after acidic workup, 5-chloro-2-nitrobenzyl *p*-tolylsulfone (**3**). The reagent **2** is typical of a number of carbanions which can undergo nucleophilic substitution *ortho* or *para* to nitro-

arenes. They require an α-stabilizing group such as $SO_2C_6H_5$, CN, COOR, $PO(OC_2H_5)_2$ and an α-leaving group such as halo, OC_6H_5, CH_3S. The value of vicarious substitution is that this reaction occurs more readily than classical nucleophilic substitution of nitroarenes.

[1] M. Makosza, A. Kinowski, and K. Sienkiewicz, *Org. Syn.* submitted (1987).

***m*-Chloroperbenzoic acid.**

Diastereoselective epoxidation.[1] *cis*-4-Benzyloxycarbonylamino allylic alcohols (**1**) are oxidized by $ClC_6H_4CO_3H$ almost exclusively to the *syn*-epoxide **2**, probably because the Cbo group can form a hydrogen bond with the reagent. The oxides are reduced by SMEAH (Red-Al®) selectively to *syn*-amino diols (**3**). Se-

lective oxidation of the primary alcohol group provides β-hydroxy-γ-amino acids such as statine (**4**, R = *i*-Bu).

***Epoxidative lactonization.*[2]** This reaction can be used for a diastereoselective synthesis of 5-hydroxyalkylbutanolides (**3**) from (E)- and (Z)-4-alkenoic acids (**1**). Thus epoxidation of (E)-**1** results in an epoxide (**2**) that in the presence of acid is converted into *erythro*-**3**. In contrast, (Z)-**1** is converted into *threo*-**3**. Yields are essentially quantitative.

***Oxidation of β-hydroxy selenides.*[3]** β-Hydroxy selenides (**1**), obtained by reaction of a dialkyl ketone with phenylselenenylmethyllithium, are oxidized by *m*-chloroperbenzoic acid in methanol to epoxides (equation I). The same sequence when applied to an alkyl aryl ketone results in one-carbon homologation. This

(I) $\begin{array}{c} R^1 \\ \diagdown \\ R^2 \end{array} C{=}O \xrightarrow[\text{THF}]{C_6H_5SeCH_2Li} \begin{array}{c} R^1 \quad OH \\ \diagdown \diagup \\ R^2 \diagup C \diagdown \\ CH_2SeC_6H_5 \end{array} \xrightarrow[75-95\%]{\substack{ClC_6H_4CO_3H \\ CH_3OH}} \begin{array}{c} R^1 \\ R^2 \end{array} \overset{\triangle}{\underset{O}{\diagdown}}$

1

(II) $C_6H_5COCH_3 \longrightarrow \text{adduct} \xrightarrow{ClC_6H_4CO_3H} CH_3\underset{\underset{O}{\parallel}}{C}CH_2C_6H_5$

oxidation provides a method for ring enlargement of a cyclic ketone fused to a benzene ring (equation III).

(III)

Nef reaction.[4]   Nitro compounds, primary or secondary, are converted to trialkylsilyl nitronates in greater than 90% yield by reaction with a trialkylsilyl chloride and DBU in CH_2Cl_2. The silyl nitronates derived from secondary nitro compounds are oxidized by $ClC_6H_4CO_3H$ at 25° to ketones in high yield. This sequence is not useful for conversion of primary nitro compounds to aldehydes.

Example:

Corticoid side chain.[5]   The final step in a recent total synthesis of cortisone (**2**) employed a novel double hydroxylation of an enol silyl ether (**1**) for construction of the side chain. Thus treatment of **1** with 3 equiv. of the peracid and $KHCO_3$ in CH_2Cl_2 at 0° results in hydroxylation at both C_{17} and C_{21}.

$$\text{Pr}_3\text{SiO} \quad \text{CH}_3 \quad \xrightarrow[\substack{\text{KHCO}_3 \\ 83\%}]{\text{ClC}_6\text{H}_4\text{CO}_3\text{H}}$$

1   **2**

[1] H. Kogen and T. Nishi, *J.C.S. Chem. Comm.*, 311 (1987).
[2] C. W. Jefford and Y. Wang, *ibid.*, 1513 (1987).
[3] S. Uemura, K. Ohe, and N. Sugita, *ibid.*, 111, (1988).
[4] J. M. Aizpurua, M. Oiarbide, and C. Palomo, *Tetrahedron Letters*, **28**, 5361 (1987).
[5] Y. Horiguchi, E. Nakamura, and I. Kuwajima, *J. Org.*, **51**, 4323 (1986).

N-Chlorosuccinimide–Dimethyl sulfide (1).

Oxidation of β-hydroxy ketones.[1]   Reaction of the Corey–Kim reagent with these substrates can result in dimethylsulfonium dicarbonylmethylides in 80–98% yield. These S-ylides are desulfurized to β-diketones by zinc in acetic acid. β-

Hydroxy ketones that are mono- or disubstituted at the α-position are oxidized directly to 1,3-diketones by the reagent in moderate to good yield.

[1] S. Katayama, K. Fukuda, T. Watanabe, and M. Yamauchi, *Synthesis*, 178 (1988).

Chlorotriisopropoxytitanium (1).

Aldol-type reaction of zinc esters.[1]   This titanium reagent promotes condensation of (ethoxycarbonylalkyl)iodozinc compounds (**13**, 220) with aldehydes or ketones to provide hydroxy esters and/or lactones. The active reagent may be (ethoxycarbonyl)alkyltriisopropoxytitanium.

Examples:

$$\text{IZn(CH}_2)_3\text{COOC}_2\text{H}_5 + \text{C}_6\text{H}_5\text{CHO} \xrightarrow{1, 0°}$$

$$\text{CH}_3\text{-CH(ZnI)-COOC}_2\text{H}_5 + \text{C}_6\text{H}_5\text{CHO} \xrightarrow{94\%}$$

(*cis*, 99%)

[1] H. Ochiai, T. Nishihara, Y. Tamaru, and Z. Yoshida, *J. Org.*, **53**, 1343 (1988).

Chlorotrimethylsilane–Hexamethylphosphoric triamide.

Conjugate additions to enals and enones.[1]   ClSi(CH$_3$)$_3$/HMPT markedly accelerates the Cu(I)-catalyzed conjugate addition of Grignard reagents to enals, enones, and unsaturated esters. The conjugate addition of organocopper reagents is also facilitated.

Examples:

(91% E)

(97% Z)

[1] Y. Horiguchi, S. Matsuzawa, E. Nakamura, and I. Kuwajima, *Tetrahedron Letters*, **27**, 4025, 4029 (1986).

Chlorotrimethylsilane–Phenol.

Deprotection of t-*butoxycarbonylamino acids.*[1]   This reaction is generally effected with 50% CF$_3$COOH/CH$_2$Cl$_2$, with slight, if any, effect on other common protecting groups. Deprotection can also be effected by ClSi(CH$_3$)$_3$ in combination with phenol, with formation of phenyl trimethylsily ether as the co-product. This cleavage occurs rapidly at room temperature with essentially no loss of benzyl protecting groups. Reactions with ClSi(CH$_3$)$_3$ alone are markedly slower.

[1] E. Kaiser, Sr., J. P. Tam, T. M. Kubiak, and R. B. Merrifield, *Tetrahedron Letters*, **29**, 303 (1988).

Chlorotrimethylsilane–Sodium iodide.

Silyl enol ethers.[1]   Aldehydes, ketones, α,β-enals, and α,β-enones are converted into the silyl enol ethers in moderate to high yield by reaction with iodotrimethylsilane, generated *in situ*, and $N(C_2H_5)_3$ at 25° in acetonitrile. In some cases intermediate stable 1,2- or 1,4- adducts can be isolated. Thus the 1,2-adduct **a** has been isolated as an intermediate in the reaction of some aldehydes and shown to decompose to the silyl enol ether.

$$RCH_2CHO \xrightarrow[N(C_2H_5)_3]{ClSi(CH_3)_3,\ NaI,} \left[\begin{array}{c} RCH_2 \quad OSi(CH_3)_3 \\ \diagdown C \diagup \\ \diagup \quad \overset{+}{\diagdown} \\ H \quad N(C_2H_5)_3I^- \end{array} \right] \xrightarrow{25°} RCH{=}CHOSi(CH_3)_3$$

a

[1] P. Cazeau, F. Duboudin, F. Moulines, O. Babot, and J. Dunogues, *Tetrahedron*, **43**, 2075, 2089 (1987).

Chlorotris(triphenylphosphine)cobalt.

Benzoquinones (**12**, 130).[1]   The maleoyl cobalt complex **a** shows excellent regioselectivity (about 15–20:1) in thermal reactions (80°) with electron-rich al-

a (X = Cl)

(R = CH₃, Bu, CH₂SiR₃
R¹ = H, COOC₂H₅, COCH₃)

kynes; only marginal regioselectivity is observed in thermal reactions with other alkynes. However, use of Lewis acid catalysts (SnCl₄, BF₃ etherate, or AgBF₄) allows the reaction to proceed at 25° and can improve the regioselectivity of reactions with electron-deficient alkynes.

This quinone synthesis was used to obtain the isoquinoline quinone **2**, a unit of some saframycin antibiotics (equation I).

(I)   **a** + (structure with Cbo, N, Bzl, C, COCH₃) $\xrightarrow[43\%]{\text{SnCl}_4, 25°}$

(X = OCOCF₃)

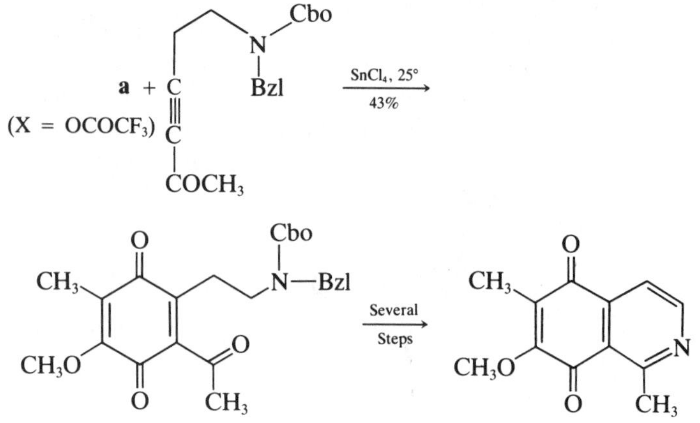

2

5-Alkylidenecyclopentene-1,4-diones (*cf.* **12**, 130).[2] The cobaltacyclopentenedione (**2**) formed on reaction of a cyclobutenedione with **1**, when complexed with dimethylglyoxime, reacts with 1-alkynes to form 5-alkylidenecyclopentene-1,4-diones (**3**). The reaction involves an alkyne–vinylidene rearrangement. Benzoquinones are usually formed in minor amounts.

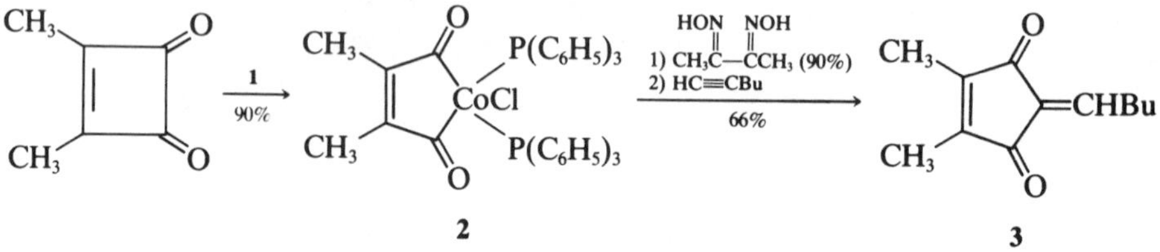

2   **3**

(Z)-Ketene silyl acetals. Hydrosilylation of acrylates with any trialkylsilane catalyzed with Wilkinson's rhodium catalyst results in (Z)-ketene silyl acetals (Z/E ≥ 98:2).[3]

Example:

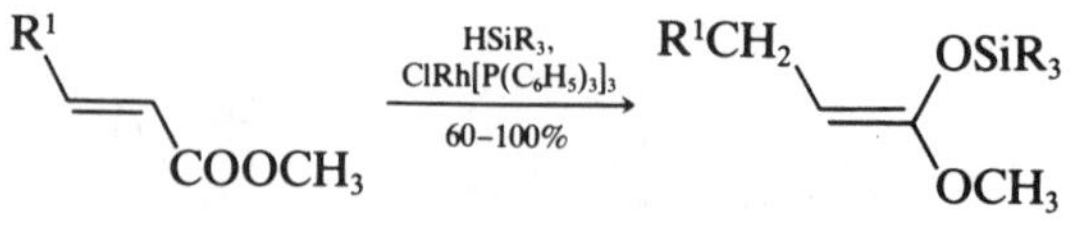

(Z/E ≥ 98:2)

[1] S. Iyer and L. S. Liebeskind, *Am. Soc.*, **109**, 2759 (1987).
[2] L. S. Liebeskind and R. Chidambaram, *ibid.*, **109**, 5025 (1987).
[3] N. Slougui and G. Rousseau, *Syn. Comm.*, **17**, 1 (1987).

Chromium carbene complexes.

Cyclopropanation of 1,3-dienes.[1] α,β-Unsaturated carbenes can undergo [4 + 2]cycloaddition with 1,3-dienes (12, 134), but they can also transfer the carbene ligand to an isolated double bond to form cyclopropanes. Exclusive cyclopropanation of a 1,3-diene is observed in the reaction of the α,β-unsaturated chromium carbene 1 with the diene 2, which results in a *trans*-divinylcyclopropane (3) and a seven-membered silyl enol ether (4), which can be formed from 3 by a Cope rearrangement. However, the tungsten carbene corresponding to 1 undergoes exclusive [4 + 2]cycloaddition with the diene 2.

Bicyclic cyclopropanes.[2] Reaction of the Fisher carbene 1 with the 1,6-enyne 2 results in the bicyclic cyclopropane 3 (a bicyclo[3.1.0]hexane) in 69% yield. The 1,7-enyne homolog of 2 reacts with 1 in the same way to form a bicyclo[4.1.0]-

heptane (46% yield). An intramolecular version of this reaction results in a tricyclic enol ether (equation I).

Anthracyclinone synthesis.   Wulff and Xu[3] have reported a high-yield formal synthesis of 11-deoxydaunomycinone (**5**) in which the first step is a benzannelation of the chromium carbene **1** with the acetylene **2** to provide the naphthol **3**, which is not isolated but treated with TFA to induce cleavage of the *t*-butyl ester and to initiate Friedel–Crafts cyclization to the tetracyclic naphthol (**4**). The synthesis of **5** is completed by oxidative demethylation to a naphthoquinone and air oxidation of ring C.

Cyclopentenones.[4] Reaction of the cyclopropylcarbene complex **1** with diphenylacetylene in 1% aqueous dioxane results in the cyclopentenone **2** as the major product. The carbonyl group of **2** is derived from a CO ligand of **1**, and C_1

1   **2** (*trans/cis* = 24:1)

and C_2 of **2** are derived from C_1 and C_2 of **1**. The remaining carbon atoms of **1** are lost as ethylene. The two hydrogens at C_4 and C_5 of **2** may be derived from water, and indeed yields of **2** are lower when the reaction is carried out in an anhydrous medium. In reactions with monosubstituted alkynes, the substituent is always attached to C_4 of **2**, α to the carbonyl group, and the other possible isomer is not observed. No reaction obtains with an alkyne such as ethyl propiolate.

$(CO)_5Cr=CHNBzl_2$. (*cf.* **11**, 400–401). This carbene, like typical Fischer carbenes, undergoes a photolytic addition to imines to give mixtures of *cis*- and *trans*-3-dibenzylamino-β-lactams in fair to good yield. The products are convertible into 3-amino-2-azetidinones.[5]

Example:

[1] W. D. Wulff, D. C. Yang, and C. K. Murray, *Am. Soc.*, **110**, 2653 (1988).
[2] P. F. Korkowski, T. R. Hoye, and D. B. Rydberg, *ibid.*, **110**, 2676 (1988).
[3] W. D. Wulff and Y.-C. Xu, *ibid.*, **110**, 2312 (1988).
[4] J. W. Herndon, S. U. Tumer, and W. F. K. Schnatter, *ibid.*, **110**, 3334 (1988).
[5] C. Borel, L. S. Hegedus, J. Krebs, and Y. Satoh, *ibid.*, **109**, 1101 (1987).

Chromium(II) chloride.

*Olefination of aldehydes with **gem-dichromium reagents.***[1]   Reduction of 1,1-diiodoethane with $CrCl_2$ in THF provides a 1,1-dichromioethane reagent (**a**) that reacts with aldehydes to furnish products of ethylidenation in 80–99% yield with high (E)-selectivity (equation I).

(I) $CH_3CHI_2 \xrightarrow{\text{CrCl}_2,\ \text{THF}} CH_3CH[Cr(III)]_2 \xrightarrow[80-99\%]{\text{RCHO}}$

$$\underset{H}{\overset{R}{>}}C=C\underset{CH_3}{\overset{H}{<}}$$

(E/Z = 84–98:22–2)

Reactions with other *gem*-diiodoalkanes under these conditions proceed in low yield; however, addition of DMF (and ultrasonic irradiation) provides a very general method for alkylidenation of aldehydes with good (E)-selectivity (equation II).

(II) $R^1CHI_2 \xrightarrow[\text{THF}]{\text{CrCl}_2,\ \text{DMF}} R^1CH[Cr(III)]_2 \xrightarrow[\substack{75-95\% \\ \text{overall}}]{R^2CHO}$

$$\underset{H}{\overset{R^1}{>}}C=C\underset{R^2}{\overset{H}{<}}$$

(E/Z = 88–99:12–1)

(E)-Alkenylsilanes and -sulfides.[2]   Dibromomethyltrimethylsilane (**1**), prepared from $ClSi(CH_3)_3$ and CH_2Br_2, is reduced by commercial $CrCl_2$ in THF to a reagent that converts aldehydes into (E)-vinylsilanes. A related reaction provides (E)-vinyl sulfides selectively from $C_6H_5SCHCl_2$.

$(CH_3)_3SiCHBr_2 \xrightarrow{\text{CrCl}_2,\ \text{THF}} (CH_3)_3SiCH[Cr(III)]_2 \xrightarrow[80-85\%]{\text{RCHO}}$

1

$$\underset{H}{\overset{R}{>}}C=C\underset{Si(CH_3)_3}{\overset{H}{<}}$$

$C_6H_5CHO \xrightarrow[83\%]{C_6H_5SCHCl_2,\ CrCl_2}$

$$\underset{H}{\overset{C_6H_5}{>}}C=C\underset{SC_6H_5}{\overset{H}{<}}$$

(E/Z = 82:18)

Reduction of α-halo sulfides.[3]   In combination with LiI, $CrCl_2$ reduces α-halo sulfides to (α-alkylthio)chromium compounds, which undergo selective 1,2-addition to aldehydes (equation I). The [1-(phenylthio)ethyl]chromium(III) reagent ob-

$$
\text{(I)}\quad CH_3SCH_2Cl \xrightarrow[\text{THF}]{CrCl_2,\ LiI,} [CH_3SCH_2Cr(III)] \xrightarrow[\text{50–90\%}]{RCHO} \overset{\displaystyle \overset{OH}{|}}{R}CHCH_2SCH_3
$$

tained in this way can undergo diastereoselective addition to aldehydes in the presence of suitable ligands, particularly 1,2-bis(diphenylphosphine)ethane.
   Example:

$$
C_6H_5SCHCH_3 \underset{\displaystyle \underset{Cl}{|}}{} \xrightarrow[\text{2) RCHO}]{\text{1) CrCl}_2,\ \text{LiI}} R \diagdown \underset{OH}{} \diagup \overset{SC_6H_5}{\underset{CH_3}{}}
$$

(*syn/anti* =
80–98:20–2)

α-Methylene-γ-lactones.   Two laboratories[4,5] have reported syntheses of α-methylene-γ-lactones by reaction of an allylic bromide with an aldehyde mediated by $CrCl_2$ to furnish a homoallylic alcohol, which cyclizes to an α-methylene-γ-lactone (equation I).

$$
\text{(I)}\quad CH_2{=}C \diagup^{CH_2Br}_{\diagdown COOC_2H_5} + C_6H_5CHO \xrightarrow{CrCl_2} \left[CH_2{=}C \diagup^{CH_2CHOH\,(C_6H_5)}_{\diagdown COOC_2H_5} \right]
$$

94% ↓

A general route to the precursors, α-bromomethyl-α,β-unsaturated esters (**1**), involves condensation of an aldehyde with methyl acrylate mediated by DABCO followed by brominative rearrangement with 48% hydrogen bromide.[5]

$$CH_2=CHCOOCH_3 + CH_3CHO \xrightarrow[80\%]{\underset{25°}{DABCO,}} CH_2=C\underset{COOCH_3}{\overset{\overset{CH_3}{|}}{\overset{CHOH}{\diagup}}} \xrightarrow[53\%]{\underset{2) H_2SO_4}{1) HBr}}$$

$$\underset{H}{\overset{CH_3}{\diagdown}}C=C\underset{COOCH_3}{\overset{CH_2Br}{\diagup}} + CH_3CHO \xrightarrow{CrCl_2} \left[\underset{H}{\overset{CH_3}{\diagdown}}C=C\underset{COOCH_3}{\overset{CH_2\overset{\overset{CH_3}{|}}{CHOH}}{\diagup}} \right]$$

1

$$\downarrow 85\%$$

cis-2

Vinylchromium compounds (12, 137).[6]   These compounds can be obtained by reaction of vinyl triflates with $CrCl_2$ catalyzed by $NiCl_2$ in DMF. They undergo selective Grignard-type addition to aldehydes. Some commercial sources of $CrCl_2$ do not require a catalyst, presumably because they contain a metal contaminant.

$$\underset{R^1}{\overset{CH_2}{\diagup}}\overset{\|}{OTf} + R^2CHO \xrightarrow[64-83\%]{CrCl_2} R^1\overset{\overset{CH_2}{\|}}{\underset{OH}{\diagup}}R^2$$

$$\overset{\diagup\diagup}{\underset{OTf}{\bigcirc}} + R^2CHO \xrightarrow[41-83\%]{} \overset{\diagup\diagup}{\bigcirc}\underset{OH}{R^2}$$

Homoallylic alcohols (8, 111–112). $CrCl_2$, prepared *in situ* by reduction of $CrCl_3$ (Strem) in THF with Na/Hg, is superior to $CrCl_2$ prepared by reduction of $CrCl_3$ with $LiAlH_4$ for the Cr(II)-mediated addition of crotyl halides to aldehydes. The homoallylic alcohols are formed in good yield and with high *anti*-selectivity.[7]
   Example:

$$n\text{-}C_6H_{11}CHO + CH_3\diagdown\diagup\diagdown Br \xrightarrow[77\%]{CrCl_2} C_6H_{11}\underset{\underset{CH_3}{\vdots}}{\overset{\overset{OH}{|}}{\diagup\diagdown}}\diagup\diagdown CH_2 + \quad syn\text{-isomer} \atop 7.4:1$$

[1] T. Okazoe, K. Takai, and K. Utimoto, *Am. Soc.*, **109**, 951 (1987).

[2] K. Takai, Y. Kataoka, T. Okazoe, and K. Utimoto, *Tetrahedron Letters*, **28**, 1443 (1987).

[3] S. Nakatsukasa, K. Takai, and K. Utimoto, *J. Org.*, **51**, 5045 (1986).

[4] Y. Okuda, S. Nakatsukasa, K. Oshima, and H. Nozaki, *Chem. Letters*, 481 (1985).

[5] S. E. Drewes and R. F. A. Hoole, *Syn. Comm.*, **15**, 1067 (1985).

[6] K. Takai, M. Tagashira, T. Kuroda, K. Oshima, K. Utimoto, and H. Nozaki, *Am. Soc.*, **108**, 6048 (1986).

[7] P. G. M. Wuts and G. R. Callen, *Syn. Comm.*, **16**, 1833 (1986).

Chromium(II) chloride–Haloform, $CrCl_2$–CHX_3.

E-*Alkenyl halides*.[1]   Iodoform in combination with $CrCl_2$ reacts with an aldehyde at 0° to form an (E)-alkenyl iodide in 75–90% yield. Replacement of CHI_3 by $CHCl_3$ and use of higher temperatures results in (E)-alkenyl chlorides. (E)-Alkenyl bromides can be obtained by use of $CHBr_3$ and $CrBr_2$. The E/Z ratio increases in the order I < Br < Cl, but the rate is in the order I > Br > Cl. Ketones undergo this reaction very slowly.

Examples:

[1] K. Takai, K. Nitta, and K. Utimoto, *Am. Soc.*, **108**, 7408 (1986).

Chromium(II) chloride–Nickel(II) chloride.

Coupling of vinyl iodides with aldehydes (**12**, 137).   Further study[1] of this 1,2-addition of alkenylchromium compounds to aldehydes to form allylic alcohols indicates that the reaction is applicable to α-alkoxy and α,β-bisalkoxy aldehydes by use of a solvent other than DMF, which can promote elimination to an enal. A wide number of other functional groups can also be accommodated. Both vinyl iodides and β-iodo enones can be used as precursors to the alkenylchromium reagent. The reaction is only modestly diastereoselective, but the stereochemistry of a disubstituted vinyl iodide is retained.

This reaction is very dependent on the source and particular batch of $CrCl_2$. This difficulty can be overcome by addition of a trace (0.1–1%) of $NiCl_2$ or $Pd(OAc)_2$ before the reaction.

$$(\alpha:\beta = 2:1)$$

This $CrCl_2/NiCl_2$-mediated coupling can also be used to obtain a propargylic alcohol with high diastereoselectivity (equation I). The aldehyde in this case under-

$$(8.3:1)$$

goes ready epimerization and dehydration, but this coupling proceeds in satisfactory yield.[2]

[1] H. Jin, J. Uenishi, W. J. Christ, and Y. Kishi, *Am. Soc.*, **108**, 5644 (1986).
[2] T. O. Aicher and Y. Kishi, *Tetrahedron Letters*, **28**, 3463 (1987).

Cobaloxime(I), (1), 11, 135–136.

   Radical cyclization of N-alkenylamino acid derivatives.[1]   Proline derivatives can be obtained by cyclization of N-alkenyl amino acid derivatives. Thus the β-iodo allylic amine **2**, prepared in 54% yield from threonine, cyclizes in the presence of cobaloxime(I) to *cis*- and *trans*-derivatives of (S)-proline, **3** and **4**. On deprotection and oxidation they afford the amino acids **5** and **6**, related to the natural kainic acids.

[1] J. E. Baldwin and C.-S. Li, *J.C.S. Chem. Comm.*, 166 (1987).

Cobalt(II) chloride.

Acylation catalyst.[1]   CoCl$_2$ in CH$_3$CN is recommended as a neutral catalyst for acetylation of β-hydroxy ketones, which can undergo elimination in the presence of DMAP–N(C$_2$H$_5$)$_3$. It is also effective for selective acylation of a primary or secondary hydroxyl group in the presence of a tertiary one.

1,2-Diones.[2]   CoCl$_2$ also catalyzes a reaction between acetic anhydride and aldehydes to provide 1,2-diones by coupling of acetyl and acyl radicals. The unsymmetrical diones are obtained as the major products by use of excess acetic anhydride.

(1) (CH$_3$CO)$_2$O + PrCHO $\xrightarrow[\text{CH}_3\text{CN}]{\text{CoCl}_2,}$ CH$_3$C–CPr + PrC—CPr + CH$_3$C—CCH$_3$

(excess)

(76–80%)   (7–8%)   (5–8%)

[1] S. Ahmad and J. Iqbal, *J.C.S. Chem. Comm.*, 114 (1987).
[2] *Idem, ibid.*, 692 (1987).

Cobalt(II) chloride–Triphenylphosphine–Sodium borohydride.

(E,E)-1,3-Dienes.[1]   A cobalt hydride prepared from these three reagents (1:4:1) in THF at −20° converts 1-alkynes into (E,E)-1,3-dienes:

2RC≡CH $\xrightarrow[65-85\%]{}$

[1] N. Satyanarayana and M. Periasamy, *Tetrahedron Letters*, **27**, 6253 (1986).

Copper(I) bromide.

Addition of RMgBr to nitriles.[1]   Grignard reagents react with nitriles slowly if at all, but even *t*-butylmagnesium chloride will add to nitriles in refluxing THF when catalyzed by a copper(I) salt. The adduct can be converted to a ketimine by anhydrous protonation, to a primary amine by reduction (Li/NH$_3$), or to a ketone by hydrolysis. The actual reagent may be a cuprate such as R$_3$Cu(MgX)$_2$.

$$C_6H_5C{\equiv}N \ + \ (CH_3)_3CMgCl \ \xrightarrow[99\%]{\underset{THF, \Delta}{CuBr,}} \ C_6H_5\overset{\overset{\displaystyle MgBr}{\underset{\displaystyle \|}{N}}}{C}C(CH_3)_3 \ \xrightarrow[95\%]{NH_3} \ C_6H_5\overset{\overset{\displaystyle NH}{\underset{\displaystyle \|}{}}}{C}C(CH_3)_3$$

$$99\% \ \Big\downarrow \ Li/NH_3$$

$$C_6H_5\overset{\overset{\displaystyle NH_2}{\underset{\displaystyle |}{}}}{C}HC(CH_3)_3$$

[1] F. J. Weiberth and S. S. Hall, *J. Org.*, **52**, 3901 (1987).

Copper(II) bromide.

Desilylbromination,[1] This reaction was first used by Fleming *et al.* (**8**, 196; **11**, 75) in connection with protection of enones, but it is also useful for synthesis of chiral 5-alkylcyclohexenones. Thus reaction of (R)-(−)-**1** with lithium dialkyl-cuprates gives the *trans*-adduct **2** as the only product. Of several bromination reagents, only CuBr$_2$ in DMF is useful for conversion of **2** to optically active **3**.

(R)-**1** + R$_2$CuLi $\xrightarrow{86-94\%}$ **2** (96–98% ee) $\xrightarrow[DMF, 60°]{CuBr_2,}$ **3**

82–92%

[1] M. Asaoka, K. Shima, and H. Takei, *J.C.S. Chem. Comm.*, 430 (1988).

Copper(II) chloride.

Carbodiimide peptide synthesis.[1] 1-Hydroxybenzotriazole (HOBt) is frequently used to suppress racemization in peptide syntheses using carbodiimides. Copper(II) chloride decreases racemization, but also depresses the yield markedly. By judicious use of both additives, racemization can be prevented with only slight effect on the yield.

[1] T. Miyazawa, T. Otomatsu, Y. Fukui, T. Yamada, and S. Kuwata, *J.C.S. Chem. Comm.*, 419 (1988).

Copper(I) chloride–Copper(II) chloride.

Indolizidines.[1] The N-chloro derivative (**1**) of a *cis*-2,6-dialkylsubstituted piperidine in the presence of this redox system undergoes homolytic cyclization with the alkenyl side chain at −45° to provide two isomeric indolizidines (**2** and **3**).

Dechlorination of the mixture provides the alkaloid gephyrotoxin-223AB (**4**) and an isomer (**5**) in the ratio 4.8:1.

[1] C. A. Broka and K. K. Eng, *J. Org.*, **51**, 5043 (1986).

Copper(I) trifluoromethanesulfonate, CuOTf.

Cyclopropanation of enones.[1]   Reaction of an enone with tris(phenylthio)-methyllithium and then with CuOTf (1 equiv.) results in a bis(phenylthio)-cyclopropane (equation I).

This activation of a C_6H_5S group as a leaving group can also effect vinylcyclo-propanation.

Example:

[1] T. Cohen and M. Myers, *J. Org.*, **53**, 457 (1988).

Copper(II) trifluoromethanesulfonate, $Cu(OTf)_2$.

Dehydration.[1] $Cu(OTf)_2$ (0.1 equiv.) is an effective catalyst for dehydration of primary, secondary, or tertiary alcohols and of diols at 25°. The alcohol can be used neat or as a suspension in decalin or heptane. Yields are generally higher than those obtained with H_2SO_4 or $POCl_3$/pyridine. Saytzeff (E)-alkenes are formed predominantly.

$Cu(OTf)_2$	92%	9:1
H_2SO_4	48%	18:1

[1] K. Laali, R. J. Gerzina, C. M. Flajnik, C. M. Geric, and A. M. Dombroski, *Helv.*, **70**, 607 (1987).

Crotyltrichlorosilane, $CH_3CH{=}CHCH_2SiCl_3$ (1).

Stereoselective reaction with aldehydes.[1] Both (E)- and (Z)-**1** react with aldehydes to give *syn*-adducts as the major products (**12**, 146). In an effort to improve the diastereoselectivity, (E)- and (Z)-**1** have been converted into pentacoordinated allylsilicates (**2**) by reaction with dilithium catecholate. Reaction of an aryl aldehyde with both (E)- and (Z)-**2** shows essentially complete diastereoselectivity.

E/Z = 88:12	82%	12:88	
E/Z = 21:79	91%	78:22	

[1] M. Kira, K. Sato, and H. Sakurai, *Am. Soc.*, **110**, 4599 (1988).

(Z)- and (E)-Crotyldiisopinocampheylboranes, $Ipc_2BCH_2CH\!=\!CHCH_3$ (1). Brown and Bhat[1] have prepared the two (Z)-crotyldiisopinocampheylboranes [derived from (−)- and (+)-pinene] by reaction of (Z)-crotylpotassium with the methoxy-diisopinocampheylboranes. The (E)-crotyldiisopinocampheylboranes are prepared in the same way from (E)-crotylpotassium.

The two (Z)-crotylboranes react with typical aldehydes to give *syn*-β-methyl-homoallyl alcohols with high diastereo- and enantioselectivity (equation I), whereas the (E)-crotylboranes provide a useful route to the *anti*-homoallylic alcohols (equation II). Thus by proper choice of the crotylborane, any one of the four possible

(I)

from (+)-α-pinene   75%
from (−)-α-pinene   72%

95:5
4:96

(II)

from (+)-α-pinene   78%   95:5
from (−)-α-pinene   76%   4:96

homoallylic alcohols can be prepared with high diastereo- and enantioselectivity from an aldehyde.

Addition to chiral aldehydes. Brown et al.[2] have examined the reaction of the chiral aldehyde 2 with the (E)- and (Z)-isomers of (+)- and (−)-1. The stereochemistry at the newly formed bond between C_4 and C_5 is controlled by the chirality of 1. The reaction affords two of the possible diastereomeric products in 64% de. Similar results have been reported by Roush et al.[3] for reaction of a β-alkoxy-α-methyl aldehyde.

+ (−)-(E)-1   75%
+ (+)-(E)-1   70%
96:4
9:91

(S)-2

+ (−)-(Z)-1   73%
+ (+)-(Z)-1   79%
82:18
4:96

[1] H. C. Brown and K. S. Bhat, *Am. Soc.*, **108**, 293, 5919 (1986).
[2] H. C. Brown, K. S. Bhat, and R. S. Randad, *J. Org.*, **52**, 3701 (1987).
[3] W. R. Roush, A. D. Palkowitz, and M. A. J. Palmer, *ibid.*, **52**, 316 (1987).

Crotyl-*trans*-2,5-dimethylborolanes (2).
(E)- and (Z)-Crotyl-*trans*-2,5-dimethylborolanes (**2**) are prepared by reaction of B-methoxy-(R,R)-2,5-dimethylborolane (**1**) with (E)- and (Z)-crotylpotassium cat-

alyzed by BF₃ etherate. (E)- and (Z)-Crotyl-(S,S)-2,5-dimethylborolanes can also be prepared in the same way from (S,S)-1.

Asymmetric crotylboration.[1]   The reaction of an achiral, unhindered aldehyde with (E)-(R,R)-**2** gives *anti*- and *syn*-homoallylic alcohols in the ratio of about 20:1;

the major *anti*-product is obtained in 95–97% ee. The reaction with (Z)-**2** results in *syn*- and *anti*-homoallylic alcohols also in the ratio of about 20:1, and the major product is formed in 86–97% ee, with the highest enantioselectivity obtained with hindered aldehydes.

The diastereofacial selectivity of (E)- and (Z)-(R,R)-**2** with the chiral aldehyde (R)-2,3-O-isopropylideneglyceraldehyde (**3**) is approximately 35:1 and 50:1, re-

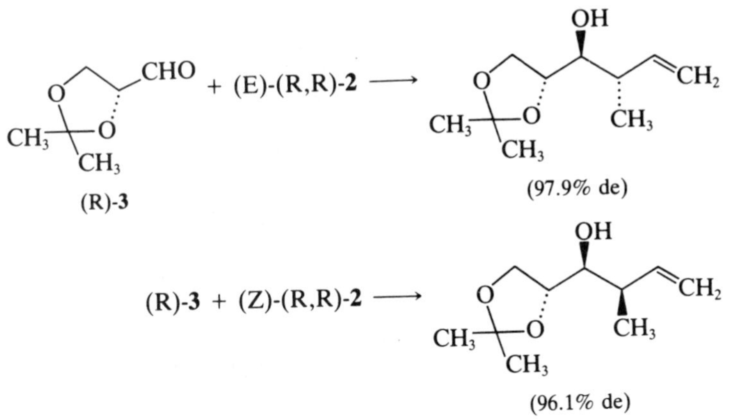

(R)-**3** + (E)-(R,R)-**2** ⟶

(97.9% de)

(R)-**3** + (Z)-(R,R)-**2** ⟶

(96.1% de)

spectively. These diastereoselectivities are among the highest reported for crotylboranes. The diastereoselectivity is markedly lower in reactions of (R)-**3** with both (E)- and (Z)-(S,S)-**2** because of mismatching.

[1] J. Garcia, B. M. Kim, and S. Masamune, *J. Org.*, **52**, 4831 (1987).

Crotyltrimethylsilane (1).

Reaction with glycals.[1]   (E)-**1** reacts with glycals in the presence of BF_3 etherate to give pyrans substituted at C_2 with a 3-methylpropenyl group that is *trans* to a substituent at C_6.

Example:

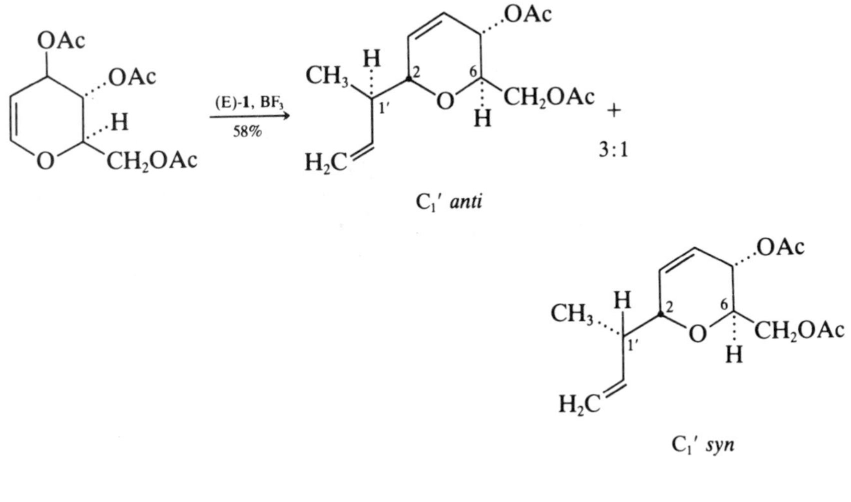

$\xrightarrow[58\%]{\text{(E)-1, BF}_3}$

C_1' *anti*

3:1

C_1' *syn*

A refined version of this reaction was used to obtain optically active **3** from glycal **2**. This product is readily converted to the tetrahydropyran **4**, a subunit of the antibiotic indanomycin.

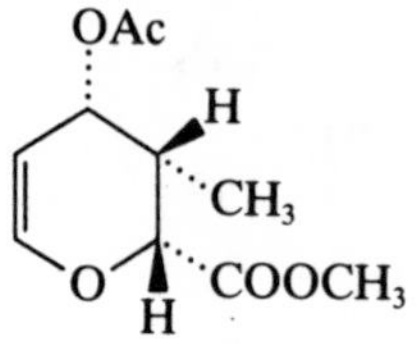

2

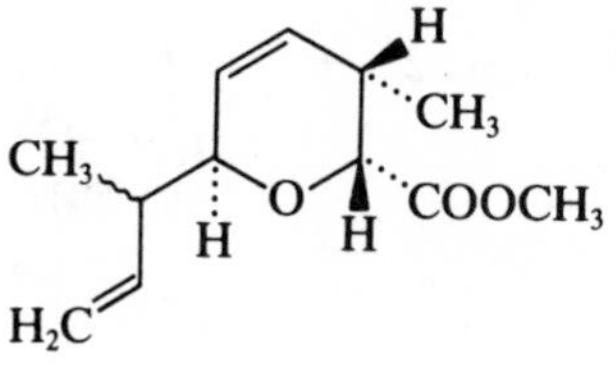

3

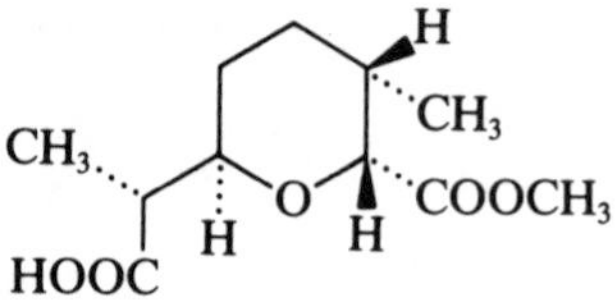

4

[1] S. J. Danishefsky, S. DeNinno, and P. Lartey, *Am. Soc.*, **109**, 2082 (1987).

Cyanomethylenetriphenylphosphorane.

Nitriles.[1] Cyanomethylenetriphenylphosphorane (**1**) can be deprotonated with sodium bis(trimethylsilyl)amide to give sodium cyanotriphenylphosphoran-ylidenemethanide (**2**). This ylide anion can be alkylated to give a wide variety of

$$(C_6H_5)_3P{=}CHCN \xrightarrow{\text{NaN[Si(CH_3)_3]_2}} Na^+[(C_6H_5)_3\overset{+}{P}\text{-}\overset{-}{C}CN]$$

1   2

$$\downarrow RX$$

$$\overset{\displaystyle R}{\underset{\displaystyle 3}{\overset{|}{(C_6H_5)_3\overset{+}{P}\text{-}\overset{-}{C}CN}}}$$

cyanomethylenetriphenylphosphoranes (**3**), which react as expected with aldehydes to give α,β-unsaturated nitriles. The ylide anion **2** is also useful for synthesis of

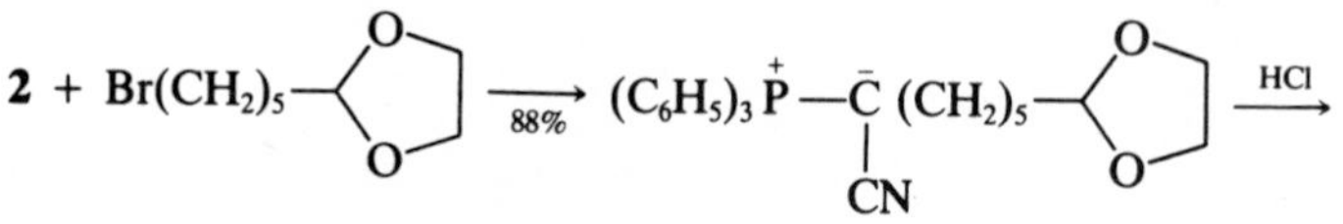

$$\left[\begin{array}{c} (C_6H_5)_3\overset{+}{P} \ \underset{\underset{CN}{|}}{\overset{\overset{H}{|}}{-C}}(CH_2)_5-CHO \\ Cl^- \end{array} \right] \xrightarrow[52\%]{} $$

4

cyclic unsaturated nitriles such as **4**. Reaction of **2** with carboxylic esters provides a synthesis of acetylenic nitriles (equation I).

(I) $2 + RCOOR^1 \longrightarrow (C_6H_5)_3\overset{+}{P}-\underset{\underset{CN}{|}}{\bar{C}}-COR \xrightarrow[65-78\%]{\Delta} RC{\equiv}CCN + (C_6H_5)_3P{=}O$

[1] H. J. Bestmann and M. Schmidt, *Angew. Chem. Int. Ed.*, **26**, 79 (1987).

Cyanotrimethylsilane, $(CH_3)_3SiCN$.

Preparation:[1]

$$2(CH_3)_3SiCl + H_2SO_4 \xrightarrow[-2HCl]{} [(CH_3)_3SiO]_2SO_2 \xrightarrow[65-83\%]{KCN} 2(CH_3)_3SiCN$$

Substitution of β-nitro sulfides.[2] β-Nitro sulfides in the presence of a Lewis acid undergo displacement reactions with either allyl- or cyanotrimethylsilane. The reaction is considered to involve an episulfonium intermediate, which is then substituted at the more positive carbon (equation I). The reaction proceeds with retention of configuration, and *anti* β-nitro sulfides react much more rapidly than

(I) $(CH_3)_2\underset{\underset{SC_6H_5}{|}}{C}-CH_2NO_2 \xrightarrow[64\%]{\overset{NCSi(CH_3)_3}{SnCl_4}} (CH_3)_2\underset{\underset{CN}{|}}{C}-CH_2SC_6H_5 \quad + \quad (CH_3)_2\underset{\underset{SC_6H_5}{|}}{C}-CH_2CN$
$ 73{:}27$

the *syn*-isomer. This reaction can be used to effect cyclization of β-nitro sulfides with an aromatic ring (equation II).

(II)

(*trans*, 100%)

[1] J.-P. Picard, R. Calas, and J. Dunoques, *Org. Syn.*, submitted (1988).
[2] N. Ono, A. Kamimura, H. Sasatani, and A. Kaji, *J. Org.*, **52**, 4133 (1987).

Cyclopropanone ethyl hemiacetal (1). Preparation.[1]

Tricyclic heptanones.[2]   The reaction of 1 with CH_3MgBr and then with lithium cyclohexenolate results in 2 and the tricyclic product 3. The minor product 2 can be converted into 3 in high yield by further reaction with CH_3MgBr and lithium

cyclohexenolate. The formation of 3 is believed to involve generation of cyclopropanone, which undergoes two consecutive reactions with the enolate.

[1] J. Salaün and J. Marguerite, *Org. Syn.*, **63**, 147 (1985).
[2] J. T. Carey, C. Knors, and P. Helquist, *Am. Soc.*, **108**, 8313 (1986).

D

1,8-Diazabicyclo[5.4.0]undecene-7 (DBU).

1,3-Elimination.[1] The dehydroiodination of γ-iodo ketones or γ-iodo esters with DBU at 25° to form cyclopropanes occurs readily only when the proton and the iodo group can adopt a W-shaped geometry. Thus the cyclopropane **1** is formed

readily from **2a** in which the hydrogen and iodine have the *syn*-orientation, but is formed slowly and in lower yield from the *anti*-isomer **2b**. The cyclopropane **1** is also formed from **3**, but only with a strong base; reaction of **3** with DBU results in a 1,2-elimination.

[1] M. Mori, N. Kanda, Y. Ban, and K. Aoe, *J.C.S. Chem. Comm.*, 12 (1988).

Diazomethane.

Insertion into Si—I bond.[1] Reaction of trimethylsilyl iodide or triflate with diazomethane results in insertion of a methylene group into the Si—I or Si—OTf bond (equation I). Similar insertion into $(CH_3)_3SiBr$ requires $ZnBr_2$ catalysis.

ether
(I) $(CH_3)_3SiX$ + CH_2N_2 $\xrightarrow[65\%]{-78\rightarrow -10°}$ $(CH_3)_3SiCH_2X$ + N_2

X = I   65%
X = OTf   81%

[1] J. G. Lee and D. S. Ha, *Synthesis*, 318 (1988).

Dibenzyl peroxydicarbonate, $(C_6H_5CH_2OCO_2)_2$ **(1).** The reagent is obtained by reaction of benzyl chloroformate with H_2O_2 in a basic medium. It is stable to water and is not readily detonated.

α-Hydroxy ketones.[1] The reagent oxidizes enolates of ketones to α-hydroxy ketones, isolated as the carbonates in moderate to good yield. It is comparable to 2-(phenylsulfonyl)-3-phenyloxaziridine for asymmetric oxidation of oxazolidinone carboximides.

Example:

[1] M. P. Gore and J. C. Vederas, *J. Org.*, **51**, 3700 (1986).

Diborane.

Chiral oxazaborolidines.[1] Enantioselective reduction of ketones with a reagent prepared from BH_3 and the chiral *vic*-amino alcohol **1** (**12**, 31) is now known to involve an oxazaborolidine. Thus BH_3 and (S)-**1**, derived from valine, react rapidly in THF to form **2**, m.p. 105–110°, which can serve as an efficient catalyst

for reductions with BH_3. Thus BH_3 (1.2 equiv.) in combination with **2** (0.025 equiv.) can reduce a typical ketone to the alcohol in 95% ee, presumably via **3**.

A more efficient bicyclic oxazaborolidine (**5**) is obtained by reaction of BH_3 with (S)-diphenylprolinol (**4**). The combination of **5** (0.05 equiv.) and BH_3 (0.6 equiv.)

$$(S)\text{-}4 \xrightarrow{BH_3} 5 \xrightarrow{BH_3} 6$$

(R)-7
(88–97% ee)

effects rapid reduction of a variety of ketones in high yield and >90% enantiose-lectivity. The enhanced reactivity of **5** is associated with the strain of a double bond at a bridge head, which is relieved on formation of the complex **6**. The enantio-selectivity of reduction can be explained on the basis of a six-membered cyclic transition state (as formulated in A).

[1] E. J. Corey, R. K. Bakshi, and S. Shibata, *Am. Soc.*, **109**, 5551 (1987).

1,1-Dibromoalkanes–Zinc–Titanium(IV) chloride–Tetramethylethylenediamine.

Alkylidenation. These four reagents in the approximate ratio of $RCHBr_2$/ $Zn/TiCl_4/TMEDA = 1:4:2:4$ react in THF to form a metal carbene complex that reacts with the carbonyl group of esters to form alkenyl ethers with (Z)-selectivity, which increases as the size of the R group of the reagent decreases. Alkylidenation of ketones with these reagents shows low Z/E selectivity. Although the reagent prepared from CH_2Br_2–Zn–$TiCl_4$ is useful for methylenation of ketones (**12**, 339), it reacts in low yield with esters. These complexes also react with lactones, but γ-hydroxy ketones are formed as well.[1]

Examples:

$$C_6H_5\overset{O}{\overset{\|}{C}}OCH(CH_3)_2 + CH_3CHBr_2 \xrightarrow[\substack{TMEDA \\ 88\%}]{Zn, TiCl_4}$$

(Z/E = 92:8)

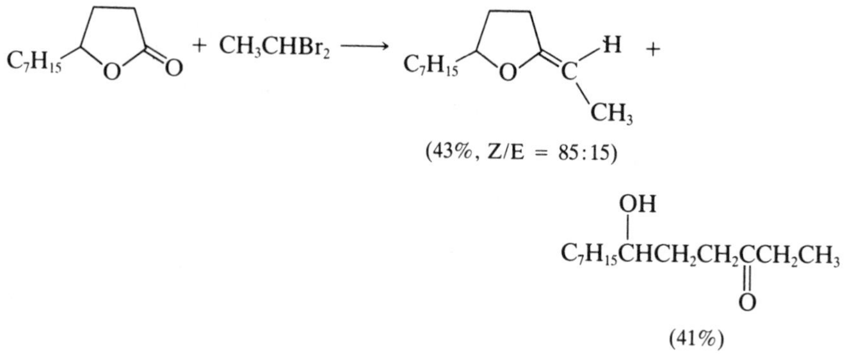

(43%, Z/E = 85:15)

$$\underset{\underset{O}{\|}}{C_7H_{15}\overset{OH}{\underset{|}{C}}HCH_2CH_2CCH_2CH_3}$$

(41%)

Alkylidenation of silyl esters.[2] Application of this alkylidenation process to silyl esters provides (Z-)silyl enol ethers stereoselectively.

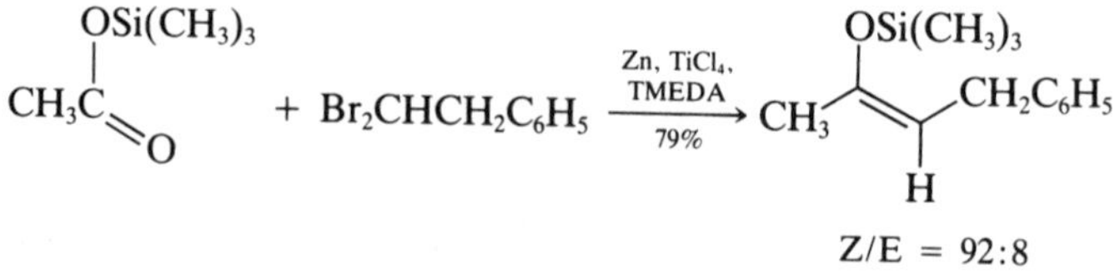

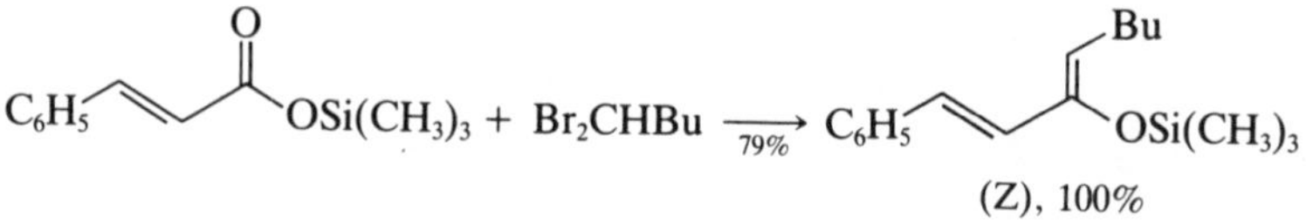

Z/E = 92:8

(Z), 100%

[1] T. Okazoe, K. Takai, K. Oshima, and K. Utimoto, *J. Org.*, **52**, 4410 (1987); K. Takai, Y. Kataoka, T. Okazoe, K. Ushihma, and K. Utimoto, *Org. Syn.* submitted (1988).
[2] K. Takai, Y. Kataoka, T. Okazoe, and K. Utimoto, *Tetrahedron Letters*, **29**, 1065 (1988).

Dibromomethyllithium,LiCHBr₂, **dichloromethyllithium.** Dibromomethyllithium can be generated from CH_2Br_2 with **LDA** as base rather than BuLi, used for generation of dichloromethyllithium from CH_2Cl_2.

Diastereoselective homologation of chiral alkylboronates (*cf.* **12**, 80–81). Investigations of this reaction have been carried out mainly on boronic esters (**1**)

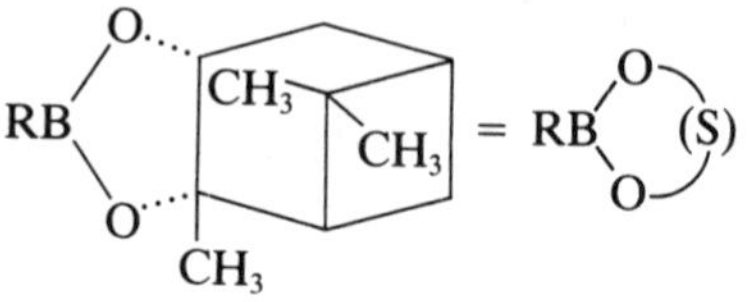

1

derived from (+)-pinanediol (from α-pinene) and formulated for convenience as indicated.[1] Reaction of **1** with LiCHCl$_2$ at $-100°$ gives a borate complex (**2**) that is rearranged by ZnCl$_2$ (1–1.5 equiv.) to a (1S)-1-chloroalkylboronic ester (**3**) with 98–99% diastereoselectivity. Nucleophilic displacements (RMgX, ROLi, RLi) on **3** yield a new chiral boronic ester (**4**) which can be further homologated. Application to the synthesis of a chiral *vic*-diol (**5**) is shown in equation (I). The configuration of the new chiral centers is determined by the pinanediol and by the order in which the substituent groups are introduced; but the unique feature of boronic ester homologation is the possibility of unlimited repetition without removal of the chiral

directing group. And indeed (L)-(+)-ribose has been synthesized from **1** (R = BzlOCH$_2$) in 13% overall yield by homologations with dibromomethyllithium and displacement by benzyl oxide.[2] Introduction of the first four carbons is efficient, but introduction of the fifth carbon is difficult because of steric hindrance.

This methodology provides a general synthesis of L-amino acids in 92–96% ee and in chemical yields of about 40–60%.[3] Thus reaction of **3** (X = Br) with NaN$_3$ under phase-transfer conditions provides **6**, which is homologated to the 1-chloro-2-azidoboronate **7**. This product is oxidized by sodium chlorite directly to an azido carboxylic acid (**8**). Hydrogenation of **8** provides L-amino acids (**9**).

$$3 \xrightarrow{\text{NaN}_3} \quad -R\blacktriangleright\overset{\displaystyle H}{\underset{\displaystyle N_3}{C}}\blacktriangleleft B\underset{O}{\overset{O}{\diagup}}(S) \xrightarrow{\text{LiCHCl}_2} \quad R\blacktriangleright\overset{\displaystyle H}{\underset{\displaystyle N_3}{C}}-\overset{\displaystyle Cl}{\underset{\displaystyle H}{C}}\blacktriangleleft B\underset{O}{\overset{O}{\diagup}}(S)$$

$$\qquad\qquad\qquad\qquad 6 \qquad\qquad\qquad\qquad\qquad\qquad 7$$

$$\downarrow \text{NaClO}_2$$

$$R\blacktriangleright\overset{\displaystyle H}{\underset{\displaystyle N_3}{C}}\blacktriangleleft COOH \xrightarrow{\text{H}_2} R\blacktriangleright\overset{\displaystyle H}{\underset{\displaystyle NH_2}{C}}\blacktriangleleft COOH$$

$$\qquad\qquad 8 \qquad\qquad\qquad\qquad 9\ (25\text{--}50\!:\!1)$$

Homologation of RCHO to α-bromo-α,β-enones.[4] The initial steps of this homologation involve addition of dibromomethyllithium (**1**) to an aldehyde followed by oxidation of the adduct to a dibromomethyl ketone (**2**). The aluminum enolate of **2** undergoes an aldol reaction to provide an α-bromo-β-hydroxy ketone

$$R^1CHO + 1 \longrightarrow \left[\overset{\displaystyle OH}{\underset{\displaystyle }{R^1CHCHBr_2}}\right] \xrightarrow[45\text{--}55\%]{PCC} \overset{\displaystyle O}{\overset{\displaystyle \|}{R^1CCHBr_2}} \xrightarrow{\substack{1)\ \text{Zn, CuBr, } (C_2H_5)_2AlCl \\ 2)\ R^2CHO}}$$

$$\qquad\qquad\qquad\qquad\qquad a \qquad\qquad\qquad\qquad\qquad 2$$

$$\left[\overset{\displaystyle OH}{R^2}\overset{\displaystyle }{\underset{\displaystyle Br}{}}\overset{\displaystyle O}{\underset{\displaystyle }{\|}}R^1\right] \xrightarrow[50\text{--}65\%]{\substack{CH_3SO_2Cl \\ N(C_2H_5)_3}} R^2\overset{\displaystyle O}{\overset{\displaystyle \|}{\underset{\displaystyle Br}{\diagdown}}}R^1 \xrightarrow{\substack{1)\ H^- \\ 2)\ -HBr}} R^2C\equiv C\overset{\displaystyle OH}{\underset{\displaystyle }{CHR^1}}$$

$$\qquad\qquad b \qquad\qquad\qquad\qquad\qquad 3 \qquad\qquad\qquad\qquad 4$$

(**b**), which on dehydration via the mesylate provides an α-bromo-α,β-enone (**3**). Hydride reduction of **3** followed by dehydrobromination provides an acetylenic alcohol (**4**). The intermediate dibromomethyl ketone (**2**) can also be prepared directly by reaction of a methyl ester with **1** (68–78% yield).

[1] D. S. Matteson, K. M. Sadhu, and M. L. Peterson, *Am. Soc.*, **108**, 810 (1986).
[2] D. S. Matteson and M. L. Peterson, *J. Org.*, **52**, 5116 (1987).
[3] D. S. Matteson and E. C. Beedle, *Tetrahedron Letters*, **28**, 4499 (1987).
[4] A. Takahashi and M. Shibasaki, *J. Org.*, **53**, 1227 (1988).

(1S,2R)-(−)-2-(N,N-Dibutylamino)-1-phenylpropanol-1 (1).

$$
\begin{array}{c}
\text{H} \\
\text{CH}_3 \diagdown \vdots \diagup \text{NBu}_2 \\
| \\
\text{C}_6\text{H}_5 \diagup \vdots \diagdown \text{OH} \\
\text{H}
\end{array}
$$

(1S, 2R)-**1**, α_D − 15.5°

This α-amino alcohol is obtained by dibutylation of (1S,2R)-norephedrine.

Enantioselective addition of $(C_2H_5)_2Zn$ *to aldehydes.*[1]   Addition of diethylzinc to either aromatic or aliphatic aldehydes catalyzed by **1** (6 mole %) results in (S)-secondary alcohols in generally 90–95% ee. Although several chiral amino alcohols are known to effect enantioselective addition of R_2Zn to aromatic aldehydes, this one is the first catalyst to be effective for aliphatic aldehydes. The dibutylamino group of **1** is essential for the high enantioselectivity; the dimethylamino analog of **1**, (1S,2R)-N-methylephedrine, effects this addition in only about 60% ee.

[1] K. Soai, S. Yokoyama, K. Ebihara, and T. Hayasaka, *J.C.S. Chem. Comm.*, 1690 (1987).

Di-*t*-butyl azodicarboxylate, BocN=NBoc (**1**). Preparation.[1] Supplier: Fluka.

Amination.   Three laboratories[2–4] have reported use of esters of azodicarboxylic acid for amination of chiral substrates to provide a synthesis of optically active α-hydrazino and α-amino acids. The di-*t*-butyl ester is particularly useful because the diastereoselectivity improves with increasing size of the ester group, and in addition these esters are hydrolyzed by TFA at 25°. Two laboratories[2,3] used the lithium enolates of chiral N-acyloxazolidones (**2**) as the chiral precursors. A typical procedure is outlined in equation (I). Thus reaction of the lithium enolate of **2**

(I)

with **1** proceeds with high diastereoselectivity to give **3**. Chiral α-hydrazino acids (**4**) are obtained by transesterification, acid hydrolysis, and catalytic hydrogenation.

These are converted into chiral α-amino acids (5) on hydrogenation with Raney nickel.

The third group[4] used (1R,2S)-N-methylephedrine as the chiral auxiliary. Thus the derived silylketene acetal (6) reacts with 1 in the presence of TiCl₄ to give 7 in 45–70% yield with ~90% stereoselectivity. The products are converted by TFA and LiOH to (R)-α-hydrazino acids (8), which are obtained in ≥98% ee after one crystallization.

[1] W. J. Paleveda, F. W. Holly, and D. F. Veber, Org. Syn., 63, 171 (1984).
[2] D. A. Evans, T. C. Britton, R. L. Dorow, and J. F. Dellaria, Am. Soc., 108, 6395 (1986).
[3] L. A. Trimble and J. C. Vederas, ibid., 108, 6397 (1986).
[4] C. Gennari, L. Colombo, and G. Bertolini, ibid., 108, 6394 (1986).

Dicarbonylcyclopentadienylcobalt.

Dihydroindoles.[1] A novel synthesis of fused dihydroindoles involves [2 + 2 + 2]cycloaddition of alkynes with the 2,3-double bond of N-alkynoylated pyrroles. The reaction of 1 with bis(trimethylsilyl)ethyne results in two diaster-

eomeric cobalt diene complexes (3), which are converted to the same free diene on oxidaton with CAN (1 equiv.) or to the aromatized indole by excess CAN.

[1] G. S. Sheppard and K. P. C. Vollhardt, J. Org., 51, 5496 (1986).

Di-μ-carbonylhexacarbonyldicobalt, $Co_2(CO)_8$.

Medium-size cycloalkynes.[1] In the presence of BF_3 etherate, a cobalt complexed propargylic ether can undergo an intramolecular alkylation with an allylic silane to provide six-, seven-, and eight-membered complexed cycloalkynes. This reaction is an extension of propargylation of allylsilanes to provide 1,5-enynes (**10**, 129–130).

1, n = 1,2   55%
      n = 3   67%

Oxidative removal of the $Co_2(CO)_6$ ligand of **2** does not result in a cycloalkyne, but **2** can be used as such for annelation to complex products.

This reaction can also be used for exocyclic intramolecular alkylation of an allylic silane, such as the cyclization of **3** to *trans*-**4** with complete stereocontrol.

Medium-ring acetylenic lactones.[2] Cyclization to medium size acetylenic lactones is difficult because of geometric constraint imposed by the triple bond, and has been considered practicable only for at least fifteen-membered rings. Surprisingly, the acetylenic ω-hydroxy acid **1** when complexed with $Co_2(CO)_6$ is cyclized in 28% yield to the seven-membered complexed acetylenic lactone **2** by Mukaiyama's reagent, 2-chloro-1-methylpyridinium chloride (**8**, 95–96). The yield is essentially the same as that observed in lactonizaton to a complexed 10-membered acetylenic lactone. Evidently a cobalt–alkyne group has less geometric constraints than an alkyne group.

The same paper reports another route to 11-membered acetylenic lactones via an intramolecular retro-Dieckmann reaction (equation I). Thus treatment of **3** with

(I)

NaH (1 equiv.) at 25° results in the complexed lactone **4** in 71% yield. The uncomplexed lactone is a stable, isolable compound.

Pauson–Khand cyclopentenone synthesis.[3] The cycloaddition of an alkene with an alkyne complexed with $Co_2(CO)_8$ usually furnishes a mixture of two cyclopentenones when the alkene is unsymmetrical. The regioselectivity can be improved markedly if the alkene bears a heteroatom that can coordinate with the cobalt complex. Both sulfur and nitrogen ligands can improve the yield and regiocontrol of this reaction.

Examples:

[1] S. L. Schreiber, T. Sammakia, and W. E. Crowe, *Am. Soc.*, **108**, 3128 (1986).
[2] N. E. Schore and S. D. Najdi, *J. Org.*, **52**, 5296 (1987).
[3] M. E. Krafft, *Am. Soc.*, **110**, 968 (1988).

Dichloroalane, Cl_2AlH.

Hydroalumination. Monosubstituted alkenes undergo hydroalumination with dichloroalane in the presence of $(C_2H_5)_3B$ or $C_6H_5B(OH)_2$. Hydroalumination of more highly substituted alkenes proceeds reluctantly. The reaction is regioselective, and the products react with a variety of electrophiles under mild conditions. Example:

$$C_{10}H_{21}CH{=}CH_2 + Cl_2AlH \xrightarrow{\text{R}_3\text{B}} [C_{10}H_{21}CH_2CH_2AlCl_2] \xrightarrow[91\%]{\text{O}_2} C_{12}H_{25}OH$$

$$71\% \Big\downarrow (CH_3)_2CHCOCl$$

$$C_{12}H_{25}COCH(CH_3)_2$$

[1] K. Maruoka, H. Sano, K. Shinoda, S. Nakai, and H. Yamamoto, *Am. Soc.*, **108**, 6036 (1986).

Dichloroaluminum phenoxide, $Cl_2AlOC_6H_5$ (1).

This aluminum reagent is prepared by reaction of CH_3AlCl_2 with phenol in CH_2Cl_2.

Ring enlargement.[1] Treatment of an α-(silylmethyl)cycloalkanecarbaldehyde (2) with a Lewis acid results in a one-carbon ring enlargement to a cycloalkanone

	3	4
2 a, n = 3	100%	—
b, n = 9	85%	—

(3) and small amounts of the 2-methylenealkanol 4. Dichloroaluminum phenoxide is preferred over CH_3AlCl_2 for selective rearrangement of 2 to 3. Ring enlargement of 2 to the methyl ether (5) of 4 is best effected with methoxytrimethylsilane (2 equiv.) and trimethylsilyl triflate (1 equiv.). The dimethyl acetals (6) of 2 also

6, n = 3 − 9   5

rearrange to **5** on treatment with zinc bromide in refluxing CH_2Cl_2. These rearrangements all involve stabilization of a cationic center at the position β to the silicon atom.

[1] K. Tanino, T. Katoh, and I. Kuwajima, *Tetrahedron Letters*, **29**, 1815, 1819 (1988).

Dichlorobis(cyclopentadienyl)titanium.
Hydromagnesiation (**11**, 163–164; **12**, 168–169). Cp_2TiCl_2-catalyzed *syn*-addition of isobutylmagnesium chloride to propargylic alcohols affords a stereoselective route to vinylmagnesium chlorides for further synthetic use.[1]

Example:

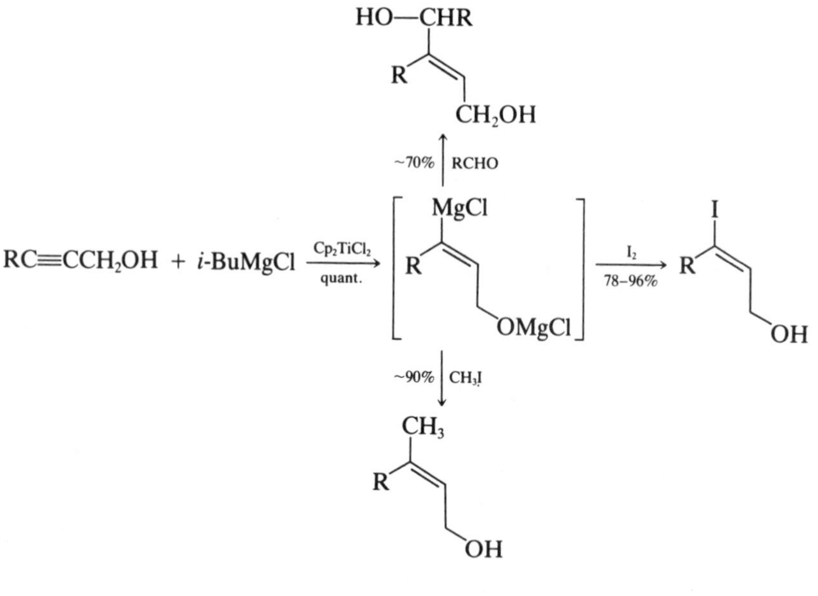

(E,E)-*Exocyclic dienes* (**12**, 170).   Reaction of 2,8-decadiyne with a reagent prepared from Cp_2TiCl_2, $CH_3P(C_6H_5)_2$, and sodium amalgam results in a bicyclic titanacyclopentadiene (**a**), which is hydrolyzed by dilute acid to 1,2-(E,E)-bis(ethylidene)cyclohexane (**2**). This bicyclization reaction can also be used to convert enynes to cycloalkanes with one exocyclic double bond, but fails when the diyne has bulky terminal groups such as $Si(CH_3)_3$.[2] For such hindered diynes a

$$CH_3C{\equiv}C(CH_2)_nC{\equiv}CCH_3 \xrightarrow[\text{Na/Hg, THF}]{\text{Cp}_2\text{TiCl}_2,\ \text{CH}_3\text{P(C}_6\text{H}_5)_2}$$

1 (n = 4)

a

76–80% | H_3O^+

2

zirconium-mediated reaction[3] can be used (equation I).

(I) $(CH_3)_3SiC{\equiv}C(CH_2)_4C{\equiv}CSi(CH_3)_3 \xrightarrow[82\%]{\substack{\text{Cl}_2\text{ZrCp}_2, \\ \text{Mg, HgCl}_2}}$

Bicyclization of **1** (n = 3 or 5) by this method is also successful, but yields are lower (60%, 27%). The reaction fails when the acetylene groups are separated by six methylene groups.

[1] F. Sato and Y. Kobayashi, *Org. Syn.*, submitted (1988).
[2] W. A. Nugent, D. L. Thorn, and R. L. Harlow, *Am. Soc.*, **109**, 2788 (1987); W. A. Nugent, *Org. Syn.*, submitted (1988).
[3] E. Negishi, S. J. Holmes, J. M. Tour, and J. A. Miller, *Am. Soc.*, **107**, 2568 (1985).

Dichlorobis(cyclopentadienyl)titanium–*sec*-Butylmagnesium chloride, Cp_2TiCl_2-*sec*-BuMgCl (**1**).

syn-**Selective pinacol reduction.**[1] Reduction of Cp_2TiCl_2 with this Grignard reagent provides a Ti(III)–Mg(II) complex (**1**) that reduces aromatic aldehydes to pinacols with high *syn*-selectivity (60–100:1).

$$2 \; ArCHO \xrightarrow[80-95\%]{1} \underset{\underset{OH}{}}{\overset{\overset{OH}{}}{Ar}} \overset{Ar}{\underset{}{}} + \underset{\underset{OH}{}}{Ar} \overset{\overset{OH}{}}{\underset{}{}} Ar$$
$$60-100:1$$

[1] Y. Handa and J. Inanaga, *Tetrahedron Letters*, **28**, 5717 (1987).

Dichlorobis(cyclopentadienyl)zirconium.

Stereoselective aldol condensation.[1] The stereoselectivity of the reaction of **1** with the ester **2** can be controlled by the choice of the metal enolate. The products are intermediates to 1-methylcarbapenems.

LDA, ClSi(CH₃)₃	87%
LDA, Cp₂ZrCl₂	52%

	95:5
	3:97

[1] C. U. Kim, B. Luh, and R. A. Partyka, *Tetrahedron Letters*, **28**, 507 (1987).

Dichlorobis(cyclopentadienyl)zirconium–Butyllithium.

Bicyclocyclization of enynes.[1] This combination, in the ratio 1:2, results in a reagent tentatively formulated as "ZrCp₂" (**1**), used to promote bicyclization of unsaturated trimethylsilylalkynes (equation I). This reaction fails with terminal enynes, as does replacement of the silyl group of **2** with various electrophiles.

[1] E. Negishi, S. J. Holmes, J. M. Tour, and J. A. Miller, *Am. Soc.*, **107**, 2568 (1985).

Di-μ-chlorobis[1,2-bis(dicyclohexylphosphine)ethane]dirhodium (1).

1, m.p. > 210° dec.

The Rh(I) complex is prepared by reaction of $[(COD)_2RhCl]_2$ with 1,2-bis(dicyclohexylphosphine)ethane (Strem).

Alkylidene lactones.[1]   In the presence of this catalyst, alkynoic acids (**2**) can cyclize to exocyclic enol lactones (**3**) with marked preference for the Z-isomer. The reaction in CH_2Cl_2 proceeds at room temperature. Cyclization to five-mem-

2, n = 1, 2   **3**
R = H, CH_3

bered lactones is strongly favored over cyclization to six-membered lactones, as are cyclizations catalyzed by Lewis acids [HgO, $Hg(TFA)_2$].
    Example:

$$RC{\equiv}C(CH_2)_2COOH \xrightarrow{1, CH_2Cl_2}$$

R = H	93%	—
R = CH_3	76% (Z only)	15%

$$HC{\equiv}C(CH_2)_3COOH \xrightarrow[85\%]{}$$

[1] D. M. T. Chan, T. B. Marder, D. Milstein, and N. J. Taylor, *Am. Soc.*, **109**, 6385 (1987).

Dichlorobis(1,4-diphenylphosphinobutane)palladium(II), $Cl_2Pd(dppb)_2$.
Coupling of vinyl dichlorides.[1]   This catalyst effects selective coupling of Grignard reagents or organozinc chlorides with only one of the chlorine atoms of 1,1-dichloro-1-alkenes (**2**) to give (Z)-vinyl chlorides (**3**). The selective coupling involves the chlorine that is *trans* to the R substituent, probably because of a steric

$$
\underset{\substack{\textbf{2}\ (R\ =\ Cl,\\ CH_3,\ Ar)}}{\overset{H}{\underset{R}{>}}C=C\overset{Cl}{\underset{Cl}{<}}}\ +\ C_6H_5MgBr\ \xrightarrow[80-95\%]{\textbf{1}}\ \underset{\textbf{3}}{\overset{H}{\underset{R}{>}}C=C\overset{C_6H_5}{\underset{Cl}{<}}}
$$

effect. The products can undergo a second coupling with Grignard reagents in the presence of the conventional catalyst $Cl_2Pd[P(C_6H_5)_3]_2$ to provide alkenes substituted by three different groups.

[1] A. Minato, K. Suzuki, and K. Tamao, *Am. Soc.*, **109**, 1257 (1987).

Dichloro[1,1'-bis(diphenylphosphino)ferrocene]palladium(II). $PdCl_2(dppf)$ (**1, 10**, 37–38)
Aryl ketones.[1]   Although several palladium complexes can catalyze carbonylative coupling of aryl triflates with organostannanes to form ketones, this ferrocenylphosphine complex (**1**) is preferred because of consistently high yields. A variety of functional groups on either partner is tolerated (aldehyde, ester, hydroxyl). Transfer of vinyl, acetylenic, alkyl, and aryl groups from tin to form the corresponding ketones proceeds in high yield.

Coupling of aryl or alkenyl halides with trialkylboranes.[2]   This reaction can be carried out in refluxing THF with a base (NaOH or $NaOCH_3$, 1 equiv.) catalyzed by $PdCl_2(dppf)$. A B-alkyl-9-BBN is more useful than a trialkylborane, since only one alkyl group is utilized.

Example:

$$C_6H_5 \diagdown \diagup_{Br} + CH_3(CH_2)_7B \diagdown \diagup \xrightarrow[\substack{PdCl_2(dppf) \\ NaOH \\ 85\%}]{} C_6H_5 \diagdown \diagup (CH_2)_7CH_3$$

[1] A. M. Echavarren and J. K. Stille, *Am. Soc.*, **110**, 1557 (1988).
[2] N. Miyaura, T. Ishiyama, M. Ishikawa, and A. Suzuki, *Tetrahedron Letters*, **27**, 6369 (1986).

Dichlorobis(triphenylphosphine)nickel(II).
Homoallylic alcohols. The formation of homoallylic alcohols by Ni-catalyzed reaction of Grignard reagents with 5-alkyl-2,3-dihydrofurans[1] has been shown to provide a stereoselective route to homoallylic alcohols with a trisubstituted double bond.[2] Thus reaction of **1** with Grignard reagents that lack β-hydrogens gives coupled products with ≥95% retention of configuration. Vinyl- or allylmagnesium

$$\underset{\substack{(1, R^1 = CH_3, Pr, \\ C_5H_{11}\text{-}n)}}{\overset{}{\text{(furan structure)}}} + R^2MgX \xrightarrow[75-95\%]{\substack{NiCl_2L_2, \\ C_6H_6}} \underset{\mathbf{2}}{HOCH_2 \diagdown \diagup \underset{R^2 \quad R^1}{}}$$

halides fail to couple or give complex mixtures. Grignard reagents that are branched in the α position react inefficiently.

Dehydrohalogenation of RCH_2CH_2X.[3] In the presence of a low-valent Ni catalyst prepared from this Ni(II) complex, $P(C_6H_5)_3$, and butyllithium, primary alkyl bromides or iodides undergo dehydrohalogenation with DBU (2 equiv.) in THF at 25° to give 1-alkenes in about 50–80% yield. Hydroxyl groups and THP ethers are not affected.

[1] E. Wenkert, E. L. Michelotti, C. S. Swindell, and M. Tingoli, *J. Org.*, **49**, 4894 (1984).
[2] S. Wadman, R. Whitby, C. Yeates, P. Kocienski, and K. Cooper, *J.C.S. Chem. Comm.*, 241 (1987).
[3] S. Jeropoulos and E. H. Smith, *ibid.*, 1621 (1986).

Dichlorobis(triphenylphosphine)nickel–Chromium(II) chloride.
Cyclization of 1,6- and 1,7-enynes.[1] This combination produces an unstable Ni–Cr complex that can catalyze cyclization of enynes. The usefulness is markedly improved by coordination to a cross-linked polystyrene. The polymeric Ni–Cr complex (**1**) has several advantages over Pd(II) catalysts used previously for this cyclization, particularly the ability to effect cyclization of enynes to both five- and six-membered rings.

Examples:

(E = COOCH$_3$)

    n = 1   82%
    n = 2   60%

[1] B. M. Trost and J. M. Tour, *Am. Soc.*, **109**, 5268 (1987).

2,3-Dichloro-5,6-dicyano-1,4-benzoquinone (DDQ, **1**).

Silylation-promoted oxidation.[1] A new method for oxidation of 4-aza-3-ke-
tosteroids (**3**) to the unsaturated lactams **4** involves DDQ (1 equiv.) as the oxidant
and a silylating reagent (4 equiv.) such as bis(trimethylsilyl)trifluoroacetamide (**2**,
BSTFA).[2] An initial reaction at 20° results in an adduct of DDQ and **3**, which on

further reaction at 110° decomposes slowly to form the lactam **4** and the silylated
hydroquinone **5**.

Acetoxylation. Although furans are readily oxidized, furans substituted by a
triisopropylsilyl (TIPS) group when treated with DDQ in toluene-acetic acid at 0°
undergo acetoxylation at an adjacent α-methylene group.[3]

Example:

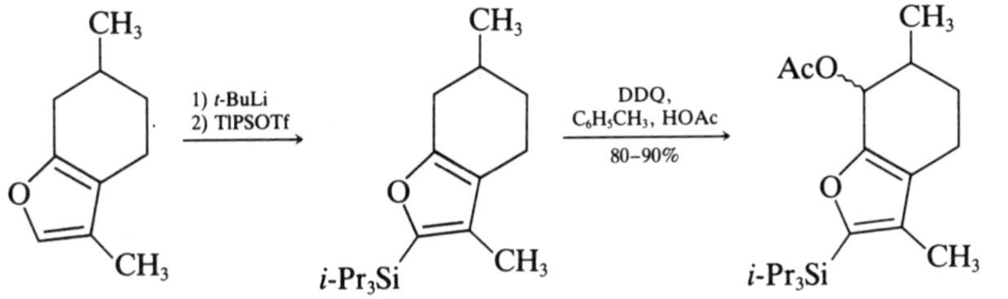

[1] A. Bhattacharya, L. M. DiMichele, U.-H. Dolling, A. W. Douglas, and E. J. J. Grabowski, *Am. Soc.*, **110**, 3318 (1988).

[2] The formula is CF₃C=NSi(CH₃)₃. This moisture sensitive, flammable reagent is available from Aldrich.

[3] E. J. Corey and Y. B. Xiang, *Tetrahedron Letters*, **28**, 5403 (1987).

(Z)-1,2-Dichloroethylene. Supplier: Aldrich.

1,3-*Diynes*.[1] In the presence of tetrakis(triphenylphosphine)palladium(0), (Z)-1,2-dichloroethylene couples with a terminal alkyne to form a (Z)-chloroenyne, which undergoes *anti*-elimination of HCl to give a 1,3-diyne on treatment with Bu₄NF.

Example:

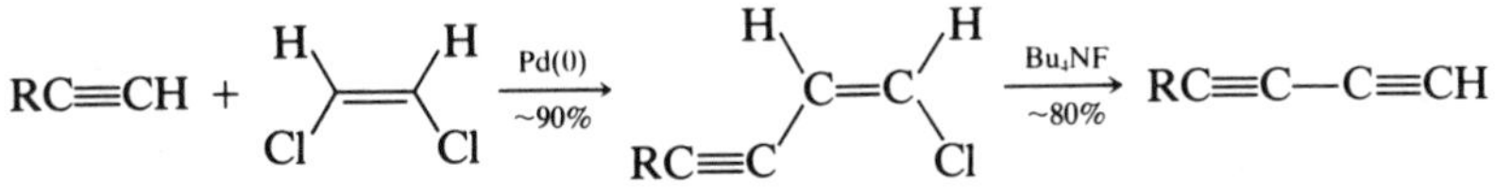

[1] A. S. Kende and C. A. Smith, *J. Org.*, **53**, 2655 (1988).

Dichloroketene.

Bicyclo-γ-butyrolactones.[1] The reaction of ketenes with chiral vinyl sulfoxides to obtain optically pure γ-arylsulfanylbutyrolactones (**12**, 177) can be extended to a synthesis of bicyclic butyrolactones. Thus the arylsulfanyl group of **1** undergoes

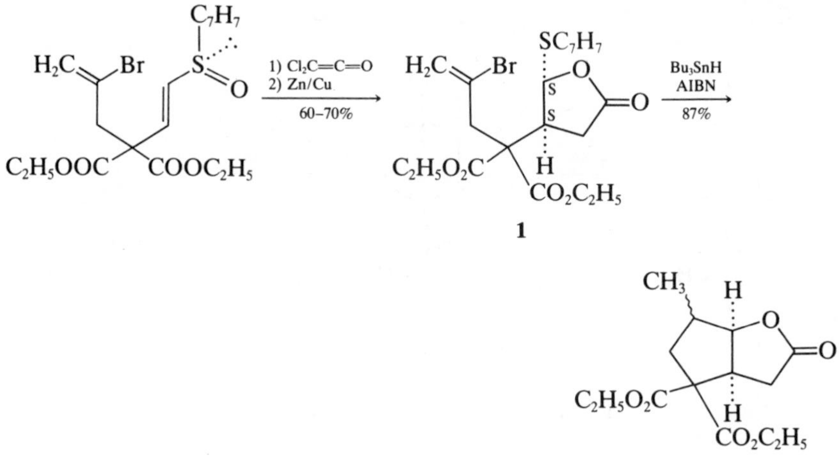

1

2 (α/β = 1:2)

displacement on radical cyclization (Bu₃SnH and AIBN) to form two epimeric *cis*-fused bicyclic butyrolactones (**2**). Lewis acid-catalyzed cyclization of the γ-aryl-sulfanylbutyrolactone **3** results in a single isomer of the *cis*-fused lactone **4**.

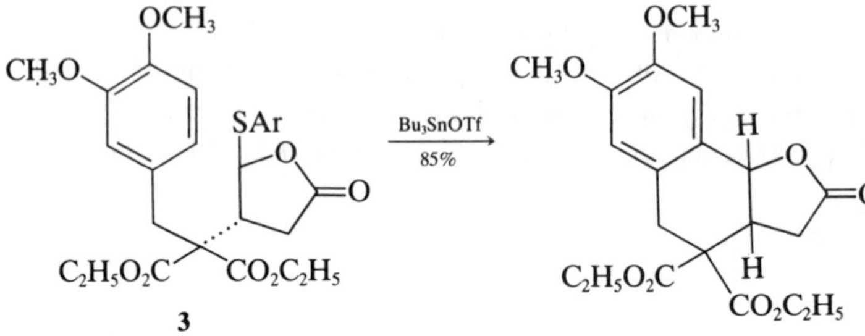

3   *cis*-**4**

Enantioselective [2 + 2]cycloaddition.[2]   The chiral allylic ether (**1**), prepared from (1S,2R)-(+)-2-phenylcyclohexanol, undergoes enantioselective cycloaddition with dichloroketene to furnish, after one crystallization, optically pure (−)-**2**. This cyclobutanone after ring expansion and exposure to chromium(II) perchlorate gives

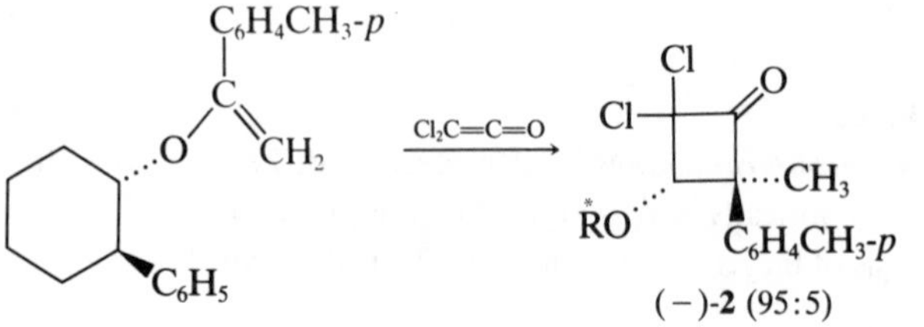

1   (−)-**2** (95:5)

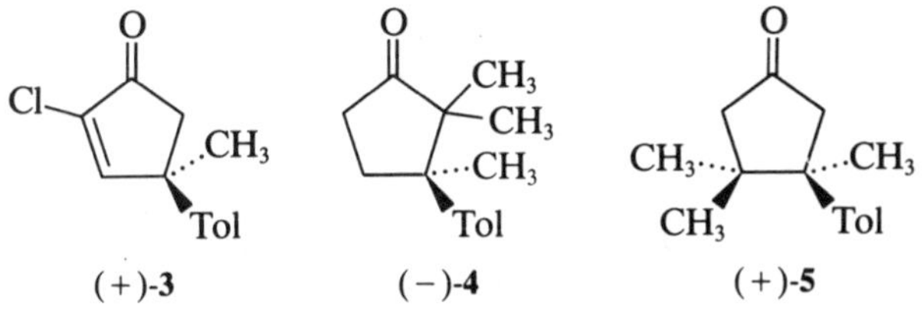

(+)-3   (−)-4   (+)-5

the α-chloroenone **3**, which can be converted to (−)-α-cuparenone (**4**) or (+)-β-cuparenone (**5**).

Cycloaddition to alkynes; cyclobutenones. This ketene when formed *in situ* from CCl₃COCl and Zn/Cu, reacts with alkynes to form 4,4-dichlorocyclobutenones,[3] which can rearrange in part to 2,4-dichlorocyclobutenones.[4] Both products are reduced to the same cyclobutenone by Zn(Cu) in HOAc/pyridine (4:1) or by zinc and acetic acid/TMEDA.[5]

Example:

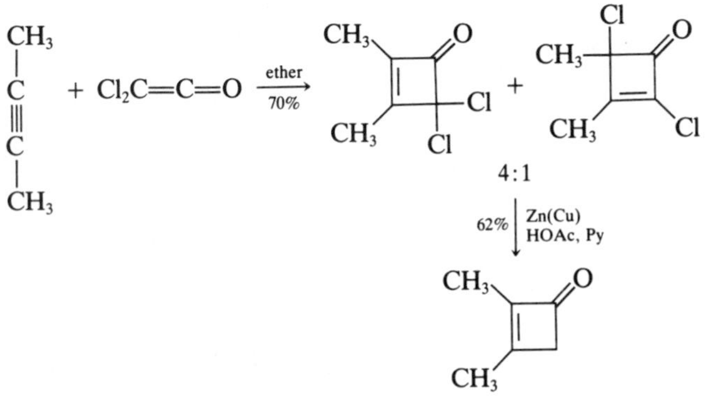

[1] J. P. Marino, E. Laborde, and R. S. Paley, *Am. Soc.*, **110**, 966 (1988).
[2] A. E. Greene, F. Charbonnier, M.-J. Luche, and A. Moyano, *ibid.*, **109**, 4752 (1987).
[3] A. Hassner and J. L. Dillon, Jr., *J. Org.*, **48**, 3382 (1983).
[4] A. A. Ammann, M. Rey, and A. S. Dreiding, *Helv.*, **70**, 321 (1987).
[5] R. L. Danheiser, S. Savariar, and D. D. Cha, *Org. Syn.*, submitted (1988).

2,3-Dichloropropene.

Biaryl synthesis.[1] This reagent promotes coupling of aryl Grignard reagents to symmetrical biaryls in 70–95% yield with formation of allene and 3-aryl-2-chloropropene as co-products. The reaction is retarded by galvinoxyl and evidently involves an electron-transfer from the Grignard reagent to the dichloropropene.

$$2\text{ArMgBr} + \text{CH}_2\!\!=\!\!\text{C}\!\!\begin{array}{c}\diagup\text{Cl}\\\diagdown\text{CH}_2\text{Cl}\end{array} \xrightarrow{\text{THF, 25°}}$$

$$\text{Ar—Ar} + \text{MgBrCl} + \text{CH}_2\!\!=\!\!\text{C}\!\!=\!\!\text{CH}_2 + \text{CH}_2\!\!=\!\!\text{C}\!\!\begin{array}{c}\diagup\text{Cl}\\\diagdown\text{CH}_2\text{Ar}\end{array}$$

(70–95%)

[1] J.-W. Cheng and F.-T. Luo, *Tetrahedron Letters*, **29**, 1293 (1988).

1,4-Dichloro-1,1,4,4-tetramethyldisilylethylene, 10, 140; **12**, 179.

Allylic amines.[1]   The stabase adduct (**1**) of 2-chloroallylamine couples with Grignard reagents in the presence of $\text{NiCl}_2(\text{dppp})$, dppp = $(\text{C}_6\text{H}_5)_2\text{P}(\text{CH}_2)_3\text{P}(\text{C}_6\text{H}_5)_2$, to form, after deprotection, allylic amines.

Example:

1

55–90% $\big|$ H_3O^+

$$\begin{array}{c}\text{CH}_2\\\|\\\text{RCCH}_2\text{NH}_2\end{array}$$

[1] T. M. Bargar, J. R. McCowan, J. R. McCarthy, and E. R. Wagner, *J. Org.*, **52**, 678 (1987).

Dichlorotris(triphenylphosphine)ruthenium(II).

Intramolecular cyclization of unsaturated α,α-dichloro esters[1] and acids.
Addition of halocarbons to alkenes in the presence of transition metals is a well-known radical reaction. Weinreb *et al.*[1] have now reported an intramolecular version leading to cyclic esters or bicyclic lactones. Typical substrates are the α,α-dichloro ester **1** or the α,α-dichloro acid **2**, readily available by reaction of ethyl lithiodichloroacetate with 5-bromo-1-pentene. When **1** is heated in benzene at 160° with a metal catalyst, mixtures of epimeric α,ω-dichloro esters **3** and **4** are obtained. The ratio and yields of **3** and **4** are dependent on the catalyst and concentration of **1**, but **3** and **4** are the major products formed in the presence of Ru(II) and Fe(II) catalysts. In contrast cyclization of **2** under the same conditions gives the bicyclic γ-lactone **5** in high yield.

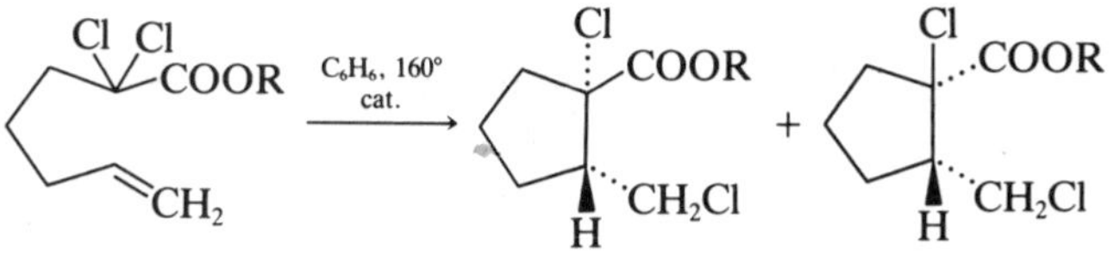

3 41%   **4** 36%

1, R = C₂H₅   Ru(II)

2, R = H $\xrightarrow{\text{Ru(II)}}$ —   —

5, 94%

Indole synthesis.[2] 2-Aminophenethyl alcohols cyclize to indoles in refluxing toluene in the presence of this Ru(II) catalyst in 75–100% yield.

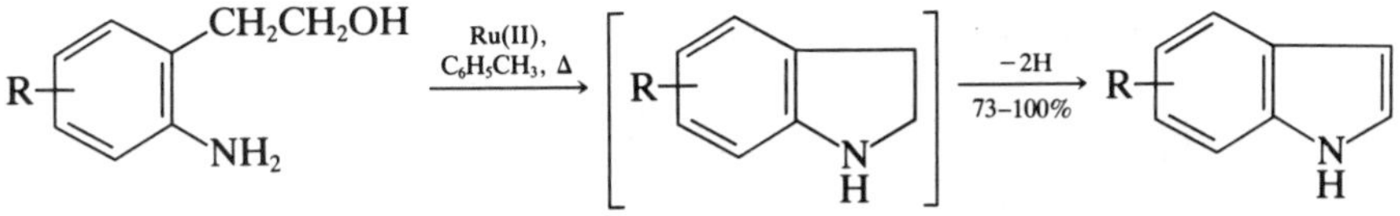

[1] T. K. Hayes, A. J. Freyer, M. Parvez, and S. M. Weinreb, *J. Org.*, **51**, 5501 (1986).
[2] Y. Tsuji, K.-T. Huh, Y. Yokoyama, and Y. Watanabe, *J.C.S. Chem. Comm.*, 1575 (1986).

Dicyclohexylcarbodiimide (DCC).
Inversion of configuration.[1] The configuration of a secondary alcohol (**2**) can be inverted by reaction with dicyclohexylcarbodiimide (**1**) to form an isourea ether (**3**), which is allowed to react, without isolation, with formic acid with formation of the ester **4** with inverted configuration.

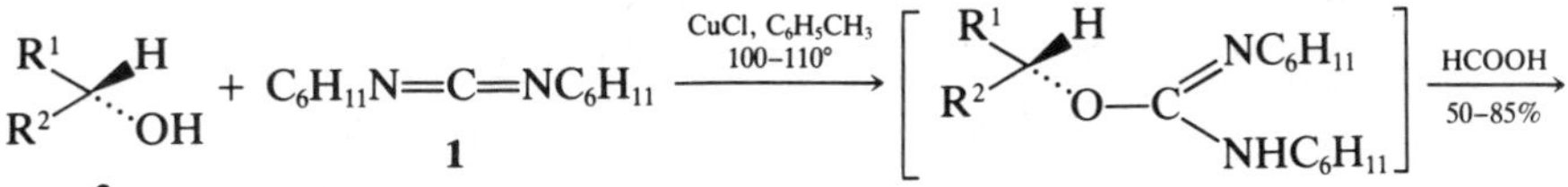

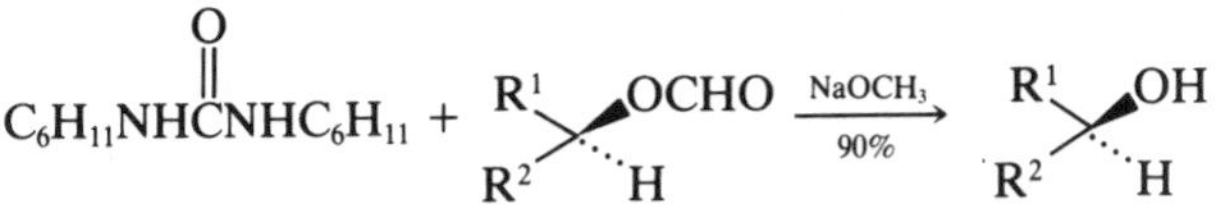

N-*Tosyl amides and lactams*.[2]   DCC in combination with 4-pyrrolidinopyridine (4-PPy) effects condensation of carboxylic acids with secondary sulfonamides to provide N-tosyl amides in 75–90% yield. The intramolecular version of this reaction provides 4-, 5-, and 6-membered N-tosyl lactams in 60–90% yield.
Example:

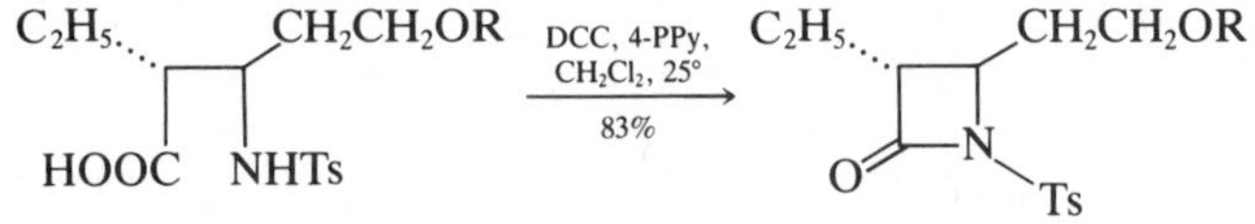

[1] J. Kaulen, *Angew. Chem. Int. Ed.*, **26**, 773 (1987).
[2] D. Tanner and P. Somfai, *Tetrahedron*, **44**, 613, 619 (1988).

Dicyclohexylcarbodiimide–4-Dimethylaminopyridine.
Macrolactonization (**13**, 107–108).   The final steps in a total synthesis of (+)-(9S)-dihydroerthythronolide (**1**), the aglycone of erythromycin A (**2**), involve cy-

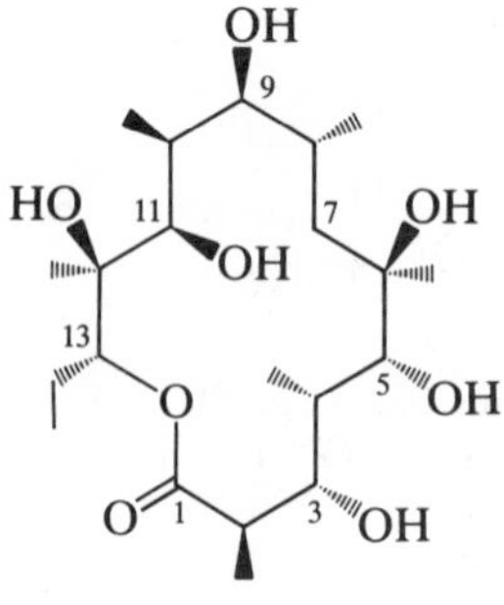

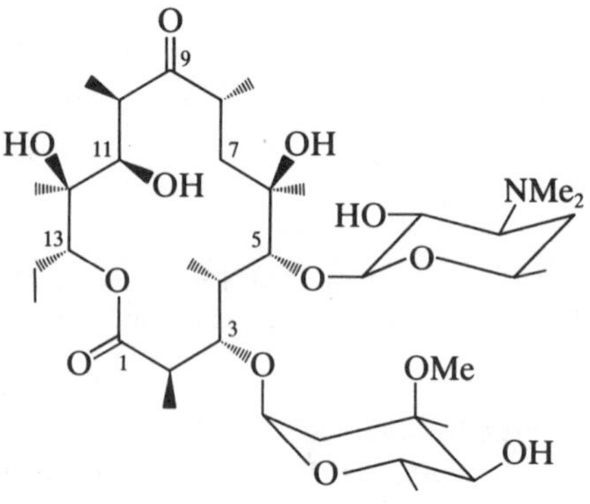

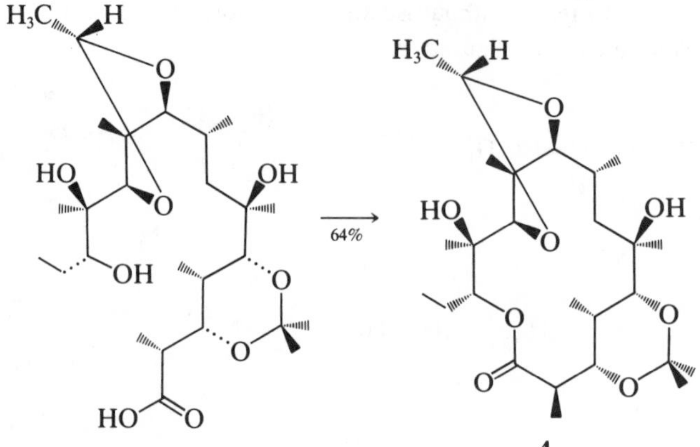

clization of the seco acid (3), which proceded in 64% yield using DCC, DMAP, and the trifluoroacetate of DMAP. Deprotection of the resulting lactone (4) provides 1 in 56% yield.

[1] G. Stork and S. D. Rychnovsky, *Am. Soc.*, **109**, 1565 (1987).

Dicyclopentadienylmethylzirconium chloride, $Cp_2Zr(CH_3)Cl$ (1).

The reagent is prepared by reaction of $(Cp_2ZrCl)_2O$ with $Al(CH_3)_3$ (75% yield).[1]

Benzyne zirconocene complexes.[2] This reagent reacts with an aryllithium at $-50°$ to form an intermediate **a** that loses methane at $25°$ to give a zirconocene

complex (2) of a benzyne. A nitrile reacts with 2 with insertion into a C—Zr bond to form metallacycles **3a** and/or **3b**, depending on steric factors. When the benzyne has only one *ortho* substituent, the corresponding **3a** is formed preferentially. Acid hydrolysis of **3a** provides a substituted acetophenone **4a**. Reaction of **3a** with iodine before hydrolysis provides **5a**.

The complex 2 also reacts with ethylene to form the metallacycle **6**, which is converted into a benzocyclobutane (**8**) in two steps.

$$2 \quad \xrightarrow{\text{CH}_2=\text{CH}_2} \quad \mathbf{6} \quad \xrightarrow{\text{I}_2} \quad \mathbf{7} \quad \xrightarrow[50\%]{\text{BuLi}} \quad \mathbf{8}$$

[1] P. C. Wailes, H. Weigold, and A. P. Bell, *J. Organometal Chem.*, **33**, 181 (1971).
[2] S. L. Buchwald, B. T. Watson, R. T. Lum, and W. A. Nugent, *Am. Soc.*, **109**, 7137 (1987).

Diethoxytriphenylphosphorane, $(C_6H_5)_3P(OC_2H_5)_2$ **(1).** The phosphorane is obtained by reaction of diethyl peroxide (caution) with triphenylphosphine at 0–70°.
Cyclodehydration of diols to ethers. 1,3-, 1,4-, and 1,5-Diols react with **1** to

$$\begin{array}{c} \text{HO(CH}_2)_n\text{OH} + \mathbf{1} \\ n = 3\text{-}5 \end{array} \xrightarrow[-\text{C}_2\text{H}_5\text{OH}]{\text{CH}_2\text{Cl}_2} \left[\begin{array}{c} \text{OC}_2\text{H}_5 \\ | \\ \text{C}_6\text{H}_5-\text{P(C}_6\text{H}_5)_2 \\ | \\ \text{O(CH}_2)_n\text{OH} \end{array} \right] \xrightarrow[70\text{-}95\%]{42°} \begin{array}{c} \text{CH}_2 \\ \text{(CH}_2)_n \end{array} + \text{O=P(C}_6\text{H}_5)_3$$

$$n = 1\text{-}3$$

form ethers via acyclic phosphoranes.[1] The reaction of **1** with 1,2,4-butanetriol results in an epoxide and a hydroxyfuran:[2]

$$\text{HOCH}_2\overset{\text{OH}}{\underset{|}{\text{CH}}}\text{CH}_2\text{CH}_2\text{OH} \xrightarrow{\mathbf{1}}$$

81%   19%

1,4-Oxathianes.[3] Dehydration of 2,2'-bis(hydroxyethyl) sulfides with $(C_6H_5)_3P(OC_2H_5)_2$ provides a stereoselective route to 1,4-oxathianes in moderate (isolated) yields.
Example:

$$\xrightarrow[47\%]{(C_6H_5)_3P(OC_2H_5)_2 \atop C_6H_6}$$

[1] P. L. Robinson, C. N. Barry, J. W. Kelly, and S. A. Evans, Jr., *Am. Soc.*, **107**, 5210 (1985).
[2] J. W. Kelly and S. A. Evans, Jr., *ibid.*, **108**, 7681 (1986).
[3] W. T. Murray, J. W. Kelly, and S. A. Evans, Jr., *J. Org.*, **52**, 525 (1987).

Diethylaluminum ethoxide, $(C_2H_5)_2AlOC_2H_5$ **(1)**, available from Aldrich.

Intramolecular aldol condensation.[1] This base can be effective for intra-molecular aldol condensation of extremely hindered diketones that resist cyclization with the usual bases.

Example:

[1] J. W. ApSimon and R. F. Lawuyi, *Syn. Comm.*, **17**, 1773 (1987).

$$\overset{O}{\underset{\|}{}}$$

Diethyl diazomethylphosphonate, $(C_2H_5O)_2PCHN_2$ **(1)**.

Furans.[1] Reaction of α,α-dimethoxy ketones with **1** affords a dihydrofuran (**2**) presumably via a carbene (**a**) that inserts intramolecularly into a C—H of an adjacent methoxy group. The reaction often results directly in a furan, since the elimination of methanol from **2** is facile.

[1] S. R. Buxton, K. H. Holm, and L. Skattebøl, *Tetrahedron Letters*, **28**, 2167 (1987).

Diethyl isocyanomethylphosphonate (1).

Aldehydes.[1] This Wittig–Horner reagent (available from Fluka) converts al-dehydes or ketones into α,β-unsaturated isocyanides, which can be hydrolyzed to the one-carbon homologated aldehyde.

Example:

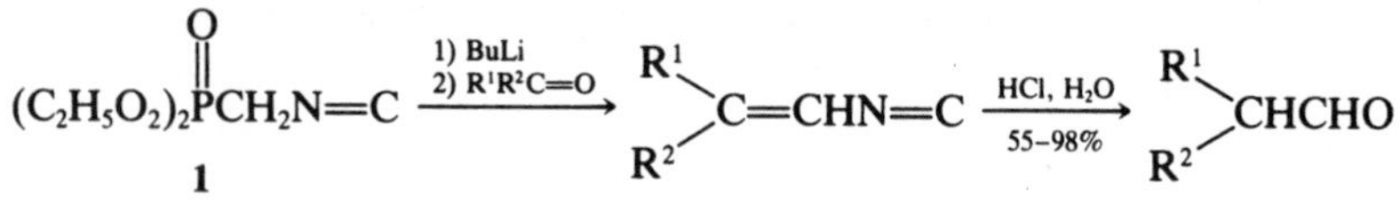

$$(C_2H_5O_2)_2\overset{\overset{O}{\|}}{P}CH_2N{=}C \xrightarrow[\text{2) } R^1R^2C{=}O]{\text{1) BuLi}} \overset{R^1}{\underset{R^2}{>}}C{=}CHN{=}C \xrightarrow[55-98\%]{HCl, H_2O} \overset{R^1}{\underset{R^2}{>}}CHCHO$$

1

[1] J. Moskal and A. M. van Leusen, *Rec. Pays.-Bas.*, **106**, 137 (1987).

Dihyridotetrakis(triphenylphosphine)ruthenium, $RuH_2[P(C_6H_5)_3]_4$ **(1).**

Amides.[1]   In the presence of this catalyst nitriles react at 160° with amines in an aqueous medium to give amides in a single step in 75–95% yield.

Example:

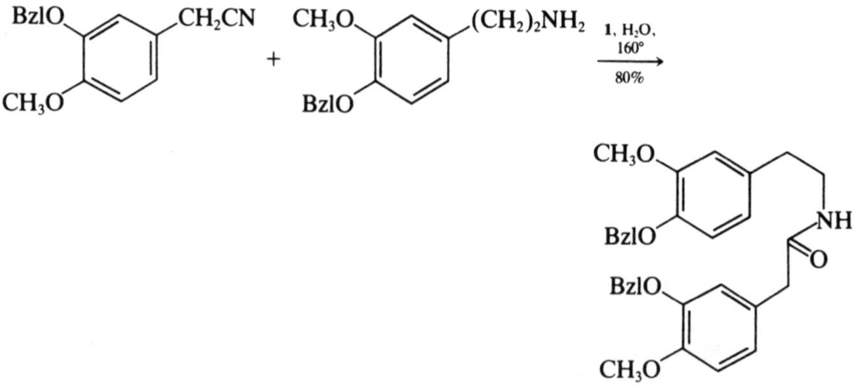

(E,E)-α,β;γ,δ-Dienones.[2]   Aryl α,β-alkynyl ketones rearrange in the presence of this ruthenium catalyst in refluxing toluene to conjugated (E,E)-dienones in 75–85% yield. A similar rearrangement with alkyl α,β-alkynyl ketones proceeds less readily.

Example:

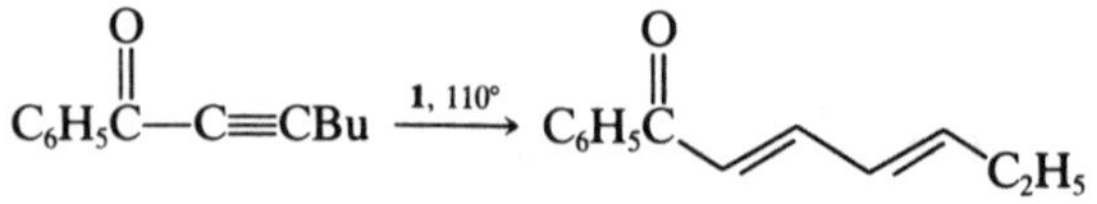

$$C_6H_5\overset{\overset{O}{\|}}{C}-C{\equiv}CBu \xrightarrow{1, 110°} C_6H_5\overset{\overset{O}{\|}}{C}\diagdown\diagup\diagdown_{C_2H_5}$$

[1] S.-I. Murahashi, T. Naota, and E. Saito, *Am. Soc.*, **108**, 7846 (1986).
[2] D. Ma, Y. Lin, X. Lu, and Y. Yu, *Tetrahedron Letters*, **29**, 1045 (1988).

1,3-Dihydrotetramethyldisilazane, $[(CH_3)_2HSi]_2NH$ **(1),** b.p. 99–100°, moisture-sensitive.

Intramolecular hydrosilylation.[1]   Hydrosilylation of internal double bonds requires drastic conditions and results in concomitant isomerization to the terminal position. However, an intramolecular hydrosilylation is possible with allylic or homoallylic alcohols under mild conditions by reaction with **1** at 25° to give a hydrosilyl ether (**a**), which then forms a cyclic ether (**2**) in the presence of $H_2PtCl_6 \cdot 6H_2O$ at 60°. Oxidative cleavage of the C—Si bond results in a 1,3-diol (**3**).

The reaction can be *anti-* or *syn-* selective.
Example:

[1] K. Tamao, T. Tanaka, T. Nakajima, R. Sumiya, H. Arai, and Y. Ito, *Tetrahedron Letters*, **27**, 3377 (1986); K. Tamao, T. Nakajima, R. Sumiya, H. Arai, N. Higuchi, and Y. Ito, *Am. Soc.*, **108**, 6090 (1986).

Diiodosilane (DIS), I_2SiH_2, b.p. 148–155°.
   The reagent is obtained as a fairly pure oil by reaction of phenylsilane with iodine catalyzed by a trace of an organic ester. After removal of volatile side products (benzene and HI), the oil can be purified by fractional distillation.
   The reagent is similar to iodotrimethylsilane in reactivity. It also converts alcohols into iodides, but in contrast to $ISi(CH_3)_3$, it reacts more rapidly with secondary alcohols (with inversion) than with primary ones. It also cleaves ethers, and again it cleaves secondary alkyl ethers more readily than primary alkyl ethers.

[1] E. Keinan and D. Perez, *J. Org.*, **52**, 4846 (1987).

Diisobutylaluminum 2,6-di-*t*-butyl-4-methylphenoxide (**1**), **9**, 171.
   Diastereoselective reduction of β-keto esters.[1]   Reduction of β-keto esters (**3**) of the chiral α-naphthylborneol (**2**)[2] is stereoselective because one face of the carbonyl group is blocked by the naphthyl group. Chelation of the keto ester with

2 = X*OH

| 1 | 80–98% | 96–99:4–1 |
| Zn(BH$_4$)$_2$/ZnCl$_2$ | 92–96% | 10–6:90–94 |

ZnCl$_2$ followed by reduction with Zn(BH$_4$)$_2$ results in **4** and **5**, in a ratio of about 8:92. Reduction of the ketone with **1** results in the opposite diastereoselectivity to give **4** as the major product. Reduction of **4** or **5** with LiAlH$_4$ liberates the optically active β-hydroxy primary alcohols.

[1] D. F. Taber, P. B. Deker, and M. D. Gaul, *Am. Soc.*, **109**, 7488 (1987).
[2] D. F. Taber, K. Raman, M. D. Gaul, *J. Org.*, **52**, 28 (1987).

Diisobutylaluminum hydride–Hexamethylphosphoric triamide.

Conjugate reductions. This combination (usually 1:3 ratio) effects conjugate reduction of α,β-acetylenic ketones or esters to α,β-enones or unsaturated esters at −50° with moderate (E)-selectivity. The HMPT is believed to function as a ligand to aluminum and thus to promote hydroalumination to give a vinylaluminum intermediate, which can be trapped by an allylic bromide (equation I).[1] The re-

(I) HC≡CCOOCH$_3$

duction can be catalyzed by CH$_3$Cu, with some increase in (Z)-selectivity. Methylcopper is a necessary catalyst for effective conjugate reduction of α,β-enones with DIBAH and HMPT.[2]

[1] T. Tsuda, T. Yoshida, T. Kawamoto, and T. Saegusa, *J. Org.*, **52**, 1624 (1987).
[2] T. Tsuda, T. Hayashi, H. Satomi, T. Kawamoto, and T. Saegusa, *ibid.*, **51**, 537 (1986).

(+)- and (−)-Diisopinocampheylborane trifluoromethanesulfonate, (+)- and (−)-(Ipc)$_2$BOTf (**1**). The triflate is prepared *in situ* from (Ipc)$_2$BH and triflic acid.

Diastereoselective aldol condensations.[1] The aldol condensation of a chiral ethyl ketone such as **2** with aldehydes catalyzed by Bu$_2$BOTf gives a mixture of all four possible diastereomeric adducts with little or no stereocontrol. In contrast, reactions catalyzed by either (+)- or (−)-**1** are highly diastereoselective. By proper choice of (+)- and (−)-**1** and of (+)- and (−)-**2**, each one of the four possible 1,2-*syn*-diastereomers can be obtained in high purity.

[1] I. Paterson and M. A. Lister, *Tetrahedron Letters*, **29**, 585 (1988).

Diisopropyl 2-allyl-1,3,2-dioxaborolane-4,5-dicarboxylate,

(R,R)-**1**

The allyl boronate esters (R,R)- and (S,S)-1 are prepared by reaction of allylboronic acid, CH_2=$CHCH_2B(OH)_2$ with L- and D-diisopropyl tartrate.[1]

Diastereoselective reaction with β-alkoxy-α-methylpropionaldehydes.[1] The reaction of (R,R)-1 with the chiral aldehyde 2a provides the *syn*-homoallylic alcohol

2a, R = SiMe$_2$-*t*-Bu   (R,R)-1 (71%)   89:11
2b, R = SiPh$_2$-*t*-Bu   (S,S)-1 (72%)   13:87
                                       3

                                       4

3 with 78% de. Although the absolute asymmetric induction is controlled by the tartrate auxiliary of 1, the extent of induction is dependent on the hydroxyl pro-tecting group. Thus the *anti*-homoallylic alcohol 4 is best obtained by reaction of 2b with (S,S)-1. The reaction of β-benzyloxy-α-methylpropionaldehyde with (R,R)- or (S,S)-1 proceeds in lower stereoselectivity than that observed with 2a or 2b.

The corresponding crotylboronates, (R,R)- and (S,S)-5 also undergo highly diastereoselective reactions with the same chiral aldehydes, and again the diaster-

(R,R)-5

2a + (R,R)-5 ⟶   6   7

              80%   97:3
2b + (S,S)-5   85%   11:88

eoselectivity is in part dependent on the hydroxyl-protecting group.[2] The highest optical yields of *anti,syn* and *anti,anti* products are obtained from reaction of 2a with (R,R)-5 and of 2b with (S,S)-5.

[1] W. R. Roush, A. E. Walts, and L. K. Hoong, *Am. Soc.*, **107**, 8186 (1985).
[2] W. R. Roush, A. D. Palkowitz, and M. A. J. Palmer, *J. Org.*, **52**, 316 (1987).

Diisopropylamine trihydrofluoride (1). The salt is obtained by addition of 3 equiv. of aqueous hydrogen fluoride to the amine.

Fluorohydrins.[1]   Both this reagent (1) and Olah's pyridinium poly(hydrogen fluoride) (2) serve as an attenuated source of hydrogen fluoride, and both convert simple epoxides into fluorohydrins, but reactions with 1 require higher temperatures than those with 2. The two reagents can differ in the regioselectivity, with steric factors being more important in reactions of 1 than those of 2.

Examples:

n = 1	1, 110°	70%	5%
	2, −5°	33%	27%
n = 2	1, 110°	69%	12%
	2, −50°	20%	78%

[1] M. Muehlbacher and C. D. Poulter, *J. Org.*, **53**, 1026 (1988).

9-O-(1,2;5,6-Di-O-isopropylidene-α-D-glucofuranosyl)-9-boratabicyclo-[3.3.1]nonane, potassium salt (K 9-O-DIPGF-9-BBNH, 1). Preparation.[1]

K 9-O-DIPGF-9-BBNH, 1

Asymmetric reductions.   The reagent can effect asymmetric reduction of alkyl aryl ketones and unhindered dialkyl ketones in high optical yield.[1] It is the most useful reagent known to date for asymmetric reduction of even hindered α-keto esters to (S)-α-hydroxy esters in >90% ee.[2] It is also effective for asymmetric reduction of phosphinyl imines of dialkyl ketones, $R^1R^2C{=}NP(O)(C_6H_5)_2$ (50–84% ee).[3]

[1] H. C. Brown, W. S. Park, and B. T. Cho, *J.Org.*, **51**, 1934 (1986).
[2] *Idem*, *ibid.*, **51**, 3396 (1986).
[3] R. O. Hutchins, A. Abdel–Magid, Y. P. Stercho, and A. Wambsgans, *ibid.*, **52**, 702 (1987).

Diketene.

β-*Lactams*. Diketene can function as an equivalent to acetylketene, $CH_3COCH=C=O$, to provide 3-acetyl-β-lactams by [2 + 2]cycloaddition with imines.[1] A stereoselective cycloaddition of this type can furnish a useful precursor (**2**) to 1β-methylcarbapenems. Thus reaction of diketene with the chiral imine **1**, prepared in a few steps from the readily available methyl (S)-3-hydroxy-2-methylpropionate (Aldrich), can provide the desired 3,4-*trans*-3-acetyl-β-lactam **2** in 84% de. This lactam is readily converted to an intermediate previously developed for synthesis of the antibacterial carbapenem **4**.

[1] T. Kawabata, Y. Kimura, Y. Ito, S. Terashima, A. Sasaki, and M. Sunagawa, *Tetrahedron*, **44**, 2149 (1988).

Dilithium tetrachloropalladate, Li_2PdCl_4.

Oxidative phenolic coupling.[1] Biosynthesis of the alkaloid narwedine (**3**) is known to involve oxidative phenolic coupling of norbelladine derivatives (**1**), but the usual oxidants for such coupling *in vitro* convert **1**(R = H) into the oxomaritidine skeleton (**4**) rather than **3**. A new biomimetic synthesis of **3** involves the palladacycle **2**, formed by reaction of **1**(R = CH$_3$) with Li_2PdCl_4, which is known to form complexes with allylic amines or sulfides (**8**, 176–177). Oxidation of **2** with thallium(III) trifluoroacetate effects the desired coupling to give **3**.

[1] R. A. Holton, M. P. Sibi, and W. S. Murphy, *Am. Soc.*, **110**, 314 (1988).

Dimanganese heptoxide, Mn_2O_7. Mn_2O_7 is a dark red oil which can explode on contact with organic compounds. However, it can be prepared and handled safely in CCl_4 solution by reaction of $KMnO_4$ with concentrated H_2SO_4 at 20–25°. Mn_2O_7 is a stronger oxidant than $KMnO_4$ and can effect oxidations at $-45°$ to $-80°$. At these temperatures, solvents such as acetone, methyl acetate, CCl_4, or Freon 113 are not attacked. Oxidations are usually complete within minutes, and the products, after filtration of MnO_2, are usually isolated by evaporation of the solvent. Aside from the usual oxidation of primary and secondary alcohols in 80–90% yield, Mn_2O_7 can cleave double bonds at $-80°$ more efficiently than ozone.

Examples:

Mn_2O_7 also effects oxidation at $-45°$ of α-methylene groups of ethers to give lactones or esters.[1]

Example:

$$(CH_3)_3COCH_2CH_3 \xrightarrow[92\%]{} (CH_3)_3CO\overset{\overset{\displaystyle O}{\|}}{C}CH_3$$

[1] M. Trömel and M. Russ, *Angew. Chem. Int. Ed.*, **26**, 1007 (1987).

(2-Dimethylaminomethylphenyl)phenylsilane,   (1).

Reductions. Silicon hydrides such as **1**, which can achieve intramolecular pentacoordination, show enhanced reducing properties. Thus they can reduce aldehydes or ketones to alcohols,[1] and reduce carboxylic acids to aldehydes via thermal decomposition of a silyl carboxylate (equation I).[2] Reaction of acid chlo-

rides with **1** proceeds at room temperature to give a chlorosilane and aldehydes directly in almost quantitative yield, with no effect on C=C double bonds or halogen and methoxy substituents.[2]

[1] J. Boyer, C. Breliere, R. J. P. Corriu, A. Kpoton, M. Poirier, and G. Royo, *J. Organometal. Chem.*, **311**, C39 (1986).

[2] R. J. P. Corriu, G. F. Lanneau, and M. Perrot, *Tetrahedron Letters*, **28**, 3941 (1987); *idem*, **29**, 1271 (1988).

(R)- and (S)-(6,6'-Dimethylbiphenyl-2,2'-diyl)bis(diphenylphosphine), BIPHEMP (1). This is the first example of an optically active bis(triarylphosphine) containing the axially disymmetric biphenyl group. The 6,6'-dimethyl groups are used to

(R)-(−)-**1**
m.p. 213°, α_D − 43°

(S)-(+)-**1**
m.p. 213°, α_D + 42°

decrease thermal racemization. (RS)-**1** is prepared by standard procedures and then resolved with a chiral Pd(II)–amine complex. These ligands are as effective as BINAP for Rh(I)-catalyzed allylic isomerization of N,N-diethylnerylamine (**12**, 56–57).[1]

[1] R. Schmid, M. Cereghetti, B. Heiser, P.Schönholzer, and H.-J. Hansen, *Helv.*, **71**, 897 (1988).

(R,R)- or (S,S)-2,5-Dimethylborolane,

BH (R,R)-**1**,   BH (S,S)-**1**, **13**, 119–120.

Asymmetric reduction of dialkyl ketones. The borohydride **1** reduces dialkyl ketones with low enantioselectivity. However, treatment of the lithium dihydridoborate **2** with methanesulfonic acid provides Reagent I, which consists of 1 equiv. of R,R-**1** and 0.2 equiv. of 2,5-dimethylborolanyl mesylate, which serves as a

Reagent I

catalyst and is essential for significant asymmetric reduction. Reagent I reduces n-alkyl methyl ketones to (R)-alcohols in about 80% ee. The percent of ee increases with branching at the β-position of the alkyl group. Thus methyl, secondary, and tertiary alkyl ketones are reduced in about 96% ee.

Asymmetric aldol reactions.[2]   The reagent **3**, prepared by reaction of (S)-3-(3-ethyl)pentyl propanethioate (**2**) and ethyldiisopropylamine with (S,S)-2,5-di-

methylborolanyl triflate (**1**), undergoes enantioselective *anti*-aldol reactions with various aldehydes (equation I). The *anti/syn* ratio of the aldol reaction is 30–33:1 and the optical yields are >98% ee.

R = Pr   91%   *anti/syn* = 33:1   98% ee

The normethyl analog (**4**) of **3**, prepared in the same way from an ethanethioate, also reacts with aldehydes to form aldols but with lower enantioselectivity (89–93%).

89–93% ee

[1] T. Imai, T. Tamura, A. Yamamuro, T. Sato, T. A. Wollmann, R. M. Kennedy, and S. Masamune, *Am. Soc.*, **108**, 7402 (1986); S. Masamune, R. M. Kennedy, J. S. Petersen, K.N. Houk, and Y. Wu, *ibid.*, **108**, 7404 (1986).

[2] S. Masamune, T. Sato, B. M. Kim, and T. A. Wollman, *Am. Soc.*, **108**, 8279 (1986).

Dimethyl chlorophosphite, $(CH_3O)_2PCl$ (**1**). This dimethyl ester of phosphorochloridous acid is obtained by reaction of $(CH_3O)_3P$ with PCl_3 in $[(CH_3)_2N]_3PO$ as solvent (74.5% yield).[1]

Inositol phosphates.[2] Phosphorylation of inositol presents certain difficulties because the usual phosphorus(V) phosphorylating reagents are relatively unreactive toward secondary hydroxyl groups, and phosphate triester intermediates are subject to cyclization and migration to neighboring hydroxyl groups. These difficulties can be circumvented by phosphitylation with a phosphorus(III) reagent such as dimethyl chlorophosphite (**1**). This phosphite is more reactive than a phosphorus(V) reagent, but can still distinguish between axial and equatorial groups. Thus, reaction of the dibenzoyl-*myo*-inositol (**2**) with **1** (3 equiv.) and an amine followed by oxidation

gives the tris(dimethyl phosphate) **3**. Cleavage of the phosphate methyl groups with HBr or BrSi(CH$_3$)$_3$ and ester hydrolysis provides the desired 1,4,5-triphosphate (**4**) of *myo*-inositol. Exhaustive phosphitylation of **2** is also possible, resulting eventually in the 1,2,4,5-tetraphosphate of *myo*-inositol. This reaction shows that **1** can be used for bisphosphorylation of a *cis,vic*-diol, and also suggests that any free hydroxyl group of a protected inositol can be converted to a phosphate mono ester.

A related procedure for phosphorylation of protected inositols uses N,N-diisopropyl dibenzyl phosphoroamidite, (BzlO)$_2$PN(*i*-Pr)$_2$, followed by *m*-chloroperbenzoic acid to form dibenzyl phosphates, ROP(OBzl)$_2$, in about 90% yield. These

$$\overset{O}{\overset{\|}{ROP}}(OBzl)_2$$

are converted to phosphate esters, $\overset{O}{\overset{\|}{ROP}}(OH)_2$ in quantitative yield by hydrogenolysis.[3]

[1] Z. Mazour, *C.A.*, **87**, 52758 (1977).
[2] J. L. Meek, F. Davidson, and F. W. Hobbs, Jr., *Am. Soc.*, **110**, 2317 (1988).
[3] K.-L. Yu and B. Fraser-Reid, *Tetrahedron Letters*, **29**, 979 (1988).

Dimethyldioxirane (1), 13, 120.

Epoxidation of allenes.[1]   The spirodioxides formed by epoxidation of allenes are unstable to acids, and only hindered ones have been obtained on epoxidation with peracids. They can be obtained, however, in 90–95% yield by epoxidation of allenes (even monosubstituted ones) with dimethyldioxirane in acetone buffered with solid K_2CO_3.

Example:

[1] J. K. Crandall and D. J. Batal, *J. Org.*, **53**, 1338 (1988).

N,N-Dimethylformamide.

Bouveault reaction.[1]   DMF is recommended for formylation of tertiary Grignard reagents. Thus pivaldehyde can be obtained in 57–61% yield by reaction of *t*-butylmagnesium chloride with twofold excess of DMF in ether.

$$(CH_3)_3CMgCl + HCON(CH_3)_2 \xrightarrow[57-61\%]{\text{ether}} (CH_3)_3CCHO$$

1,5-Diketones.   Potassium enolates of some ketones can dimerize with incorporation of a methylene group from DMF to give 1,5-diketones. This novel reaction requires a *tert*-alkyl or phenyl group attached to the carbonyl group.[2]

Example:

[1] R. B. Nazarski, M. K. Tasz, and R. Showronski, *Org. Syn.* submitted (1987).
[2] S. Kiyooka, T. Yamashita, J. Tashiro, K. Takano, and Y. Uchio, *Tetrahedron Letters*, **27**, 5629 (1986).

Dimethylhydrazine.
anti-α-*Amino alcohols*. Organolithium reagents add diastereoselectivity to the dimethylhydrazones of α-alkoxy aldehydes to form adducts that provide *anti*-α-amino alcohols on hydrogenolysis.

Examples:

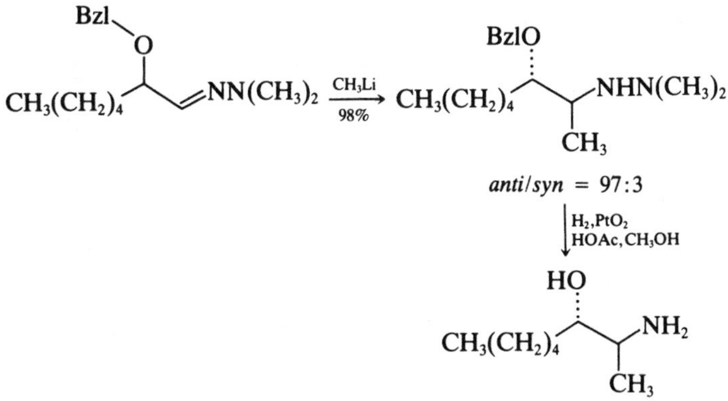

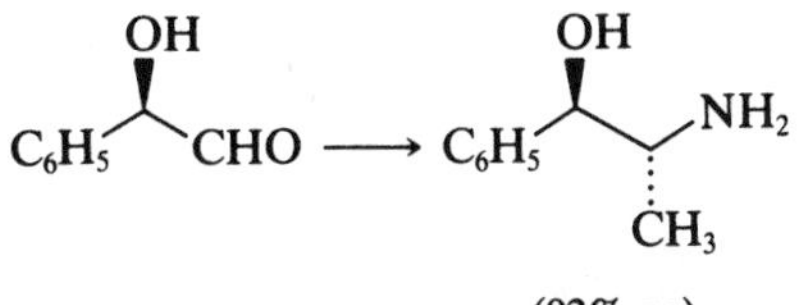

anti/syn = 97:3

(92% ee)

The reaction was used for synthesis of (−)-norpseudoephedrine from a chiral α-hydroxy aldehyde (second example). The diastereoselectivity can be reversed by addition of alkyllithiums to α-trityloxy aldehydes, presumably because chelation with the oxygen atom is no longer possible.

[1] D. A. Claremon, P. K. Lumma, and B. T. Phillips, *Am. Soc.*, **108**, 8265 (1986).

1,3-Dimethyl-2-phenylbenzimidazoline (DMBI),

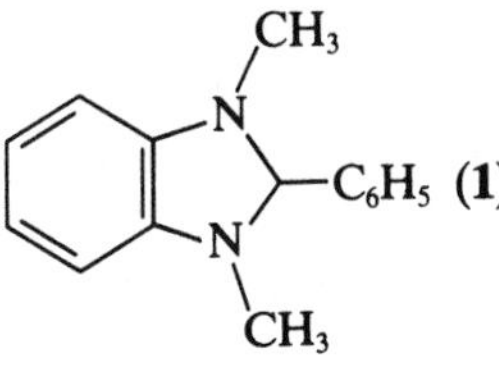

Preparation.[1]

Dehalogenation.[2] DMBI effects dehalogenation of α-halo carbonyl compounds in a variety of ethereal solvents with formation of DMBI⁺X⁻ in generally high yield. The order of relative reactivity is Br > Cl > F (halides) and primary > secondary > tertiary (for the α-substituted position). In combination with HOAc (1 equiv.) the reagent also reduces acyl chlorides to aldehydes (70–90% yield).

[1] J. C. Craig, N. N. Ekwuribe, C. C. Fu, and K. A. M. Walker, *Synthesis*, 303 (1981).
[2] H. Chikashita, H. Ide, and K. Itoh, *J. Org.*, **51**, 5400 (1986).

Dimethyl sulfoxide–Acetic anhydride.

—CHO → —COOH.[1] This oxidation can be effected in 75–100% yield by oxidation of the hydrogen sulfite adduct of the aldehyde with DMSO–Ac₂O. Esters or amides can be prepared by quenching the oxidation with an alcohol or an amine rather than water.

Example:

[1] P. G. M. Wuts and C. L. Bergh, *Tetrahedron Letters*, **27**, 3995 (1986).

Dimethyl sulfoxide–Oxalyl chloride.

Several laboratories have reported that Swern oxidation of alcohols can be accompanied of α-chlorination of keto or β-keto ester groups. Undesired electrophilic chlorination can be avoided by use of oxalyl chloride (1.05 equiv.) and DMSO (2.5 equiv.) in stoichiometric amounts or by use of acetic anhydride or trifluoroacetic anhydride in place of oxalyl chloride.[1]

Oxidation of amines to imines.[2] This Swern reagent effects oxidation of benzylamines to the corresponding Schiff bases in 40–60% yield. It also oxidizes indoline to indole in 88% yield.

Ring A diosterols.[3] The ring A diosterols (**3** and **4**) of triterpenes can be prepared from the Δ²-alkene (**1**) by osmylation to form the two possible *cis*-diols (**2**), which on Swern oxidation give the α-diketone (**3**). The same diketone is also obtained by Swern oxidation of the 2β,3α-diol, the product of peracid oxidation followed by acid cleavage. The diketone **3** rearranges to the more stable diosphenol (**4**) in the presence of base.

1　　　　2 ($\beta,\beta/\alpha,\alpha$ = 5:1)　　　　3　　　　4

[1] A. B. Smith, III, T. L. Leenay, H.-J. Liu, L. A. K. Nelson, and R. G. Ball, *Tetrahedron Letters*, **29**, 49 (1988).
[2] D. Keirs and K. Overton, *J.C.S. Chem. Comm.*, 1660 (1987).
[3] S. V. Govindan and P. L. Fuchs, *J. Org.*, **53**, 2593 (1988).

Dimethyl sulfoxide–Phenyl dichlorophosphate, $C_6H_5OP(O)Cl_2$.

Oxidation of alcohols.[1] This phosphate is apparently as efficient as oxalyl chloride for activation of DMSO for oxidation, and the derived reagent is less prone to give chlorine-containing byproducts. The reactions are rapid at 20° or below, and yields are generally 75–95%.

[1] H.-J. Liu and J. M. Nyangulu, *Tetrahedron Letters*, **29**, 3167 (1988).

Dimethyl sulfoxide–Phosphorus pentoxide.

Oxidation. DMSO activated by P_2O_5 (1 equiv.) and in combination with triethylamine is useful for oxidation of alcohols to ketones and aldehydes, particularly in cases where the Swern reagent results in chlorinated byproducts. Yields are typically 80–85%.

[1] D. F. Taber, J. C. Amedio, Jr., and K.-Y. Jung, *J. Org.*, **52**, 5621 (1987).

Dimethyl sulfoxide–Trifluoroacetic anhydride.

α-Dicarbonyl compounds.[1] This variant of the Pfitzner–Moffatt oxidant (**7**, 136; **10**, 168) can be superior to DMSO-oxalyl chloride for oxidation of acyclic and cyclic *vic*-diols to α-dicarbonyl compounds.

[1] C. M. Amon, M. G. Banwell, and G. L. Gravatt, *J. Org.*, **52**, 4851 (1987).

Dimethyl sulfoxide–Trifluoromethanesulfonic acid.

α-*Ketols from epoxides*.[1]   Epoxides can be converted to α-ketols by reaction with DMSO and a strong arenesulfonic acid (**5**, 265) but this reaction has seen little use probably because of the need for a strong acid. Actually this oxidative cleavage can be carried out at 25° with DMSO and triflic acid (1 equiv.) in CH_2Cl_2 in 55–70% yield in the case of simple epoxides. The example shown here indicates that the reaction can be stereospecific and that an acetonide group is not affected.

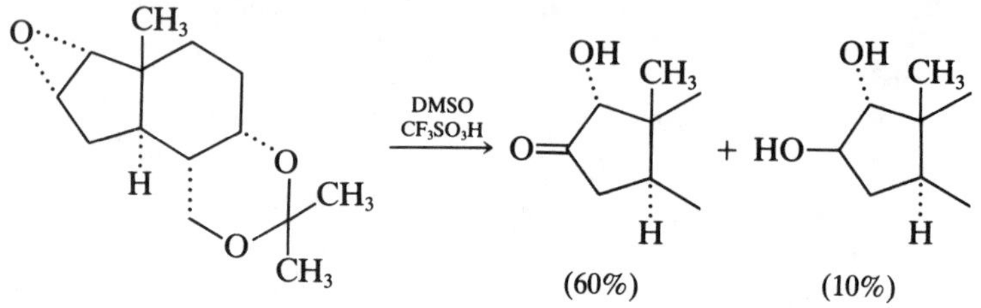

(60%)   (10%)

[1] B. M. Trost and M. J. Fray, *Tetrahedron Letters*, **29**, 2163 (1988).

Dimethylsulfoxonium methylide.

Review.[1]   This review covers synthetic uses of Corey's reagent from January 1974 to December 1985 (345 references).

[1] Yu. G. Gololobov, A. N. Nesmeyanov, V. P. Lysenko, and I. E. Boldeskul, *Tetrahedron*, **43**, 2609 (1987).

Dimethylzinc–Titanium(IV) chloride. These reagents are known to form CH_3TiCl_3 and $(CH_3)_2TiCl_2$ (**10**, 270).

Carbosulfenylation of alkenes.   This organotitanium reagent reacts with the *trans*-β-chloro sulfide **2a**, formed by addition of phenylsulfenyl chloride to cyclo-

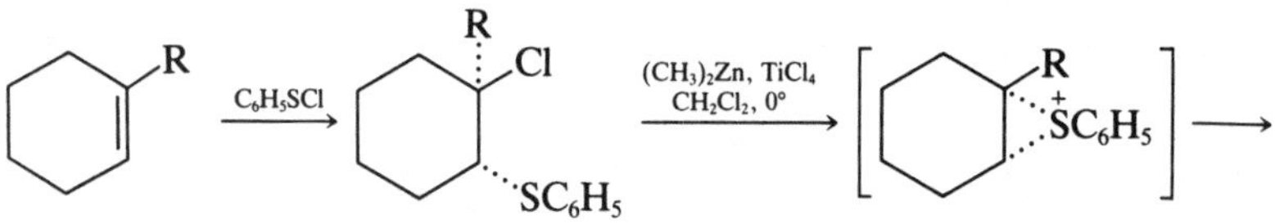

2a, R = H
   b, R = CH_3

a

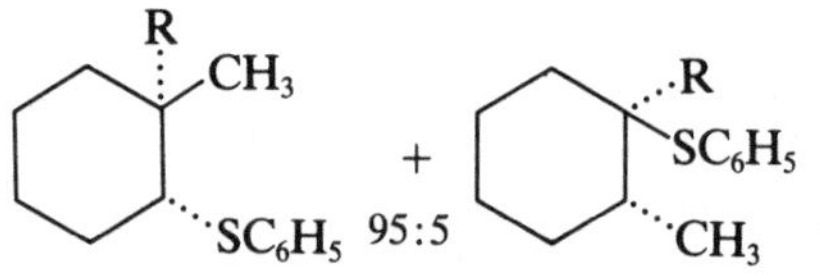

95:5

3   **4b**

hexene, to give **3a** as the major product (68% yield). The coupling with retention of configuration is ascribed to an episulfonium intermediate (**a**). A similar reaction with **2b** affords **3b** and **4b** in 59% yield in the ratio 95:5. This methylation occurs regioselectively at the more substituted position.

Methylsulfenylation of acyclic chiral alkenes such as **5** can also be regio- and stereoselective as a result of steric and electronic factors.

Example:

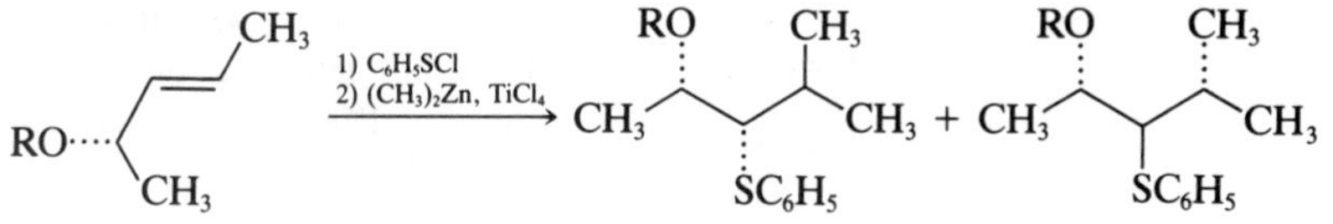

| R = CH$_3$ | 50% | 99:1 |
| = Si(CH$_3$)$_3$ | 69% | 84:16 |

[1] M. T. Reetz and T. Seitz, *Angew. Chem. Int. Ed.*, **26**, 1028 (1987).

4H-1,3-Dioxin (1).
Preparation[1]:

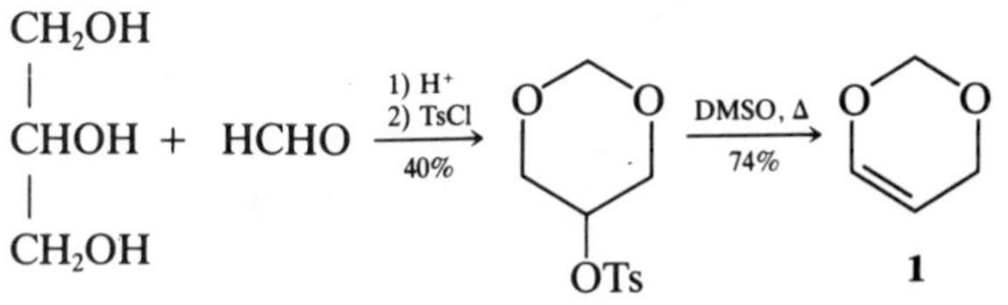

α,β-Enals.[2] The anion (*sec*-BuLi) of **1** reacts with electrophiles, including alkyl halides, carbonyl compounds, epoxides, to provide 4-substituted (4*H*)-1,3-dioxins (**2**), which on mild acid treatment (HCl,CH$_3$OH, 0°) are hydrolyzed to β-hydroxy acetals (**3**). Although the α,β-enal can be liberated by treatment with aqueous HCl, a milder conversion is achieved by retrocycloaddition in refluxing toluene (110°), which affords the (E)-α,β-enal (**4**).

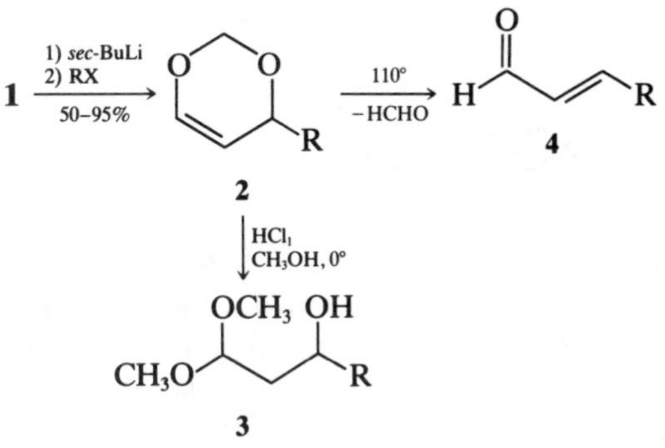

Example:

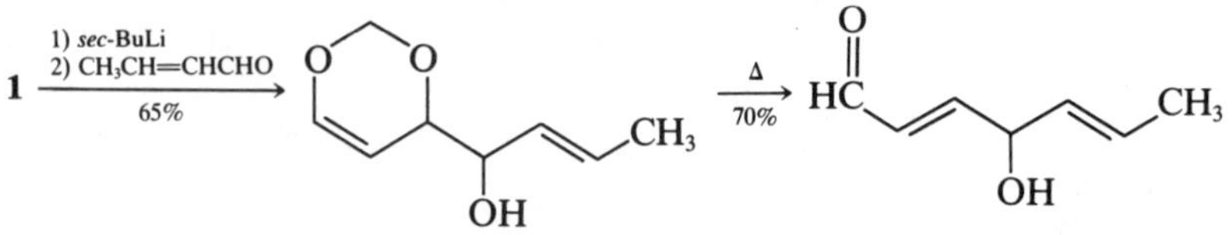

4,6-Dialkyl-1,3-dioxins.[3] Deprotonation of a 4-alkyldioxin (**2**) with *sec*-BuLi (−78°) followed by alkylation with CH_3I provides a 4,6-dialkyl-1,3-dioxin (**3**) as the major product. Such dioxins also undergo a retro Diels–Alder reaction to

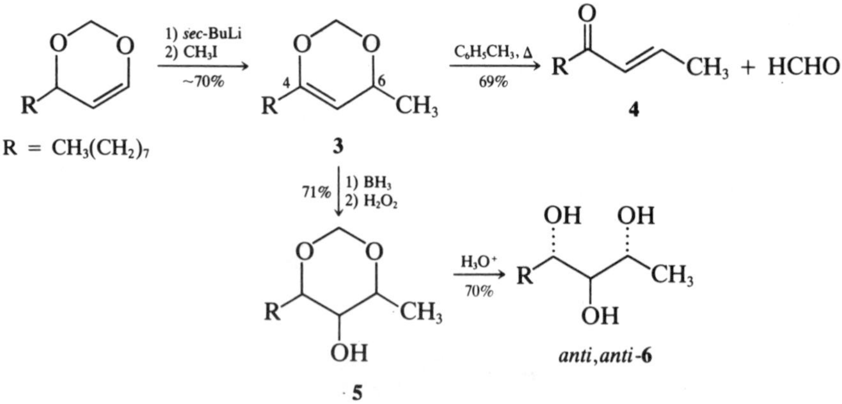

R = $CH_3(CH_2)_7$

provide α,β-enones (**4**). Hydroboration of **3** followed by oxidation results in a single 5-hydroxydioxan (**5**), which is hydrolyzed to an *anti*, *anti*-1,2,3-triol (**6**). Similarly, catalytic hydrogenation of **3** furnishes a dioxan that is hydrolyzed to a *syn*-1,3-diol.

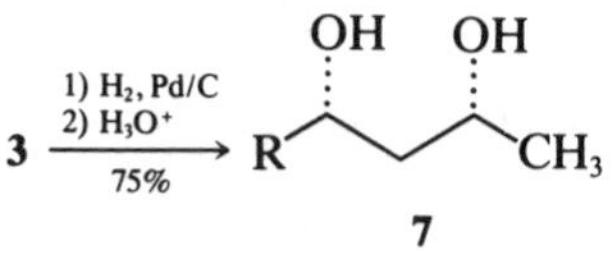

[1] R. Camerlynck and M. Anteunis, *Tetrahedron*, **31**, 1837 (1975).
[2] R. L. Funk and G. L. Bolton, *Am. Soc.*, **110**, 1290 (1988).
[3] *Idem*, *Tetrahedron Letters*, **29**, 1111 (1988).

Diphenylboryl trifluoromethanesulfonate.

Asymmetric esterification. *meso*-Cyclohexanedicarboxylic anhydride (**1**) undergoes a highly stereoselective esterification with the diphenylboric ester (**2**) of (R)-2-methoxy-1-phenylethanol in the presence of diphenylboryl triflate to provide, after reaction with diazomethane, the optically active ester **3**.[1] The asymmetric esterification can be used to resolve racemic **1**. Thus *dl*-**1** (2 equiv.) reacts with **2** in the presence of the triflate to give, after esterification with diazomethane,

meso-**1**

(1S,2R)-**3**, (99% de)

dl-**1**

(1R,2R)-**3**, $\alpha_0 - 87.6°$ (90% de)

(1R,2R)-**3** in 90% de.[2] Evidently, the (1R,2R)-enantiomer of **1** is esterified in preference to the (1S,2S)-enantiomer.

[1] M. Ohshima and T. Mukaiyama, *Chem. Letters*, 377 (1987).
[2] M. Ohshima, N. Miyoshi, and T. Mukaiyama, *ibid.*, 1233 (1987).

Diphenyl disulfide.

Radical [3 + 2]cycloaddition.[1] Cyclopentanes can be prepared by addition of alkenes across vinylcyclopropanes catalyzed by phenylthio radicals formed from $(C_6H_5S)_2$ and AIBN. A Lewis acid such as $Al(CH_3)_3$ can increase the rate and the stereoselectivity of this radical initiated cycloaddition. Thus the combination of the vinylcyclopropyl ester **1** with *t*-butyl acrylate (**2**) provides the four possible cyclo-

1

cis-**3**

R = R' = COOC(CH₃)₃   53%   3:1
        + Al(CH₃)₃   52%   6:1

trans-**3**

pentanes with some preference for the *cis*-adducts. This *cis*-selectivity is enhanced by the presence of a Lewis acid.

[1] K. S. Feldman, A. L. Romanelli, R. E. Ruckle, Jr., and R. F. Miller, *Am. Soc.*, **110**, 3300 (1988).

S-(−)-2-(Diphenylhydroxymethyl)pyrrolidine [(S)-Diphenylprolinol], 1, m.p. 75°, α_D − 68.1°.

The reagent is prepared by reaction of C_6H_5MgCl with the N-Cbz derivative of (S)-proline methyl ester.

Enantioselective reduction of ketones.[1] The 1:1 complex **2** can be isolated from the reaction of **1** with $BH_3 \cdot THF$ (3 equiv.). It forms another complex (**3**) with BH_3 in solution, which can be identified spectroscopically. A complex similar to **2**

(S)-**1**   (S)-**2**, m.p. 107–124°   **3**

is also formed on reaction of BH_3 with (S)-diphenylvalinol, and identified as an intermediate in the asymmetric reductions of that amino alcohol with BH_3. Similarly, complex **2** in combination with BH_3 can effect highly enantioselective reduction of ketones, but **2** can function catalytically. Thus optimum asymmetric reduction is effected with 0.6 equiv. of BH_3 and 0.05 equiv. of (S)-**2** in THF at 25°, and is the highest enantioselectivity observed so far for reductions effected with $LiAlH_4$ or BH_3 in combination with a chiral amino alcohol. Typical enantioselectivities are shown for some of the alcohols obtained by this procedure.

$C_6H_5CHCH_3$ $C_6H_5CHC_2H_5$ $(CH_3)_3CCHCH_3$
| | |
OH   OH   OH
(R), 97% ee   (R), 90% ee   (R), 88% ee   (R), 89% ee

Chemical yields in all cases are >99.7%, and the reductions are complete in about 1 minute. The paper includes a rational explanation for asymmetric reductions with complex **3**.

[1] E. J. Corey, R. K. Bakshi, and S. Shibata, *Am. Soc.*, **109**, 5551 (1987).

β-Diphenylphosphinopropanoic acid, $(C_6H_5)_2PCH_2CH_2COOH$ (**1**). The ester is available from Strem.

Wittig reagents.[1] Wittig reagents prepared from **1** show higher (E)-selectivity than the corresponding ones prepared from triphenylphosphine. Moreover the by-product, $(C_6H_5)_2P(O)CH_2CH_2COOH$, is soluble in water and readily separated from the alkene.

[1] H. Daniel and M. LeCorre, *Tetrahedron Letters*, **28**, 1165 (1987).

2,3-Diphenylsulfonyl-1,3-butadiene.

3-Pyrrolines.[1] This diene undergoes a [4 + 1]annelation with primary amines to form a pyrrolidine **2** that can be converted to a 3-(3-phenylsulfonyl)pyrroline (**3**) in high yield. These pyrrolines are oxidized by DDQ to pyrroles (**4**), which can be converted readily to 2,3-disubstituted pyrroles (**5**).

[1] A. Padwa and B. H. Norman, *Tetrahedron Letters*, **29**, 3041 (1988).

1,3-Dithiane.

Ring enlargement.[1] A new route to seven-membered ring systems from a cyclohexenone (**1**) involves a photocycloaddition of ethylene to provide the bicyclooctanone **2**. Addition of lithio-1,3-dithiane to **2** provides the adduct **3**, which on reaction with HgO and HBF$_4$ forms an unstable rearranged hydroxy aldehyde

(a) that undergoes retroaldol cleavage to a cycloheptane derivative that is oxidized to the dicarboxylic acid **4**.

[1] B. C. Ranu and D. C. Sarkar, *J.C.S. Chem. Comm.*, 245 (1988).

E

Ephedrine.

Enantioselective conjugate addition of cuprates. This enantioselective reaction has been demonstrated using the amino alcohol **1**, prepared by reaction of (2-chloroethyl)dimethylamine with(1R,2S)-(−)ephedrine, as a ligand.[1] The cuprates obtained from **1** by deprotonation (RLi), reaction with $CuI \cdot S(CH_3)_2$, and

1, α_D + 1.86°

then with an additional equivalent of RLi, add to cyclohexenone to give (R)-3-alkylcyclohexanones (**2**) in 85–92% ee with recovery of **1**. Optical yields are critically dependent on the purity of RLi. The enantioselectivity is explained on the basis of the cuprate model A.

A

Enantioselective addition of R_2Zn to aldehydes. Corey and Hannon[2] have prepared the diamino benzylic alcohol **1** from (S)-proline and (1S,2R)-(+)-ephedrine and report that the chelated lithium salt of **1** is an effective catalyst for enantioselective addition of diethylzinc to aromatic aldehydes. Thus benzaldehyde can be converted into (S)-(−)-**3** with 95% ee, via an intermediate tridentate lithium complex such as **2** formed from **1**. Similar reactions, but catalyzed by diastereomers of **1**, show that the chirality of addition of dialkylzincs to aldehydes is controlled by the chirality of the benzylic alcohol center of **1**.

The same laboratory has prepared three tridentate zinc chelates from chiral tertiary amino phenolic alcohols and used them for enantioselective addition of diethylzinc to aryl aldehydes in 70–87% ee. Results with the ligand **4** [from (1S,2S)-(+)-pseudoephedrine] are typical.

[1] E. J. Corey, R. Naef, and F. J. Hannon, *Am. Soc.*, **108**, 7114 (1986).
[2] E. J. Corey and F. J. Hannon, *Tetrahedron Letters*, **28**, 5233, 5237 (1987).

(Ethoxycarbonyliodomethyl)triphenylphosphonium iodide,
$(C_6H_5)_3P^+$—$CHICOOC_2H_5I^-$ (1). The salt is prepared by iodination of a two-phase system of solid K_2CO_3 and (ethoxycarbonylmethyl)triphenylphosphonium bromide in CH_3OH; m.p. 151°, 80% yield.

Propiolic acids.[1]   The ylide derived from **1** converts aldehydes into propiolic esters or the acids in 50–70% yield.

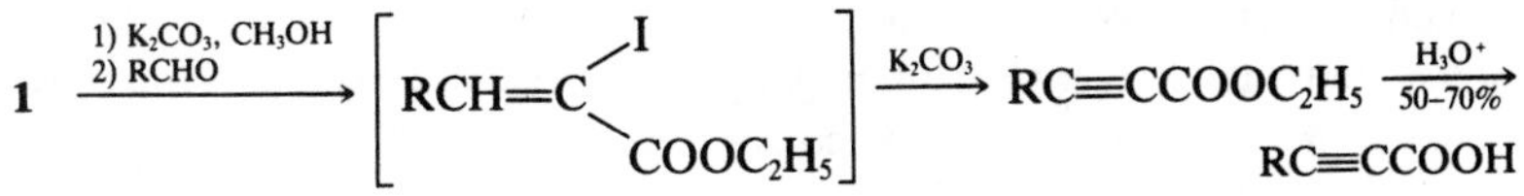

$$1 \xrightarrow[\text{2) RCHO}]{\text{1) K}_2\text{CO}_3,\ \text{CH}_3\text{OH}} \left[RCH{=}C \begin{smallmatrix} I \\ \\ COOC_2H_5 \end{smallmatrix} \right] \xrightarrow{\text{K}_2\text{CO}_3} RC{\equiv}CCOOC_2H_5 \xrightarrow[50-70\%]{\text{H}_3\text{O}^+} RC{\equiv}CCOOH$$

[1] J. Chenault and J.-F. E. Dupin, *Synthesis*, 498 (1987).

Ethyl 2-bromocrotonate, $CH_3CH{=}C(Br)COOC_2H_5$ (**1**). The reagent is prepared by bromination of ethyl crotonate followed by dehydrobromination with DBU.

Vinylcyclopropanation; cyclopentenes. The lithium dienolate of **1** adds to α,β-enones to form an α-keto vinylcyclopropane, which on pyrolysis (550°) provides a bicyclic keto acrylate (equation I).[1]

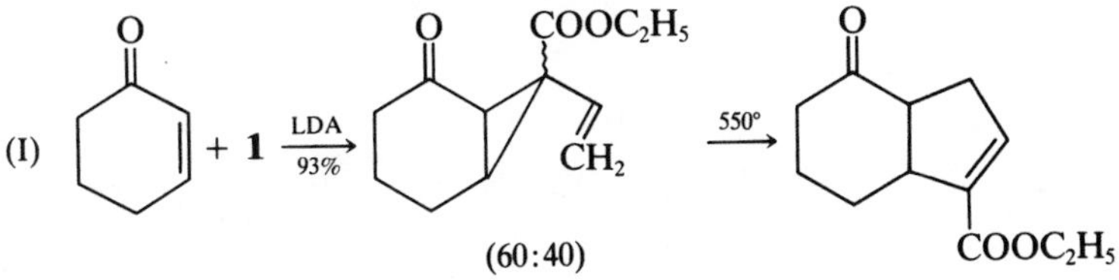

(I)

(60:40)

This [3 + 2]cyclopentene annelation has been used to obtain the sesquiterpene pentalenene (**6**) from the enone **2**.[2] Reaction of **2** with the dienolate **1** provides the

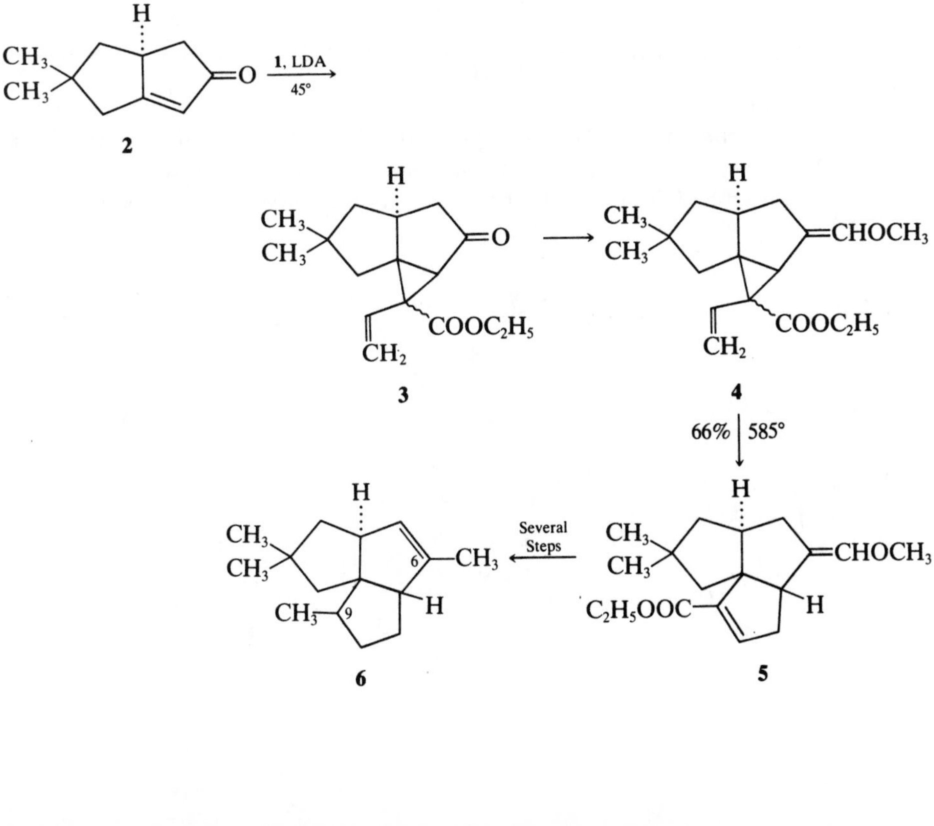

2

3

4

66% | 585°

6

5

expected adduct **3**. Pyrolysis of **3** does not result in the expected tricyclic cyclo-pentene, but the derivative **4** undergoes the expected cleavage to provide the triquinane (**5**), which is a useful precursor to **6**.

[1] T. Hudlicky, L. Radesca, H. Luna, F. E. Anderson, III, *J. Org.*, **51**, 4746 (1986).
[2] T. Hudlicky, M. G. Natchus, G. Sinai–Zingde, *ibid.*, **52**, 4641 (1987).

$$\overset{CH_2}{\underset{\|}{}}$$

Ethyl α-(bromomethyl)acrylate, $BrCH_2CCOOC_2H_5$ (**1**).

α-Methylene-γ-lactams.[1] The Reformatsky zinc reagent derived from this acrylate reacts with imines to form α-methylene-γ-lactams.

Example:

$$1 + Zn \xrightarrow[\text{17–20°}]{\text{THF,}} BrZnCH_2\overset{CH_2}{\underset{\|}{C}}COOC_2H_5 \xrightarrow[\text{75–80\%}]{ArCH=NCH_3}$$

[1] N. E. Alami, C. Belaud, and J. Villieras, *Tetrahedron Letters*, **28**, 59 (1987).

Ethyl (S)-3-hydroxybutanoate, Preparation.[1]   Suppliers: Aldrich, Fluka.

2-Azetidinones.[2] The dianion (BuLi) of **1** condenses with the N-aryl aldimine **2** to provide a 1:1 mixture of the *trans-* and *cis-*adducts **3**. The adducts (**3**) can be converted by Mitsunobu silylation (inversion), oxidative degradation of the side

1 +

2, Ar = $C_6H_4OCH_3$-*p*

*trans-***3**

*cis-***3**

4 (α$_D$ = 67°)

(+)-**5**

chain, and oxidative dearylation (CAN) into the 3-(hydroxyethyl)-2-azetidinone **4**, a known precursor to natural (+)-thienamycin (**5**). Thus use of **1** solves the difficult problem in an asymmetric synthesis of carbapenems such as **5** of construction of the three adjacent chiral centers at C_5, C_6, and C_8. Carbapenems with the (S)-configuration at C_8 (olivanic acids) can be prepared via the same sequence by omission of the Mitsunobu inversion step.

[1] D. Seebach, M. A. Sutter, R. H. Weber, and M. F. Züger, *Org. Syn.*, **63**, 1 (1984).
[2] G. I. Georg, J. Kant, and H. S. Gill, *Am. Soc.*, **109**, 1129 (1987).

F

Ferric chloride.

Ene catalyst.[1] FeCl$_3$ is superior to ZnBr$_2$ or alkylaluminum halides as a catalyst for ene cyclization of the chiral 1,7-diene **1**, the Knoevenagel adduct from citronellal and dimethyl malonate. Thermal cyclization provides the 1,2-*trans*-substituted

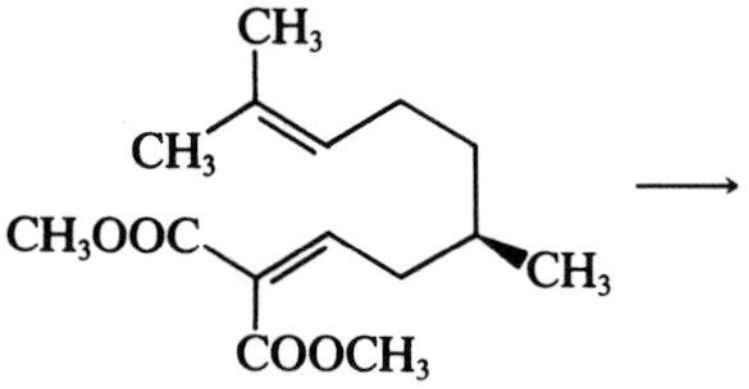

1

	180°	75%
+ FeCl$_3$,	25°	84%
+ FeCl$_3$/Al$_2$O$_3$,	−78°	92%

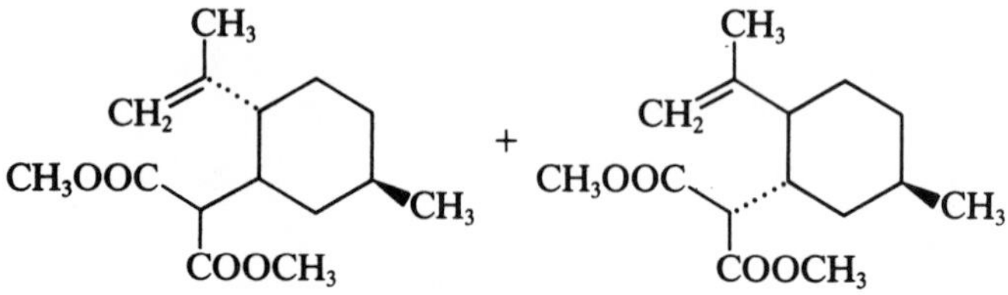

2a + **2b**

90:10
96:4
98.8:1.2

products **2a** and **2b** in the ratio 90:10. Catalysis with FeCl$_3$ (0.01 equiv.) permits a lower temperature, and increases the overall yield and the diastereoselectivity. The diastereoselectivity is further improved by use of FeCl$_3$ adsorbed in Al$_2$O$_3$ or SiO$_2$.

Si-directed Nazarov cyclization (**13**, 133–134). Denmark[2] has extended the Si-directed cyclization of β-silyl divinyl ketones to preparation of linear tricycles (triquinanes). These cyclizations proceed very readily even at low temperatures, and the position of the double bond is controlled by the silyl group. The reactions

(CH₃)₃Si

$$(CH_3)_3Si$$

1

2 (*trans/anti*)

$$FeCl_3, CH_2Cl_2 \quad -50° \quad 79\%$$

3

4a (*trans/anti*)

$$FeCl_3 \quad -20° \quad 36\%$$

5

6 (*cis/anti*)

$$FeCl_3 \quad -20° \quad 73\%$$

are generally stereoselective, resulting in the (6,5,6)- and (5,5,6)-*trans/anti*-config-uration, and the (5,5,5)-*cis-/anti*-configuration.

Cleavage of benzyl ethers.[3] Anhydrous FeCl₃ in CH₂Cl₂ cleaves carbohy-drate benzyl and *p*-phenylbenzyl ethers at 25° without effect on methyl ethers or acetate and benzoate groups. Yields are usually >70%.

Conversion of MEM ethers to esters.[4] 2-Methoxyethoxymethyl (MEM) ethers are converted into carboxylic esters by reaction with an anhydride in the presence of FeCl₃ (0.4 equiv.) (equation I). Selective cleavage is possible in the presence of a benzyl ether but not in the presence of a *t*-butyl ether. Aromatic rings, if present, can undergo acylation.

(I) $\quad ROCH_2O(CH_2)_2OCH_3 + (CH_3CO)_2O \xrightarrow[65-95\%]{FeCl_3} ROCOCH_3$

[1] L. F. Tietze and U. Beifuss, *Synthesis*, 359 (1988).
[2] S. E. Denmark and R. C. Klix, *Tetrahedron*, **44**, 4043 (1988).
[3] M. H. Park, R. Takeda, and K. Nakanishi, *Tetrahedron Letters*, **28**, 3823 (1987).
[4] R. S. Gross and D. S. Watt, *Syn. Comm.*, **17**, 1749 (1987).

Ferrocenylphosphines, chiral, **11**, 237–240.

Asymmetric aldol reaction.[1] In the presence of a gold(I) complex (**1**) and a chiral ferrocenylphosphine (**2**), various aldehydes react with methyl isocyanoacetate

(3) to form 5-alkyl-2-oxazoline-4-carboxylates (4) with high enantio- and diastereoselectivity. Example:

$[Au(c\text{-}HexNC)_2]^+BF_4^-$ (1)

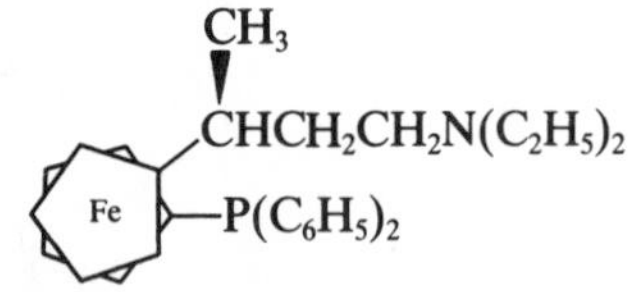

$CH_3CHO + CNCH_2COOCH_3$ $\xrightarrow[100\%]{\substack{1, 2, \\ CH_2Cl_2, 25°}}$

3

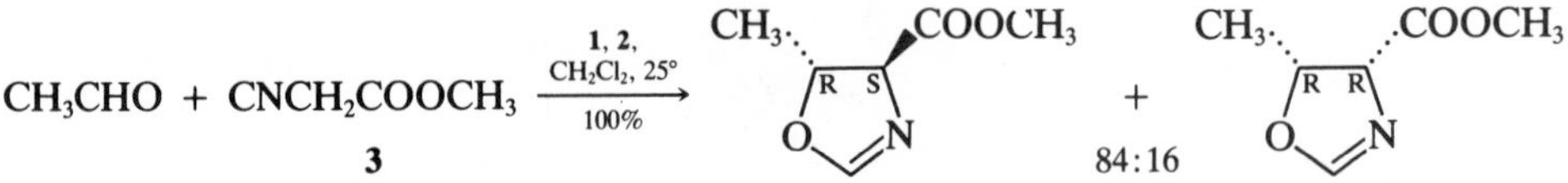

trans-**4**
(72% ee)

+

84:16

cis-**4**
(44% ee)

The same reaction but with trimethylacetaldehyde results in the *trans*-adduct exclusively in 100% yield and in 97% ee. The use of the gold catalyst is essential for the high selectivity; silver and copper catalysts are much less effective. The length of the side chain between the diethylamino group and the ferrocene nucleus is also an important factor in the selectivity.

Asymmetric hydrogenation of α,β-unsaturated acids.[2] 2-Aryl-3-methyl-2-butenoic acids (2) undergo highly stereoselective hydrogenation catalyzed by a complex of rhodium with the chiral (aminoalkyl)ferrocenylphosphine (1), but not with

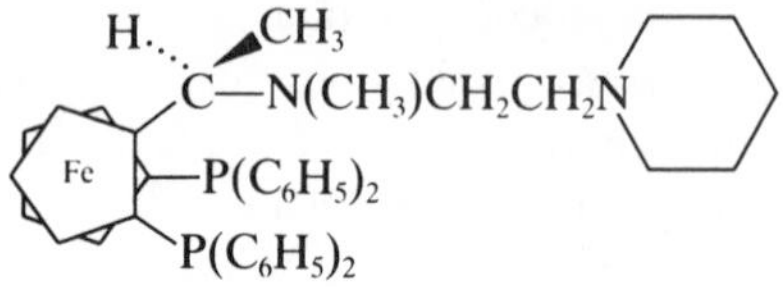

(R)-(S)-**1**

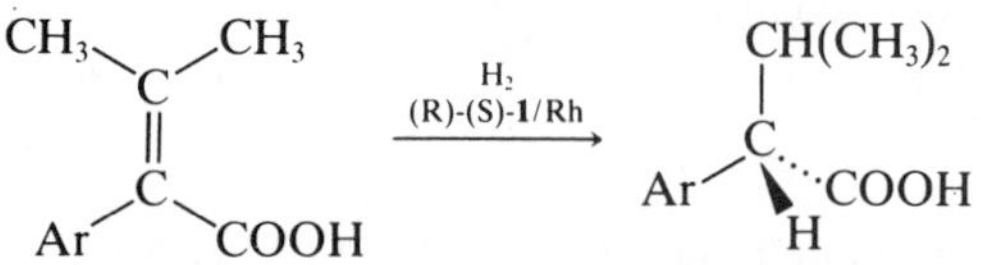

2

$\xrightarrow[(R)\text{-}(S)\text{-}1/Rh]{H_2}$

(S)-**3** (97–98.4% ee)

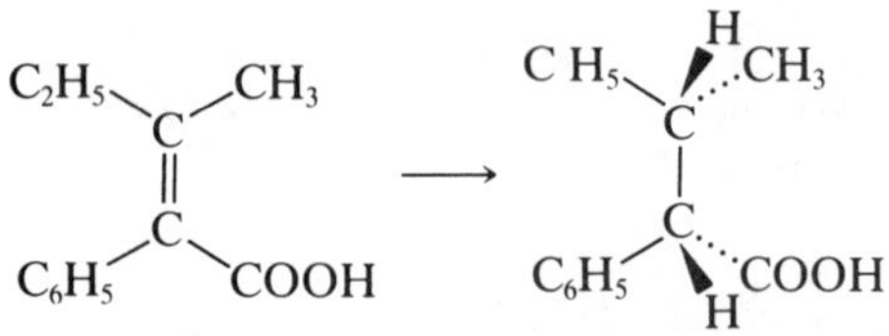

4

$\longrightarrow$

(2S,3S)-**5** (97% de)

ferrocenylphosphines lacking the terminal alkylamino side chain. This hydrogenation also provides access to acids with two vicinal chiral carbon atoms such as (2S,3S)-**5**.

[1] T. Hayashi, *Pure Appl. Chem.*, **60**, 7 (1988); Y. Ito, M. Sawamura, and T. Hayashi, *Am. Soc.*, **108**, 6405 (1986).
[2] T. Hayashi, N. Kawamura, and Y. Ito, *ibid.*, **109**, 7876 (1987).

Fluorine.

Electrophilic substitution.[1] Tertiary carbon atoms can undergo electrophilic substitution when treated in $CFCl_3/CHCl_3$ with dilute F_2 in nitrogen. The chloroform acts as a radical scavenger and stabilizer for F^-. The substitution proceeds with retention. The reaction shows some regioselectivity. Although **1** has three tertiary hydrogens, the major product of fluorination is **2**.

These products undergo dehydrofluorination when treated with BF_3 etherate or CH_3MgI.

[1] S. Rozen and C. Gal, *J. Org.*, **52**, 2769 (1987).

N-Fluorotrifluoromethylsulfonimide, $(CF_3SO_2)_2NF$ (**1**). The imide is obtained in 95% yield by reaction $(CF_3SO_2)_2NH$ with F_2 at $-196°$ with gradual warming to $22°$. The imide is stable at $22°$, but should be stored in a fluoropolymer plastic container rather than glass.

The N-fluoroimide effects direct aromatic fluorination at 22°, with high preference for *ortho*-substitution of substituted arenes. It also converts $NaC(CH_3)$-$(COOC_2H_5)_2$ into $FC(CH_3)(COOC_2H_5)_2$ in 96% yield.[1]

[1] S. Singh, D. D. DesMarteau, S. S. Zuberi, M. Witz, and H.-N. Huang, *Am. Soc.*, **109**, 7194 (1987).

Formaldehyde.

Piperidines. Grieco *et al.*[1] have described a general synthesis of piperidines by reaction of the acid salt of a primary amine with an allylsilane and 2 equiv. of formaldehyde in water. The reaction involves reaction of iminium ion (**a**), derived from the amine and formaldehyde, with the allylsilane to form a homoallylamine (**b**), which can form a second iminium ion (**c**), which cyclizes with capture of water to the piperidine.

This reaction was used to effect a polyolefin cyclization in a synthesis of the alkaloid yohimbone (**3**).[2] Thus reaction of the trifluoroacetate of **1** with HCHO provides a 63% yield of **2**, which was converted into **3** by a six-step sequence (10% overall yield).

Aminomethano destannylation; bishomoallyl amines. Methyleneimmonium trifluoroacetates, generated *in situ* from primary amines and formaldehyde, react with allyltrialkyltin compounds to form bishomoallylic amines in high yield:

Example:

The reaction can effect cyclization of the side chain of tryptamine (**1**) to form a substituted piperidine ring (**2**).[3]

[1] S. D. Larsen, P. A. Grieco, and W. F. Fobare, *Am. Soc.*, **108**, 3512 (1986).
[2] P. A. Grieco and W. F. Fobare, *J.C.S. Chem. Comm.*, 185 (1987).
[3] P. A. Grieco and A. Bahsas, *J. Org.*, **52**, 1378 (1987).

(Formylmethyl)triphenylarsonium bromide (**1**). The salt is prepared from $(C_6H_5)_3As$ and bromoacetaldehyde; m.p. 161°, yield, 90%.

(E,Z)-1,3-Dienes.[1] This unit, often present in insect pheromones, can be obtained with high stereoselectivity by reaction of an aldehyde with **1** to provide an (E)-α,β-enal (**2**), which on reaction with a phosphorane generated with BuLi and HMPT provides (E,Z)-dienes almost exclusively.

[1] Y.-Z. Huang, L. Shi, J. Yang, and Z. Cai, *J.Org.*, **52**, 3558 (1987).

G

Grignard reagents.

Monoaddition to esters. The reagent formed from RMgX and LDA (1:1) reacts with esters or amides to give enolates of ketones, which can be trapped by ClSi(CH₃)₃.[1] This technique provides a synthesis of artemisia ketone (**1**) by an aldol

reaction followed by reaction of the ester group with methallylmagnesium chloride and LDA. Actually Grignard addition/alkylation can provide complex ketones such as **2**.

Coupling of RMgX with dithioacetals.[2] In the presence of Cl₂Ni[P(C₆H₅)₃]₂, Grignard reagents react with 1,3-dithiolanes to give cross-coupled alkenes in moderate yield.

Example:

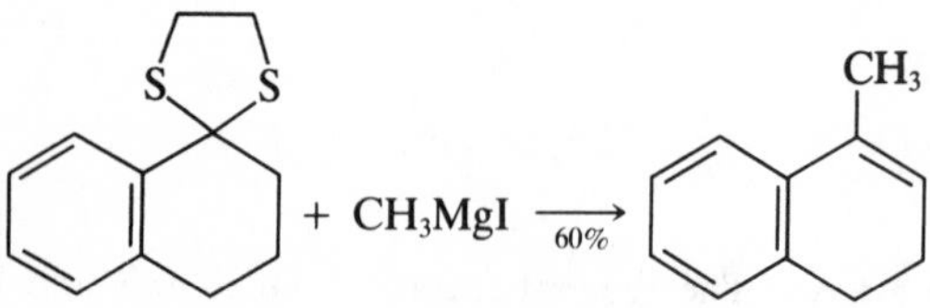

[1] C. Fehr and J. Galindo, *Helv.*, **69**, 228 (1986); C. Fehr, J. Galindo, and R. Perret, *ibid.*, **70**, 1745 (1987).
[2] Z.-J. Ni and T.-Y. Luh, *J.C.S. Chem. Comm.*, 1515 (1987).

H

N-Halosuccinimide–Sodium ethoxide.

Halodeacylation.[1] Reaction of a β-keto ester or a β-diketone with NCS or NBS and a base (alkoxide or KOH) results in an α-halo ester or an α-halo ketone by replacement of an acyl group by halogen.

Example:

$$(CH_3)_2CHCHCCH_3 \quad \xrightarrow[\substack{77\%}]{\substack{NCS,\ C_2H_5ONa, \\ C_2H_5OH,\ 20°}} \quad (CH_3)_2CHCHCHCOOC_2H_5$$

[1] G. Mignani, D. Morel, and F. Grass, *Tetrahedron Letters*, **28**, 5505 (1987).

Hexabutylditin, (Bu₃Sn)₂.

Iodine-transfer cyclization. Irradiation of unsaturated α-iodo carbonyl compounds in the presence of a hexaalkylditin (5–10%) can result in isomerization to cyclic γ-iodo carbonyls.[1] The reaction is very slow in the absence of an initiator. Thus under these conditions **1** isomerizes to a mixture of **2** and **3** in which **2** predominates. The reaction is particularly useful for formation of fused bicyclic systems (**4 → 5**).

Sunlamp irradiation of butynyl iodide (6) in the presence of hexabutylditin generates an alkyl radical that reacts with an electron-deficient alkene (7) to form an (iodomethylene)cyclopentene (8) in moderate yield. This product can be reduced by Bu₃SnH (AIBN) to the methylenecyclopentane (9).[2]

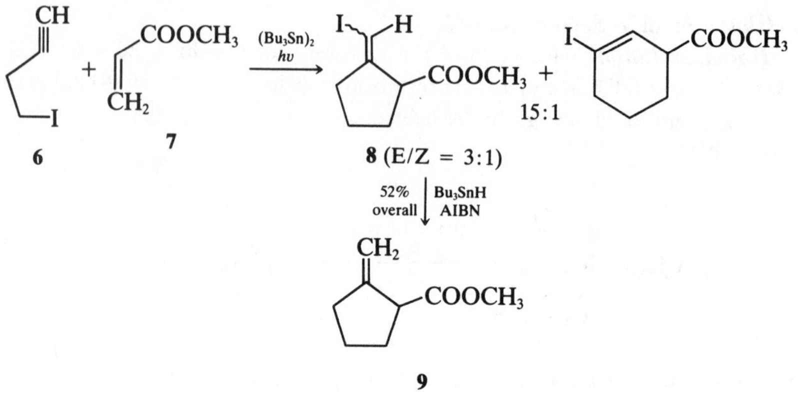

CH
‖
+   COOCH₃   (Bu₃Sn)₂   →   I   H   I   COOCH₃
CH₂   hv   COOCH₃   +
I

6   7   8 (E/Z = 3:1)   15:1

52%   Bu₃SnH
overall AIBN

CH₂
COOCH₃

9

[1] D. P. Curran and C.-T. Chang, *Tetrahedron Letters*, **28**, 2477 (1987).
[2] D. P. Curran and M.-H. Chen, *Am. Soc.*, **109**, 6558 (1987).

2,3,4,5,6,6-Hexachloro-2,4-cyclohexadiene-1-one (1), 11, 251.

Dehydrogenation.[1]   The tetrahydro-β-carboline **2** can be dehydrogenated directly to the dihydro-β-carboline **4** by two equiv. of **1**, but the conversion is effected

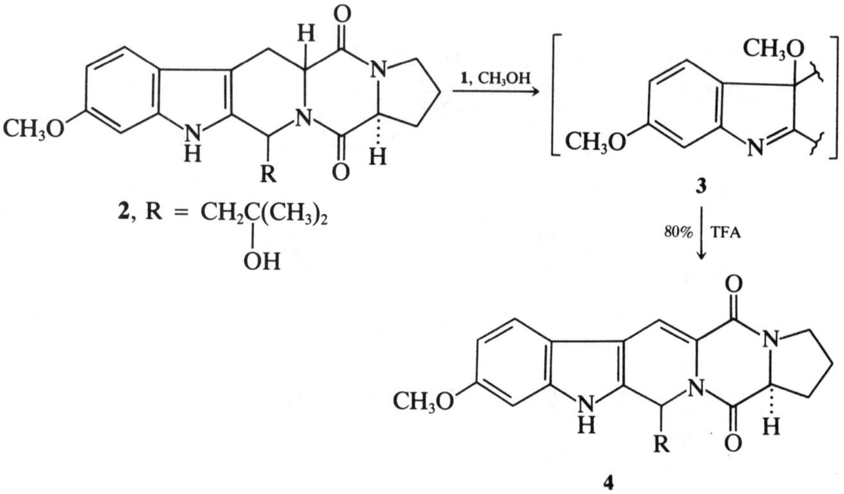

CH₃O

H   O
N

N
N
R   O   H

2, R = CH₂C(CH₃)₂
          |
          OH

1, CH₃OH   →

CH₃O   CH₃O
N

3

80%   TFA

CH₃O

O
N

N
N
H   R   O   H

4

in higher yield by a two-step reaction, as shown. Osmylation of the isolated double bond of **4** furnishes a diol, which is a known mycotoxin of some fungi.

[1] P. H. H. Hermkens, R. Plate, and H. C. J. Ottenheijm, *Tetrahedron Letters*, **29**, 1323 (1988).

Hexa-μ-hydrohexakis(triphenylphosphine)hexacopper

[hydrido(triphenylphosphine)copper(I), hexameric], $[(C_6H_5)_3PCuH]_6$ (**1**). Preparation by reaction of $[P(C_6H_5)_3CuCl]_4$ with sodium trimethoxyborohydride.[1]

Conjugate reduction.[2]   This stable copper(I) hydride cluster can effect conjugate hydride addition to α,β-unsaturated carbonyl compounds, with apparent utilization of all six hydride equivalents per cluster. No 1,2-reduction of carbonyl groups or reduction of isolated double bonds is observed. Undesirable side reactions such as aldol condensation can be suppressed by addition of water. Reactions in the presence of chlorotrimethylsilane result in silyl enol ethers. The reduction is stereoselective, resulting in hydride delivery to the less-hindered face of the substrate.

Examples:

[1] M. R. Churchill, S. A. Bezman, J. A. Osborn, and J. Wormald, *Inorg. Chem.*, **11**, 1818 (1972).
[2] W. S. Mahoney, D. M. Brestensky, and J. M. Stryker, *Am. Soc.*, **110**, 291 (1988).

Hexamethyldisilazane–Chlorotrimethylsilane.

Selective silylation.   Hexamethyldisilazane alone can effect silylation, but only at elevated temperatures. Rapid silylation of amines, alcohols, and acids can be achieved at 0° in CH_2Cl_2 if chlorotrimethylsilane is also present. Selective silylation is also possible by adjustment of the proportions of HMDS and TMSCl. Thus only the amino group of amino alcohols is silylated by HMDS (1 equiv.) and TMSCl (0.1 equiv.). The same ratio effects selective silylation of primary alcohols, but secondary alcohols are silylated by use of 1.3 equiv. of each reagent. Finally tertiary alcohols are silylated by use of 2.5 equiv. of each reagent together with DMAP as catalyst.

[1] J. Cossy and P. Pale, *Tetrahedron Letters*, **28**, 6039 (1987).

Hexamethylphosphoric triamide (HMPT).

Michael additions of ketone enolates.[1]   The stereochemistry of Michael additions of lithium enolates of ketones to α,β-enones is controlled by the geometry of the enolate. Addition of (Z)-enolates results in *anti*-products with high diastereoselectivity, which is not changed by addition of HMPT. Reaction of (E)-enolates is less stereoselective but tends to favor *syn*-selectivity, which can be enhanced by addition of HMPT.

[1] D. A. Oare and C. H. Heathcock, *Tetrahedron Letters*, **27**, 6169 (1986).

Hydrogen peroxide.

Cleavage of Si—C bonds (**12**, 243–245).   This oxidation can be used to convert vinylsilanes in three steps to *syn*- or *anti*-1,2-diols. Thus Grignard reagents cleave epoxides of vinylsilanes selectively to β-hydroxy silanes, which can be oxidized with retention of configuration to 1,2-diols. When applied to an (E)-vinylsilane, the sequence results in the *syn*-1,2-diol; the *anti*-1,2-diol is obtained from a (Z)-vinylsilane by the same reactions.

[1] K. Tamao, E. Nakajo, and Y. Ito, *J. Org.*, **52**, 4412 (1987).

Hydrogen peroxide–Benzeneseleninic acid.

RCHO $\longrightarrow$ RCOOH.[1]   This oxidation can be effected with 30% H_2O_2 (excess) in the presence of 5 mole % of $C_6H_5SeO_2H$ in refluxing THF in high yields. Both aliphatic and vinyl aldehydes undergo this reaction, but *ortho*-substituents can lower yields from aryl aldehydes.

[1] J.-K. Choi, Y.-K. Chang, and S. Y. Hong, *Tetrahedron Letters*, **29**, 1967 (1988).

Hydrogen peroxide–Diphenyl diselenide.

Oxidation of hydroquinones.[1]   Hydrogen peroxide (30%) in combination with ~0.5 mole % of diphenyl diselenide is an inexpensive but effective reagent for large-scale oxidation of hydroquinones to benzoquinones in aqueous CH_2Cl_2 containing $Bu_4N^+HSO_4^-$ as phase-transfer catalyst. Yields can range from 65 to 90%, but are poor with electron-poor substrates. The actual oxidant may be benzeneperoxyseleninic acid, $C_6H_5Se(O)OOH$, or benzeneseleninic acid, $C_6H_5Se(O)OH$, which are known to oxidize hydroquinones (**10**, 23–24). This new reagent also can effect Baeyer–Villiger oxidation, since it effects oxidation of vanillin to 2-methoxy-1,4-benzoquinone in 62% yield.

[1] D. V. Pratt, F. Ruan, and P. B. Hopkins, *J. Org.*, **52**, 5053 (1987).

Hydrogen peroxide–Selenium dioxide.

Oxidation of amines to nitrones.[1]   Secondary amines can be oxidized to nitrones by 30% H_2O_2 in the presence of a catalytic amount of SeO_2. The reaction is applicable to acyclic and cyclic amines. The products can be used without isolation in 1,3-dipolar cycloadditions.

Examples:

The actual oxidant may be peroxyseleninic acid, $HOSe(O)OOH$, which is known to oxidize secondary amines to hydroxylamines.

[1] S.-I. Murahashi and T. Shiota, *Tetrahedron Letters*, **28**, 2383 (1987).

Hydrosilanes–Tetrakis(triphenylphine)palladium(0)–Zinc chloride.

Reduction of α,β-unsaturated carbonyl compounds.[1]   Hydrosilanes, particularly $(C_6H_5)_2SiH_2$, in the presence of Pd(0), and a Lewis acid, particularly $ZnCl_2$, can effect selective conjugate reduction of unsaturated ketones, aldehydes, and carboxylic acid derivatives. Chloroform is the solvent of choice. In addition, 1 equiv. of water is required. Experiments with D_2O and $(C_6H_5)_2SiD_2$ indicate that the silane supplies the hydrogen at the β-position, while the hydrogen at the α-position is supplied by water. The reduction involves 1,4-hydrosilylation followed by hydrolysis of the intermediate silyl enol ether.

In the absence of $ZnCl_2$, the system effects chemoselective reductive cleavage of allylic acetates.[2]

[1] E. Keinan and N. Greenspoon, *Am. Soc.*, **108**, 7314 (1986).
[2] *Idem, J. Org.*, **48**, 3545 (1983).

3-Hydroxybutyric acid.

1,3-Dioxanone derivatives.[1] The optically active 1,3-dioxanone (acetal lactone) **1** is obtained by reaction of (R)-3-hydroxybutyric acid with pivalaldehyde in the presence of an acid catalyst. These derivatives can be used to effect enantioselective reactions at the 2-, 3-, and 4-positions of (R)-3-hydroxybutyric acid. Thus,

alkylation of the lithium enolate provides optically active 2-alkyl derivatives. Reaction of the enolate with C_6H_5SeCl followed by selenoxide elimination leads to the chiral acetoacetic acid derivative **3**. Addition of dialkyl cuprate to **3** provides 3-substituted derivatives with >95% de.

[1] D. Seebach and J. Zimmermann, *Helv.*, **69**, 1147 (1986).

[Hydroxy(bisphenoxyphosphoryloxy)iodo]benzene, $C_6H_5I—OP(OC_6H_5)_2$ (**1**).

(with O double bond above P and OH below I)

This iodine (III) reagent is obtained in 90% yield by reaction of $C_6H_5I(OAc)_2$ with diphenyl phosphate, $(C_6H_5O)_2PO_2H$, in aqueous CH_3CN.

Phosphorylation of ketones.[1]   Ketones bearing an adjacent methylene group react directly with **1** to form α-ketol phosphates in 60–80% yield. The reagent also effects cyclization of 4-pentenoic acid to a phosphorylated lactone.

$$CH_3COCH_3 + 1 \xrightarrow[81\%]{CH_3CN,\ \Delta} CH_3\overset{\overset{O}{\|}}{C}CH_2O\overset{\overset{O}{\|}}{P}(OC_6H_5)_2 + C_6H_5I$$

$$CH_2{=}CHCH_2CH_2COOH + 1 \xrightarrow[55\%]{CH_2Cl_2,\ 25°} (C_6H_5O)_2\overset{\overset{O}{\|}}{P}O\cdots$$

[1] G. F. Koser, J. S. Lodaya, D. G. Ray, III, and P. B. Kokil, *Am. Soc.*, **110**, 2987 (1988).

[Hydroxy(tosyloxy)iodo]benzene, $C_6H_5I(OH)OTs$ **(1),** m.p. 135–138°.

This trivalent iodine compound is obtained in about 90% yield by reaction of $C_6H_5I(OAc)_2$ with toluenesulfonic acid (2 equiv.) in CH_3CN.[1]

1,3-Enynes.   The reagent adds to terminal alkynes to form alkynylphenyliodonium tosylates **(2)** in moderate yield (equation I).[2] These salts couple with alkenylcopper reagents in ether to give 1,3-enynes, with >99% retention of the alkene

$$R'C{\equiv}CH + 1 \xrightarrow{CH_2Cl_2,\ 25°} [C_6H_5IC{\equiv}CR']^+OTs^-$$

R' = CH_3   19%   **2**
R' = *t*-Bu, C_6H_5   60–70%

geometry. Since the alkenylcopper reagents are formed by *syn*-addition to terminal alkynes, the coupling reaction provides a new route to pure 1,1-disubstituted 1,3-enynes with control of the alkene geometry.[3]

Examples:

BuC≡CH + PrCu ⟶ Bu C=C H / Pr Cu $\xrightarrow[48\%]{2 \ (R' \ = \ C_6H_5), \ -78° \rightarrow 25°}$

Bu C=C H / Pr C≡CC$_6$H$_5$

[1] G. F. Koser and R. H. Wettach, *J. Org.*, **42**, 1476 (1977).
[2] L. Rebrovic and G. F. Koser, *ibid.*, **49**, 4700 (1984); P. J. Stang, B. W. Surber, Z. C. Chen, K. A. Roberts and A. G. Anderson, *Am. Soc.*, **109**, 228 (1987).
[3] P. J. Stang and T. Kitamura, *ibid.*, **109**, 7561 (1987).

I

Indium.

β-Hydroxy esters.[1] Indium powder can promote a Reformatsky-type condensation of ethyl iodoacetate with aldehydes in THF at 25° to form β-hydroxy esters in 65–90% yield. Ketones can be used, but yields are lower.

$$RCHO + ICH_2COOC_2H_5 \xrightarrow[65-90\%]{\substack{In, THF \\ 25°}} R\overset{\overset{\displaystyle OH}{|}}{C}HCH_2COOC_2H_5$$

Allylation.[2] Indium also effects addition of allyl halides to carbonyl compounds under mild conditions.

$$RCHO + (CH_3)_2C{=}CHCH_2Br \xrightarrow[70-75\%]{\substack{In, DMF, \\ 25°}} R\overset{OH}{\underset{CH_3}{\overset{|}{C}}}{-}\underset{CH_3}{\overset{}{C}}{=}CH_2$$

[1] S. Araki, H. Ito, and Y. Butsugan, *Syn. Comm.*, **18**, 453 (1988).
[2] *Idem, J. Org.*, **53**, 1831 (1988).

Iodine.

Iodocyclization to hydroxytetrahydrofurans.[1] Cyclization of γ,δ-unsaturated alcohols to tetrahydrofurans (**12**, 254–256) can be directed by an allylic oxygen substituent, which also can increase the rate of cyclization. Thus derivatives of 4-pentene-1,3-diol undergo iodocyclization mainly to *cis*-3-hydroxy-2-iodomethyltetrahydrofurans.

R¹ = H, R² = H   15:1
R¹ = CH₃, R² = H   8.8:1
R¹ = CH₃, R² = Bzl   12.9:1

Diastereoselective iodolactonization of γ,δ-unsaturated acids.[2] Kinetic iodolactonization of the *meso*-1,6-heptadien-4-carboxylic acid (**1**) results in two prod-

181

ucts in the ratio 14:1 as a result of face selectivity for a 2,3-disubstituted 4-pentenoic acid group. Asymmetric induction is similar but somewhat lower for *meso-syn, syn-1*.

Iodolactonization of *anti,syn-1* could result in four iodolactones, two resulting from face selectivity, and two resulting from diastereotopic olefin selectivity. In practice only three lactones are formed in a 142:4.7:1 ratio, with **4** being essentially the only product. In fact this kinetic iodolactonization proceeds with 147:1 olefin selectivity and 30:1 face selectivity, considerably higher than the selectivity observed in previous iodolactonization of 3-methyl-4-pentenoic acid (**8**, 257). Lactonization of **1** also shows *cis*-C_4,C_5 selectivity.

[1] Y. G. Kim and J. K. Cha, *Tetrahedron Letters*, **29**, 2011 (1988).
[2] M. J. Kurth and E. G. Brown, *Am. Soc.*, **109**, 6844 (1987).

2-Iodomethyl-3-trimethylsilylpropene, $(CH_3)_3SiCH_2\overset{\overset{\displaystyle CH_2}{\|}}{C}CH_2I$ (**1**).

[3 + 4] *and* [3 + 5] *Annelation.*[1]   This reagent has been used as the equivalent of trimethylenemethane dianion for [3 + 2] annelation of enones to methylenecyclopentanes (**11**, 259–260). It has now been used for [3 + 4]annelation to 1,4-diketones to provide seven-membered carbocycles. Thus reaction of **1** with SnF_2 is believed to result in an allyltin trihalide, which can react with the diketone

2 to form a five-membered cyclic hemiketal (**a**), which cyclizes to carbocyclic **3**. A similar reaction of **1** with a 1,5-diketone results in an eight-membered carbocycle.

Keto aldehydes or even 1,4- and 1,5-dialdehydes can undergo this annelation, but yields are lower.

cis-1,2-Cyclohexanediols.[2] The reagent (**1**) combination with SnF_2 undergoes [3 + 3]cycloaddition with α,β-epoxy aldehydes (**2**) to furnish cis-1,2-cyclohexanediols (**3**).

When applied to an optically active substrate (**4**) derived from nerol or geraniol, optically active products (**5**) are obtained in about 50% yield with a diastereoselectivity of 5–9:1.

[1] G. A. Molander and D. C. Shubert, *Am. Soc.*, **109**, 6877 (1987).
[2] G. A. Molander and D. C. Shubert, *Am. Soc.*, **109**, 576 (1987).

Iodosylbenzene–Boron trifluoride.

RSn(CH₃)₃ ⟶ ROH. A methyl–tin bond is selectively cleaved by iodosylbenzene/BF_3 in CH_2Cl_2 at 0°. Quenching with NH_4X affords $RSn(CH_3)_2X$ in 80–

90% yield. The dimethylhalogenotin groups are converted to hydroxyl groups by oxidation with alkaline H_2O_2 (**12**, 243–245).[1]

This transformation has been used to effect stereoselective osmylation.[2] Thus reaction of **1a** with OsO_4 and hexamethylenetetramine (HMT) gives a 1:1 mixture of two *vic*-glycols (**2**). Osmylation of **1b** followed by reaction with CH_3MgBr affords mainly one *vic*-glycol. The significant effect is attributed to interaction between the tin atom and oxygen resulting in a pentacoordinated tin and a 1,3-diaxial conformation.

1a, X = CH_3	95%		49:51	
b, X = Cl	73%		94:6	

α-Hydroxy ketones.[3]   α-Hydroxylation of ketones can be effected by oxidation of the silyl enol ethers with this oxidant suspended in water. Addition of an organic solvent can decrease yields, which are generally 55–85%. The reaction is applicable to aryl or heteroaryl alkyl ketones as well as dialkyl ketones. Oxidation in methanol provides α-methoxy ketones.

[1] M. Ochiai et al., Tetrahedron Letters, **26**, 2351, 4501 (1985).
[2] M. Ochiai, S. Iwaki, T. Ukita, Y. Matsuura, M. Shiro, and Y. Nagao, Am. Soc., **110**, 4606 (1988).
[3] R. M. Moriarty, M. P. Duncan, and O. Prakash, J.C.S. Perkin I., 1781 (1987); R. M. Moriarty, O. Prakash, M. P. Duncan, R. K. Vaid, and H. A. Musallam, J. Org., **52**, 150 (1987).

Iron.

Gif catalyst.[1]   Reaction of iron dust, acetic acid, and pyridine at 30° results in a triiron cluster compound with the empirical formula $Fe_2FeO(OAc)_6Py_{3.5}$. This iron compound in the presence of a proton source (acetic acid), a reducing agent (Zn or Fe), and pyridine (essential) can catalyze air oxidation of saturated hydrocarbons. This system attacks mainly at secondary positions to give ketones as the major products. It also shows a preference for attack at the less hindered secondary positions, which may explain the lack of attack at tertiary positions. It does not epoxidize alkenes. Although oxidation by the Gif system can result in complex mixtures, some selectivity can be observed. Thus the major products of oxidation of cholestanes are derivatives of C_{20}-ketones resulting from cleavage of the side chain.

[1] D. H. R. Barton, J. Boivin, W. B. Motherwell, N. Ozbalik, K. M. Schwartzentruber, and K. Jankowski, *Nouv. de Chim.*, **10**, 387 (1986).

Iron phthalocyanine (1).

(1)

Wacker oxidation of 1-alkenes. The Wacker oxygenation of 1-alkenes to methyl ketones involves air oxidation catalyzed by $PdCl_2$ and $CuCl_2$, which is necessary for reoxidation of Pd(0) to Pd(II).[1] This oxygenation is fairly sluggish and can result in chlorinated by-products. A new system is comprised of catalytic amounts of $Pd(OAc)_2$, hydroquinone, and **1**, used as the oxygen activator.[2] The solvent is aqueous DMF, and a trace of $HClO_4$ is added to prevent precipitation of Pd(0). Oxygenation using this system of three catalysts effects Wacker oxidation of 1-alkenes in 2–8 hours and in 67–85% yield.

[1] J. Tsuji, H. Nagashima, and H. Nemoto, *Org. Syn.*, **62**, 9 (1984).
[2] J.-E. Bäckvall and R. B. Hopkins, *Tetrahedron Letters*, **29**, 2885 (1988).

Isopiperitenone (1).
Stereocontrolled oxy-Cope rearrangement. 1,2-Addition of a chiral vinyllithium reagent to a chiral β,γ-unsaturated ketone could give rise to at least eight

diastereomers. However, addition of various chiral cyclopentenyllithium reagents to (R)-(−)-1 results in only two diastereomeric alcohols, probably because of the rigid conformation of the enone. Thus addition of 2 to (R)-1 results in only two

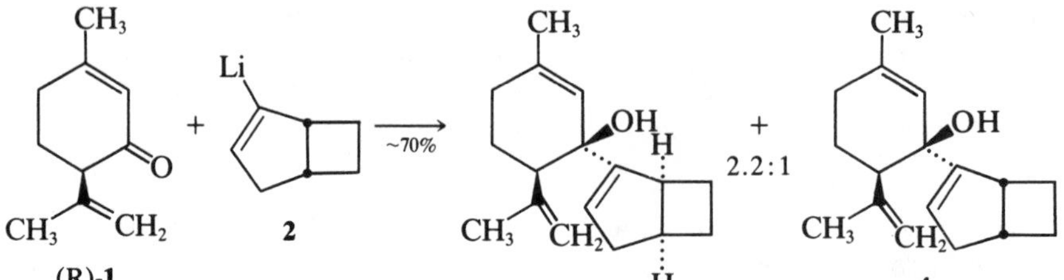

3

5 (43%)   6 (31%)

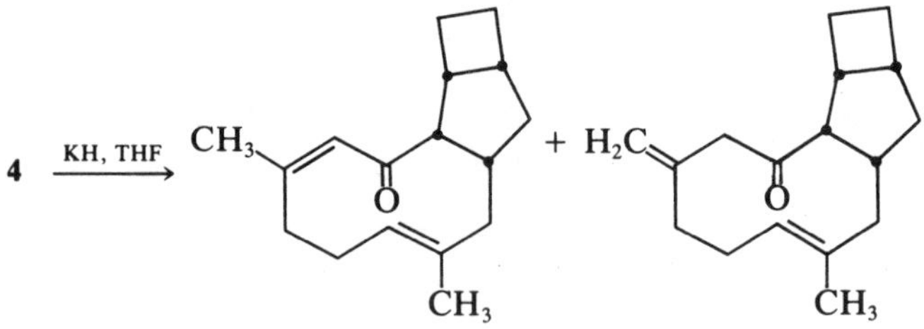

hydroxy dienes (3 and 4) of the type known to undergo oxy-Cope rearrangement (8, 412) to germacranolides. Base-catalyzed rearrangement of 4 results in the *endo*- and *exo*-isomers 5 and 6, with identical stereochemistry.

[1] L. A. Paquette, D. T. DeRussy, and C. E. Cottrell, *Am. Soc.*, **110**, 890 (1988).

(Isopropoxydimethylsilyl)methyl chloride. (i-PrO)Me$_2$SiCH$_2$Cl. This silyl chloride is obtained by reaction of i-PrOH and N(C$_2$H$_5$)$_3$ with (ClMe$_2$Si)CH$_2$Cl in refluxing ether. It is available from Aldrich.

Hydroxymethylation.[1]   The Grignard reagent (1) prepared from this silyl chloride adds to aldehydes or ketones to provide adducts (2), which undergo oxidative

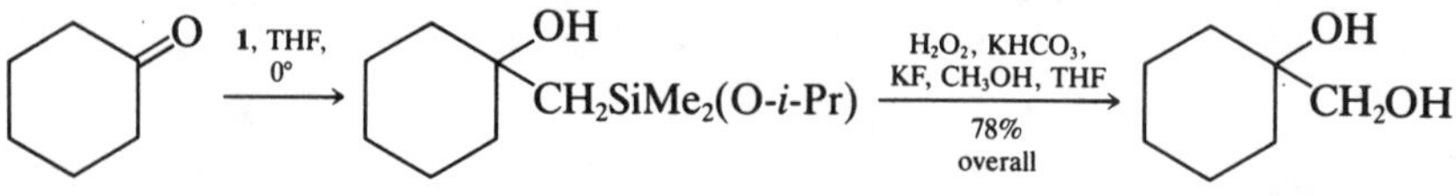

2   3

cleavage in the presence of F⁻ of the silicon—carbon bond (**12**, 243–244) to provide a hydroxymethyl group. The preferred conditions are H_2O_2 (30%, 1 equiv.), $KHCO_3$ (1 equiv.), and KF (2 equiv.), in CH_3OH/THF (1:1) at 25°. Originally (*i*-PrO)$_2$MeSiCH$_2$MgCl was preferred to **1**, because of more rapid oxidative cleavage of the adducts, but **1** is now preferred because the intermediates are more stable to mild acid or base. This nucleophilic hydroxymethylation is also applicable to halides, tosylates, and epoxides.

[1] K. Tamao, N. Ishida, Y. Ito, and M. Kumada, *Org. Syn.*, submitted (1988).

L

Lanthanum(III) trifluoromethanesulfonate is obtained by reaction of $La_2(CO_3)_3$ with triflic acid in water followed by dehydration.

Alkyllanthanum triflates, $RLa(OTf)_2$.[1] Alkyl- or aryllithium reagents are known to cleave tertiary amides to ketones, but yields are low because of a further reaction with the ketone to form a tertiary alcohol. The amides can be cleaved to ketones in high yield by reaction with alkyl- or aryllanthanum triflates, generated *in situ* by reaction of RLi or ArLi with $La(OTf)_3$ (equation I).

$$(I) \quad CH_3Li \xrightarrow{La(OTf)_3} [CH_3La(OTf)_2] \xrightarrow[90-98\%]{\overset{\overset{O}{\|}}{RCN(C_2H_5)_2}} \overset{\overset{O}{\|}}{RCCH_3}$$

[1] S. Collins and Y. Hong, *Tetrahedron Letters,* **28**, 4391 (1987).

Lead tetraacetate.

1-Alkynyllead triacetates.[1] $Pb(OAc)_4$ reacts with a 1-alkynyltrimethyltin to form trimethyltin acetate and an unstable 1-alkynyllead triacetate (**a**), which can effect alkynylation of β-dicarbonyl compounds and nitronates. In general yields are highest with $R = C_6H_5$ and lowest with $R = H$.

Example:

$$RC{\equiv}CSn(CH_3)_3 \xrightarrow[CHCl_3]{Pb(OAc)_4,} (CH_3)_3SnOAc + [RC{\equiv}CPb(OAc)_3]$$

a

[1] M. G. Moloney, J. T. Pinhey, and E. G. Roche, *Tetrahedron Letters,* **27**, 5025 (1986).

Lithioacetonitrile.

Axial addition to cyclohexanones.[1]   Addition of carbanions to cyclic ketones generally favors equatorial products. This preference may result from nonbonded interactions, since Trost *et al.*[1] now find that the addition of LiCH₂CN to cyclohexanones is axial selective (equation I). The preference for axial addition is even higher in the case of cyclohexenones (~20:1). The axial selectivity of LiCH₂CN is

(I)

evidently associated with the small size of the reagent, which allows the intrinsic bias for axial addition to predominate.

The reaction of LiCH₂CN with menthone (**4**) and pulegone (**5**) support the conclusions stated above. The addition to **4** is equatorial selective because of steric

effects, but the axial adduct (**6**) can be obtained by hydrogenation of the major adduct (**8**) of LiCH₂CN to **5**.

[1] B. M. Trost, J. Florez, and D. J. Jebaratnam, *Am. Soc.*, **109**, 613 (1987).

Lithium aluminum hydride.

anti-Selective reduction of $\searrow$=NOH **to** $\searrow$—NH$_2$.[1] A variety of metal hydrides, including LiAlH$_4$, AlH$_3$, DIBAH, reduce acyclic α-alkoxy and α,β-dialkoxy ketone oximes to primary amines with *anti*-selectivity.

Examples:

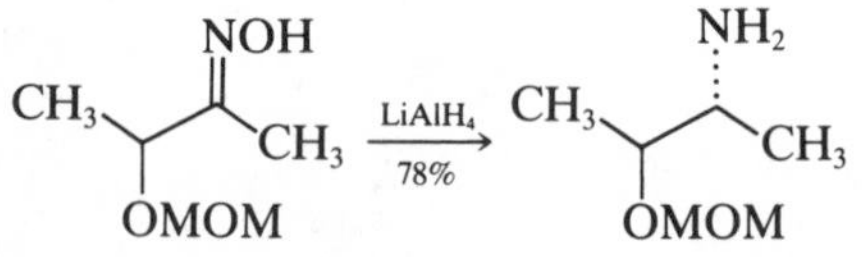

$$(MOM = CH_2OCH_3) \qquad (anti/syn = 70:30)$$

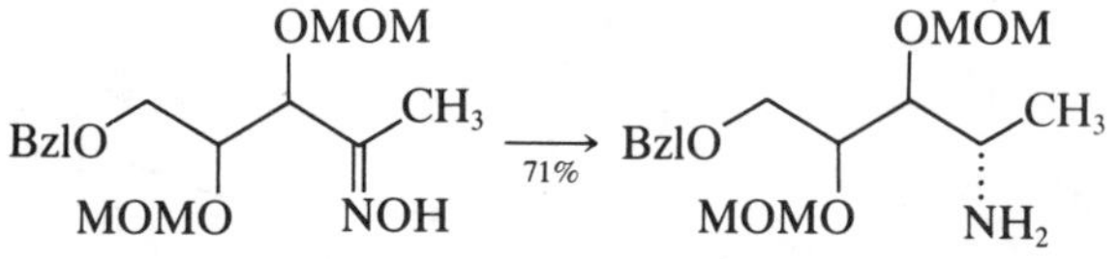

$$(anti/syn = 80:20)$$

Reduction of α-methyl-β-hydroxy ketones.[2] The *t*-butyldimethylsilyl ethers of these ketones, in which chelation is difficult, are reduced by lithium aluminum hydride with a high degree of 1,2-*anti*-selectivity. This reaction can therefore afford either *anti,anti*-1,3-diols or *anti,syn*-1,3-diols with high selectivity.

Examples:

Msp 385 — GD

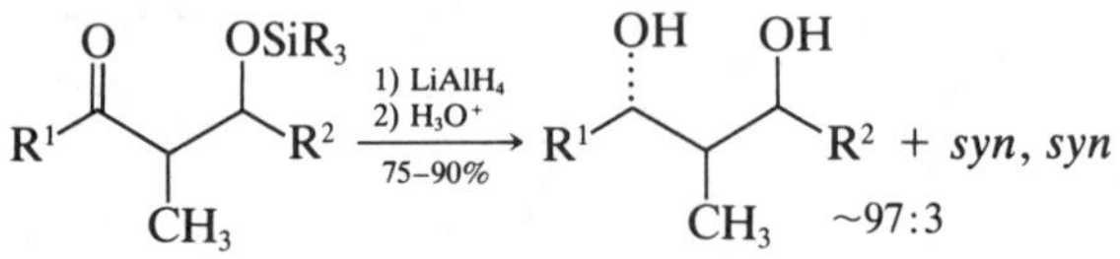

anti, syn

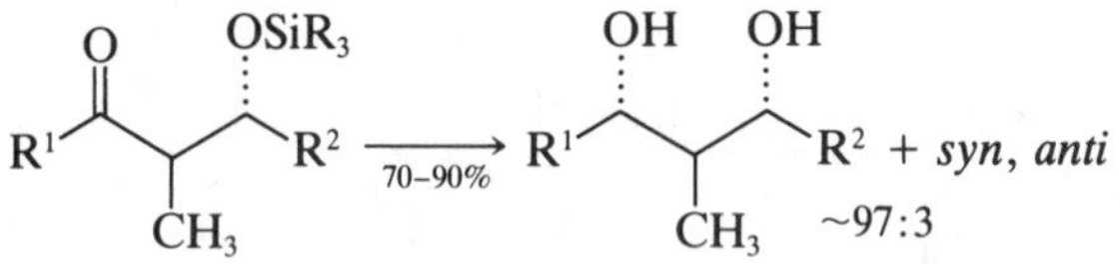

anti, anti

Substrates lacking an alkyl substituent in the β-position (R^2 = H) are reduced by LiAlH$_4$ with moderate *syn*-selectivity (60:40).

Reduction of amino acids to amino alcohols.[3] Amino acids can be reduced directly to the corresponding amino alcohols by LiAlH$_4$ in refluxing THF in 70–

90% yield. This method is now preferred to reduction of amino acid esters by borane–dimethyl sulfide (12, 64).[4]

[1] H. Iida, N. Yamazaki, and C. Kibayashi, *J.C.S. Chem. Comm.*, 746 (1987).
[2] R. Bloch, L. Gilbert, and C. Girard, *Tetrahedron Letters*, 29, 1021 (1988).
[3] G. A. Smith, G. Hart, S. Chemburkar, K. Rein, T. V. Anklekar, A. L. Smith, and R. E. Gawley, *Org. Syn.*, submitted (1988).
[4] G. A. Smith and R. E. Gawley, *Org. Syn.*, 63, 136 (1984).

Lithium aluminum hydride–Diethylamine.

RCOOR¹ ⟶ RCHO.[1] A reagent prepared as a slurry from diethylamine and LAH in a 2:1 ratio in pentane reduces esters (1 equiv.) to aldehydes at 25° in 92–94% yield. The reaction is general for aliphatic, aromatic, and α,β-unsaturated esters.

[1] J. S. Cha and S. S. Kwon, *J. Org.*, 52, 5486 (1987).

Lithium N-benzyltrimethylsilylamide, $LiN(CH_2C_6H_5)Si(CH_3)_3$ (1). Preparation.[1]

Conjugate addition.[2] This base undergoes efficient 1,4-addition to α,β-unsaturated esters to give the enolate of a β-amino ester, which can be trapped by an alkyl halide to give α-alkyl-β-amino esters (2) as a mixture of *syn-* and *anti-*isomers (about 1:1). These esters can be converted into β-lactams (3) by hydrolysis and dehydration (11, 449) or into α-alkyl-α,β-unsaturated esters (4) by N-quaternization and β-elimination on silica gel (~75% yield).

[1] J. Diekman, J. B. Thomson, and C. Djerassi, *J. Org.*, 32, 3904 (1967).
[2] N. Asao, T. Uyehara, and Y. Yamamoto, *Tetrahedron*, 44, 4137 (1988).

Lithium borohydride.

vic-*Glycols.* The one-pot conversion of esters to secondary alcohols by the combination of $LiBH_4$ and a Grignard reagent (12, 276) when applied to an α-

alkoxy ester (**1**) results in diastereoselective formation of the protected derivative of a *syn*-1,2-diol, particularly when conducted in THF at 0°.[1] The diastereoselectivity is reversed by a one-pot reduction of the ester to an aldehyde (DIBAH) followed by a Grignard addition.
Example:

(*syn*, 6–20:1)

(*anti*, 5–30:1)

[1] S. D. Burke, D. N. Deaton, R. J. Olsen, D. M. Armistead, and B. E. Blough, *Tetrahedron Letters*, **28**, 3905 (1987).

Lithium *t*-butyl(trialkylsilyl)amides, $LiN(SiR_3)C(CH_3)_3$.
The *t*-butyl(trialkylsilyl)amines are prepared by deprotonation of *t*-butylamine and reaction with a trialkylsilyl chloride; yields are 50–70%. They are converted to the corresponding lithium amides by BuLi in THF.
   Regioselective deprotonation of ketones.[1] The ability of a lithium alkyl-*t*-butylamide (LOBA, **12**, 285) to generate the kinetic (less substituted) enolate of an unsymmetrical ketone with greater selectivity than LDA has resulted in a search for related amides containing hindered groups such as these lithium trialkylsilyl-*t*-butylamides. These amides are comparable to or slightly more regioselective than LDA for deprotonation of straight-chain 2-alkanones or of 2-methylcyclohexanone. They show enhanced selectivity in the case of trimethylsilylacetone (equation I).

LDA	81:19
$LiN(t\text{-Bu})Si(CH_3)_3$	98:2
$LiN(t\text{-Bu})SiCH_3(C_6H_5)_2$	>99:1

[1] J. A. Prieto, J. Suarez, and G. L. Larson, *Syn. Comm.*, **18**, 253 (1988).

Lithium diisopropylamide–Chlorotrimethylsilane.

Claisen rearrangement of glycolates.   Two laboratories[1,2] have reported that allylic glycolate esters undergo Claisen–Ireland rearrangement (**6**, 276–277) with useful diastereoselectivity. This rearrangement was used in a synthesis of **1**, the aggregation pheromone of the European elm bark beetle.[1]

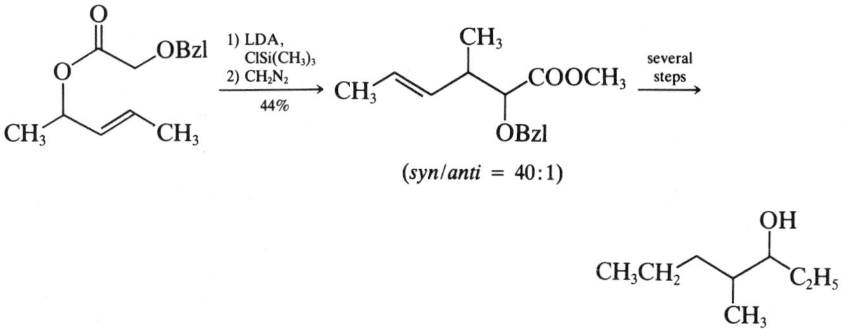

(*syn/anti* = 40:1)

1

The rearrangement provides a stereocontrolled synthesis of the C-pyranoside **2**, present in pseudomonic acids.[2]

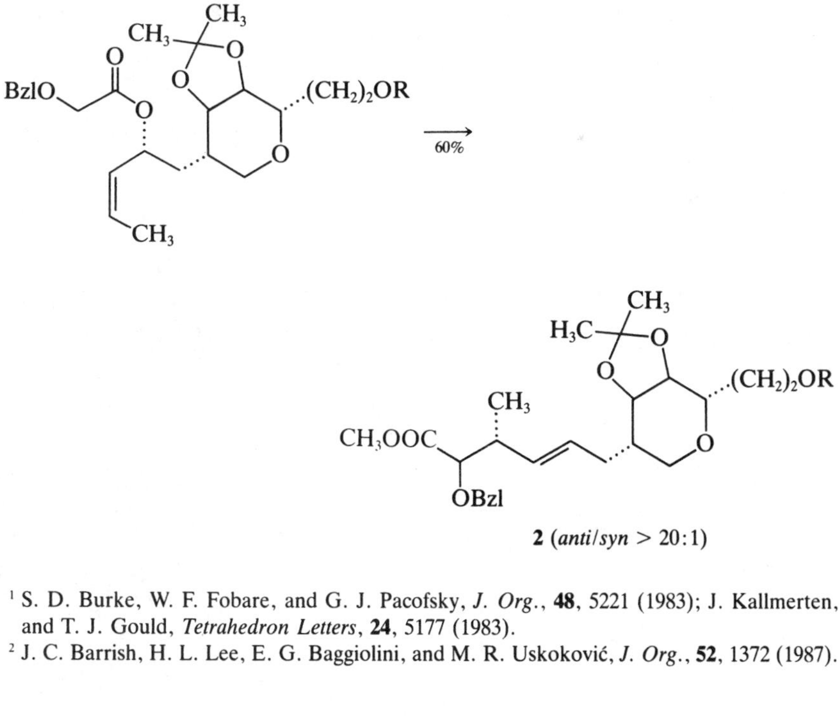

2 (*anti/syn* > 20:1)

[1] S. D. Burke, W. F. Fobare, and G. J. Pacofsky, *J. Org.*, **48**, 5221 (1983); J. Kallmerten, and T. J. Gould, *Tetrahedron Letters*, **24**, 5177 (1983).
[2] J. C. Barrish, H. L. Lee, E. G. Baggiolini, and M. R. Uskoković, *J. Org.*, **52**, 1372 (1987).

Lithium hexamethyldisilylamide.

Silylimines; β-lactams.[1]   This lithium amide converts aldehydes, even enolizable ones, into silylimines. Thus acetaldehyde can be converted into the imine (**2**), which cannot be isolated but which reacts with lithium ester enolates to form

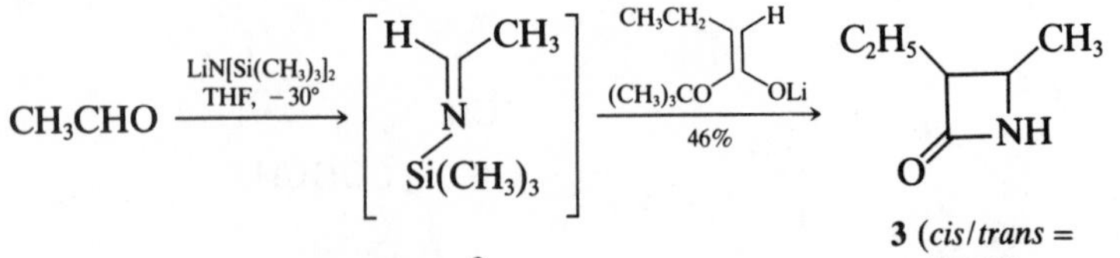

2

3 (*cis/trans* = 78:22)

β-lactams in about 45% yield. This reaction provides a general route to *cis*-3,4-disubstituted β-lactams.

[1] G. Cainelli, D. Giacomini, M. Panunzio, G. Martelli, and G. Spunta, *Tetrahedron Letters*, **28**, 5369 (1987).

Lithium tetramethylpiperidide (LTMP).

Metalation of nonenolizable carbonyl groups.[1]   This lithium amide, which lacks β-hydrogens, cannot reduce nonenolizable aldehydes or ketones but can metalate these substrates. Thus reaction of LTMP with trimethylacetaldehyde (**1**) evidently results in an acyllithium (**a**) as shown by formation of an acyloin (**2**, equation I).

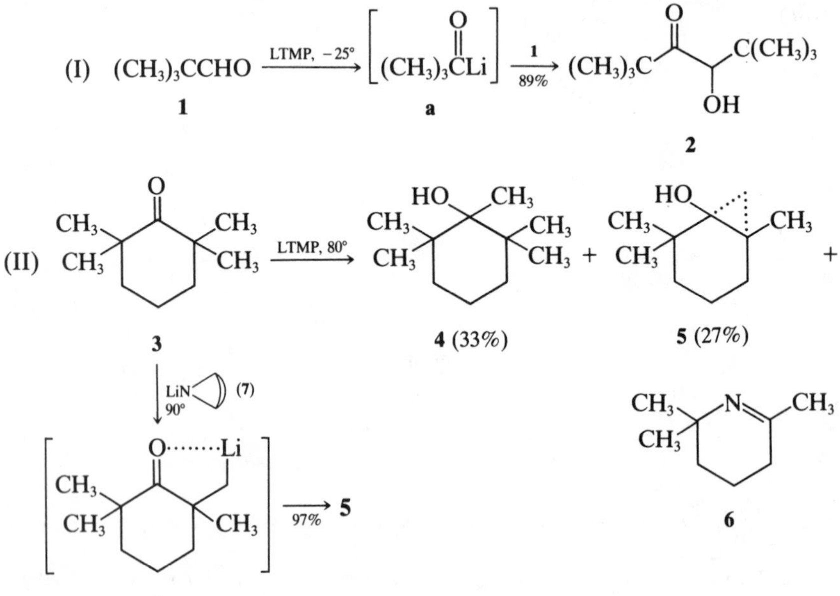

The reaction of LTMP at 80° with 2,2,6,6-tetramethylcyclohexane (**3**) to give **4** evidently involves transfer of a β-methyl of LTMP, since **6** is also formed during the reaction. The minor carbinol (**5**) of this reaction can be formed in quantitative yield when N-lithio-9-azabicyclo[3.3.1]nonane (**7**) is used as the base. This base lacks both β-hydrogens and β-methyl groups, but evidently can metalate a methyl group of **3** to form the anion **b**. The same paper also presents evidence for bridge-head metalation by LTMP.

[1] C. S. Shiner, A. H. Berks, and A. M. Fisher, *Am. Soc.*, **110**, 957 (1988).

Lithium tri-*sec*-butylborohydride.

Diastereoselective synthesis of lactones.[1]   Acylation of the enolate (LDA) of the vinylogous urethane (**1**) results in a product (**2**) that on reduction with LiBH[CH(CH₃)C₂H₅]₃ (**3**) forms the *anti*-lactone (**4**) exclusively (equation I). This two-step synthesis of lactones is the equilvalent of an aldol condensation between

(I)

1   **2a**   *anti*-**4** (99:1)

the enolate of **1** and the corresponding aldehyde (RCHO rather than RCOCl), but it is generally superior with respect to the yield and, more importantly, the dias-tereoselectivity. Another advantage of the acylation route is that reaction of the intermediate **2** with an alkyllithium is also diastereoselective and provides products

(II) **1**

2b   (99:1)

equivalent to ketone condensations (equation II). The reactions generally proceed in high yield and with high diastereoselectivity.

Moreover, this two-step equivalent of an aldol condensation can proceed with high enantioselectivity in the presence of a chiral auxiliary. Thus reaction of the enolate of chiral silyl ketene acetal (5) with isobutyryl chloride gives 6 in 89% yield and 94% ee after reduction of the intermediate.

5   6 (94% ee)

[1] R. Schlessinger, J. R. Tata, and J. P. Springer, *J. Org.*, **52**, 708 (1987).

M

Magnesium diisopropylamide, $Mg[N(CH(CH_3)_2]_2$ (**1**). The amide is obtained[1] as a pale yellow solid by reaction of $BuMg(sec\text{-}Bu)^2$ with 2 equiv. of diisopropylamine at 25° with evolution of butanes.

Reduction of carbonyl compounds.[1] The reagent reduces aldehydes or ketones to alcohols in refluxing cyclohexane in 2–5 hours; yields are 60–80%. The reduction probably involves hydride transfer from the carbon beta to the magnesium center.

---

[1] R. Sanchez and W. Scott, *Tetrahedron Letters*, **29**, 139 (1988).
[2] Lithco of America.

Magnesium monoperphthalate (MMPP, **1**). Suppliers: Aldrich, Interox Chemicals Ltd.

Alternative to **m-chloroperbenzoic acid.**[1] This oxidant has been introduced as a suitable replacement for *m*-chloroperbenzoic acid, which is no longer available from commercial sources because of hazards in the manufacture. Actually MMPP is a safer reagent than MCPBA, which is shock-sensitive and potentially explosive. MMPP is soluble in water and in low-molecular-weight alcohols. The by-product, magnesium phthalate, is water-soluble and easily removed. It is generally more stable than other percarboxylic acids. It can replace MCPBA for the usual classic oxidations: epoxidation, Baeyer–Villiger reactions, and oxidation of amines to N-oxides.

---

[1] P. Brougham, M. S. Cooper, D. A. Cummerson, H. Heaney, and N. Thompson, *Synthesis*, 1015 (1987).

Manganese(III) acetate, $Mn_3O(OAc)_7$ (**1**).

Cyclopentenones. 1,3-Dicarbonyl compounds add to enol ethers or esters (terminal) in the presence of $Mn_3O(OAc)_7$ (excess) to form 1-alkoxy-1,2-dihydrofurans. These can be converted to a 1,4-diketone, which undergoes aldol cyclization to fused (or spiro) cyclopentenones.[1]

Examples:

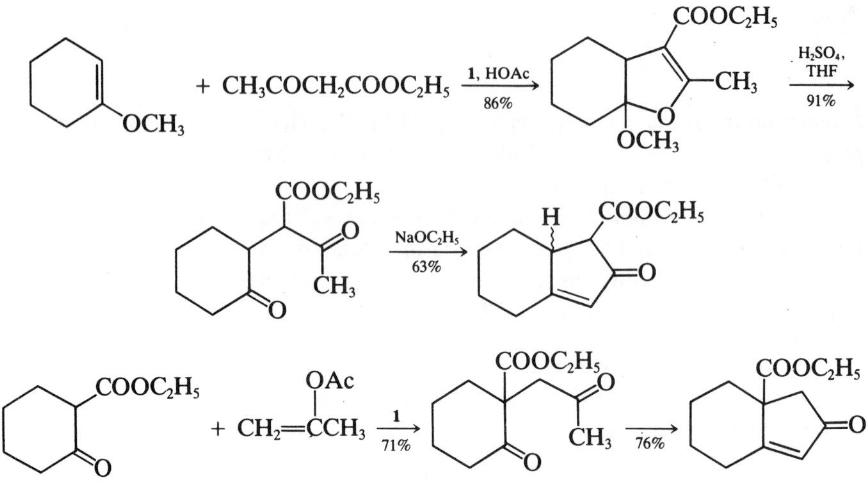

Furans.[2]   Enol ethers, β-dicarbonyl compounds, and Mn(III) acetate (2 equiv.) react in acetic acid (25°) to form 1-alkoxy-1,2-dihydrofurans, which form furans readily on acid-catalyzed elimination of ROH.

Example:

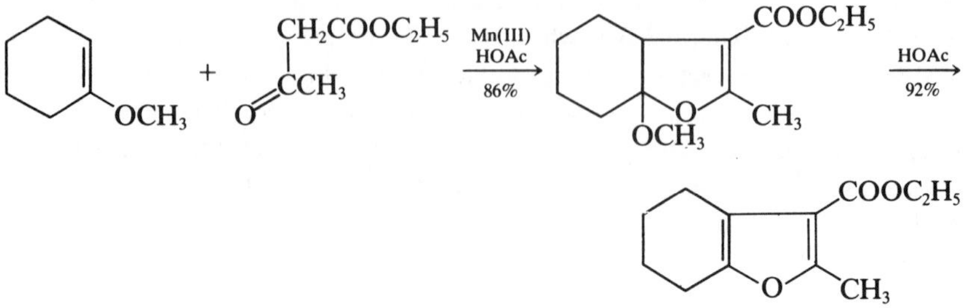

The reaction is applicable to acyclic and cyclic enol ethers and to various β-dicarbonyl compounds, but fails with silyl enol ethers and simple 1,2-disubstituted alkenes. When applicable, this route to furans is useful because the yields and regioselectivity are consistently satisfactory. The paper includes a preparation of the reagent by reaction of $Mn(NO_3)_2$ with Ac_2O at 100° to give $Mn_3O(OAc)_7 \cdot HOAc$ in 60% yield.

Oxidation of para-*methoxyphenols*.[3]   The last step in a synthesis of cyano-cycline A (**1**) requires oxidation of the *para*-methoxyphenol (**2**) to a quinone. This reaction is effected in highest yield (55%) with manganese(III) acetate $(0.3\%H_2SO_4–CH_3CN)$.

2   1

Oxidation of indolines.[4] This oxidant is superior to RuO$_4$ or CAN for oxidation of N-protected indolines to the indoles. Yields depend upon the particular protective group as well as the nature of substituent groups on the benzene ring, but are generally in the range 40–80%.

X = CON(C$_2$H$_5$)$_2$, 70°, 82%
X = COCH$_3$, 110°, 39%

[1] E. J. Corey and A. K. Ghosh, *Tetrahedron Letters*, **28**, 175 (1987).
[2] E. J. Corey and A. K. Ghosh, *Chem. Letters*, 223 (1987).
[3] T. Fukuyama, L. Li, A. A. Laird, and R. K. Frank, *Am. Soc.*, **109**, 1587 (1987).
[4] D. M. Ketcha, *Tetrahedron Letters*, **29**, 2151 (1988).

Manganese(III) acetate–Copper(II) acetate.

Oxidative radical cyclization of β-keto esters.[1] Radical cyclizations of unsaturated β-keto esters initiated by Mn(III) acetate (1) can be terminated by oxidative β-hydride elimination by Cu(OAc)$_2$ (equation I). This radical reaction can

(I)

be extended to oxidative bicyclization of diunsaturated β-keto esters (equation II),

(II)

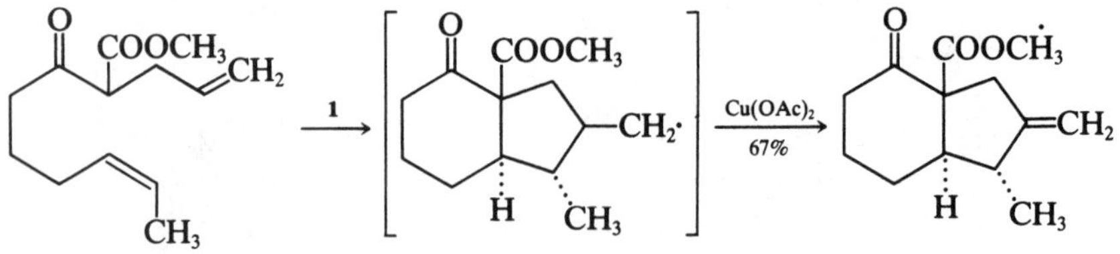

and has been used to obtain the skeleton of gibberellanes and kauranes, as in the cyclization of **2** to **3**.

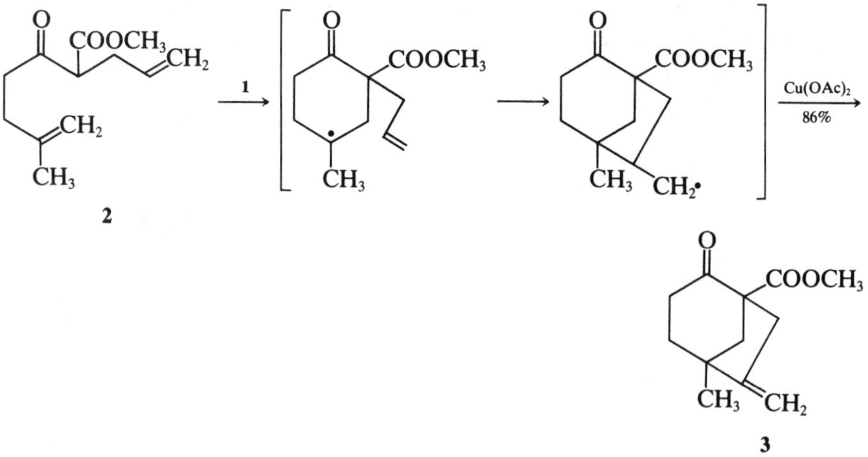

2

3

[1] B. B. Snider and M. A. Dombroski, *J. Org.*, **52**, 5487 (1987).

Manganese dioxide.

Chemical manganese dioxide (CMD). This form of MnO_2 is used for batteries; it is available from I. C. Sample office (Cleveland, Ohio, 44101). Shioiri *et al.*[1] report it is superior to commercial activated MnO_2 (Aldrich) and more convenient than freshly prepared activated MnO_2 for dehydrogenation of 2-(1-aminoalkyl)thiazolidine-4-carboxylic acids to the corresponding thiazoles (thiazole amino acids).

Example:

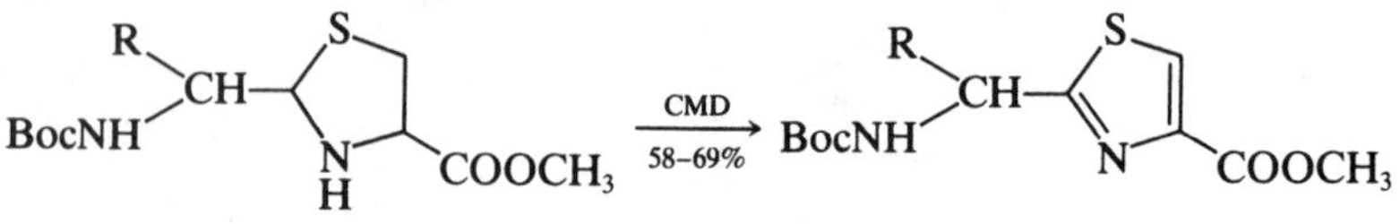

¹ Y. Hamada, M. Shibata, T. Sugiura, S. Kato, and T. Shioiri, *J. Org.*, **52**, 1252 (1987).

Menthone.

Asymmetric [2 + 2] photocycloadditions.   Acid-catalyzed condensation of *t*-butyl acetoacetate with (−)-menthone (**1**) results in two dioxacyclohexenones, **2** and **3**, which are easily separable. Photocycloaddition of (−)-**2** with methylcyclo-butene gives mainly the adduct (**4**) which has the *cis, anti, cis*-configuration at the coupling sites. The pronounced enantioselectivity in the cycloaddition (7:1) is attributed to the rigid spirocyclic structure of **2**. The major cycloadduct, (+)-**4**, from (−)-**2**, was transformed in three steps to optically active (+)-grandisol (**6**) with recovery of (−)-menthone. The same sequence when applied to (−)-**3** provides (−)-grandisol.

Deracemization of 1,3-diols.[2]   Ketalization of (−)-menthone (**1**) with a prochiral 1,3-diol (**2**) results in formation of only two of the four possible diastereomers (**3** and **4**), which can be separated by chromatography and then hydrolyzed to provide (S)- and (R)- **2** in 92–98% ee.

This ketalization reaction can also be used to convert racemic 1,3-diols to (R)-1,3-diols with high optical purity.

Chiral 2-alkyl-1,3-propanediols.[3]   Reaction of (−)-menthone (**1**) with the bis(trimethylsilyl)ether **2** catalyzed by trimethylsilyl triflate gives the more stable equatorial isomer (**3**) of a spiroketal. Ring cleavage of the equatorial bond of the ketal is effected by reaction with the enol silyl ether of acetophenone catalyzed by TiCl$_4$; after protection of the resulting hydroxyl group, the chiral auxiliary is removed with potassium *t*-butoxide to give **4** in >98% ee. The overall result is enantioselective differentiation of the hydroxymethyl groups in 2-alkyl-1,3-propanediols.

[1] M. Demuth, A. Palomer, H.-D. Sluma, A. K. Dey, C. Kruger, and Y.-H. Tsay, *Angew. Chem. Int. Ed.*, **25**, 1117 (1986).
[2] T. Harada, H. Kurokawa, and A. Oku, *Tetrahedron Letters*, **28**, 4843, 4847 (1987).
[3] T. Harada, T. Hayashiya, I. Wada, N. Iwa-ake, and A. Oku, *Am. Soc.*, **109**, 527 (1987).

(−)-Menthyl (S)-*p*-toluenesulfinate **(1).**

Esterification of *p*-toluenesulfinic acid with (−)-menthol gives a mixture of two diastereomers, which equilibrate to the pure (−)-menthyl (S)-*p*-toluenesulfinate diastereomer in the presence of hydrochloric acid (80% yield). The report includes an improved procedure for reaction of **1** with CH$_3$MgI to give (R)-(+)-methyl *p*-tolyl sulfoxide.[1]

1-Alkenyl **p-*tolyl* sulfoxides.**[2] Various 1-alkynylmagnesium bromides react stereospecifically (with inversion) with **1** in ether/toluene to give chiral 1-alkynyl sulfoxides **2**. Reduction of **2** with LiAlH$_4$ (THF, −90°) affords optically pure (E)-1-alkenyl *p*-tolyl sulfoxides (**3**). The corresponding (Z)-isomers are obtained by hydrogenation of **2** with the Wilkinson catalyst.

[1] G. Solladié, J. Hutt, and A. Girardin, *Synthesis*, 173 (1987).
[2] H. Kosugi, M. Kitaoka, K. Tagami, A. Takahashi, and H. Uda, *J. Org.*, **52**, 1078 (1987).

Mercury(II) oxide–Iodine.

Lactones by ring expansion of lactols.   Medium-sized lactones can be obtained in reasonable yield by photolysis of the hypoiodites of catacondensed lactols, which results in regioselective cleavage of the bridging bond. The substrates are available by the general route shown for the 6/6 lactol **1**, which is cleaved to the 10-membered lactone **2**. This cleavage can be used for preparation of 9-membered

lactones from 6/5 fused lactols and of 11-membered lactones from 7/6 or 8/5 fused lactols. In all cases, the bridging bond is cleaved preferentially because this cleavage results in the most substituted radical.[1]

This hypoiodite reaction can also be used for ring expansion of cyclic ketones.[2] Thus Barbier cyclization of α-(ω-iodopropyl)cyclododecanone (**3**) furnishes the bicyclic alcohol **4**, which undergoes regioselective cleavage to a 15-membered iodo ketone (**5**). The final step to exaltone (**6**) involves a free radical reduction with Bu$_3$SnH.

[1] H. Suginome and S. Yamada, *Tetrahedron*, **43**, 3371 (1987).
[2] *Idem*, *Tetrahedron Letters*, **28**, 3963 (1987).

p-Methoxybenzyl chloromethyl ether (1).
Preparation:

$$CH_3O-\text{(ring)}-CH_2ONa \xrightarrow{ClCH_2SCH_3} CH_3O-C_6H_4-CH_2OCH_2SCH_3 \xrightarrow{SO_2Cl_2}$$

$$CH_3O-\text{(ring)}-CH_2OCH_2Cl$$

1

Protection of alcohols.[1]   Even somewhat hindered secondary alcohols or tertiary alcohols are converted into (p-methoxybenzyloxy)methyl (PMBM) ethers by reaction with **1** and diisopropylethylamine in CH$_2$Cl$_2$ for 3–30 hours. Deprotection can be effected by oxidation with DDQ (65–95% yield), a method previously recommended for deprotection of p-methoxybenzyl ethers (**11**, 166–167).

[1] A. P. Kozikowski and J.-P. Wu, *Tetrahedron Letters*, **28**, 5125 (1987).

Methoxy(phenylthio)methyllithium (1), **6**, 369.
Homoallyl ethers or sulfides.[1]   gem-Methoxy(phenylthio)alkanes (**2**), prepared by reaction of **1** with alkyl halides, react with allyltributyltin compounds in the presence of a Lewis acid to form either homoallyl methyl ethers or homoallyl phenyl sulfides. Use of BF$_3$ etherate results in selective cleavage of the phenylthio group to provide homoallyl ethers, whereas TiCl$_4$ effects cleavage of the methoxy group with formation of homoallyl sulfides.

$$
\begin{array}{c}
OCH_3 \\
| \\
R\overset{|}{C}H \\
| \\
SC_6H_5 \\
\textbf{2 (R = } n\text{-C}_8H_{17}\textbf{)}
\end{array}
\quad + \quad CH_2{=}CHCH_2SnBu_3 \quad \longrightarrow
$$

| | BF$_3$·O(C$_2$H$_5$)$_2$ | 72% | 100:0 |
| | TiCl$_4$ | 79% | 6:94 |

3   **4**

[1] T. Sato, S. Okura, J. Otera, and H. Nozaki, *Tetrahedron Letters*, **28**, 6299 (1987).

$$\text{Methoxy(phenylthio)trimethylsilylmethyllithium,} \quad (CH_3)_3Si\overset{\displaystyle SC_6H_5}{\underset{\displaystyle OCH_3}{\overset{|}{\underset{|}{C}}}Li} \quad \textbf{(1)}.$$

α,β-Dialkylation of enones.[1]   Reaction of a cyclic α,β-enone with **1** in the presence of HMPT and then with an alkyl halide affords *trans*-α,β-disubstituted enones in moderate to high yield. HMPT and the trimethylsilyl group of **1** are essential for the initial 1,4-addition. This reaction was used for a synthesis of sarkomycin **(2)** from cyclopentenone, in which **1** serves as an equivalent of a carboxy group.

[1] J. Otera, Y. Niibo, and H. Aikawa, *Tetrahedron Letters*, **28**, 2147 (1987).

Methylaluminum bis(2,6-di-*t*-butyl-4-methylphenoxide), MAD (1), 13, 203.

Selective reduction of ketones.[1]   This reagent can be used to effect selective reduction of the more hindered of two ketones by DIBAH or dibromoalane. Thus treatment of a 1:1 mixture of two ketones with 1–2 equiv. of **1** results in preferential complexation of the less hindered ketone with **1**; reduction of this mixture of free and complexed ketones results in preferential reduction of the free, originally more hindered, ketone. An electronic effect of substituents on a phenyl group can also play a role in the complexation. This method is not effective for discrimination between aldehydes and ketones, because MAD-complexes are easily reduced by hydrides. MAD can also serve as a protecting group for the more reactive carbonyl group of a diketone. The selectivity can be enhanced by use of a more bulky aluminum reagent such as methylaluminum bis(2-*t*-butyl-6-(1,1-diethylpropyl)-4-methylphenoxide).

Amphiphilic alkylation (13, 203).[2] Addition of Grignard reagents to α-methyl substituted aldehydes proceeds with modest *syn*-selectivity (~2–3:1). In contrast, the addition to the same aldehydes complexed with MAD or MAT shows *anti*-selectivity (equation I). Yields are also improved by complexation.

(I)

		syn	84:16	*anti*
+ MAD	90%		25:75	
+ MAT	98%		20:80	

Complexation of enones with MAD can be used to effect conjugate additions with organolithium reagents. This unusual 1,4-selectivity is observed mainly with cyclic systems and requires 2 equiv. of the aluminum reagent as well as of the nucleophile for full effect.

Examples:

		cis/trans	
CH₃Li	69%		29:71
t-BuLi	73%		18:82
BuC≡CLi	no reaction		

[1] K. Maruoka, Y. Araki, and H. Yamamoto, *Am. Soc.*, **110**, 2650 (1988).
[2] K. Maruoka, T. Itoh, M. Sakurai, K. Nonoshita, and H. Yamamoto, *ibid.*, **110**, 3588 (1988).

N-Methylaniline.
Intramolecular [4 + 2]cycloaddition of an enamine/enal(enone).[1] Generation *in situ* of an aldehyde enamine of a substrate also containing an enal or enone group can result in a facile intramolecular [4 + 2]cycloaddition resulting in bicyclic dihydropyrans. Although several *sec*-amines can be used, N-methylaniline is particularly suitable because of the stability of the adducts.

Example:

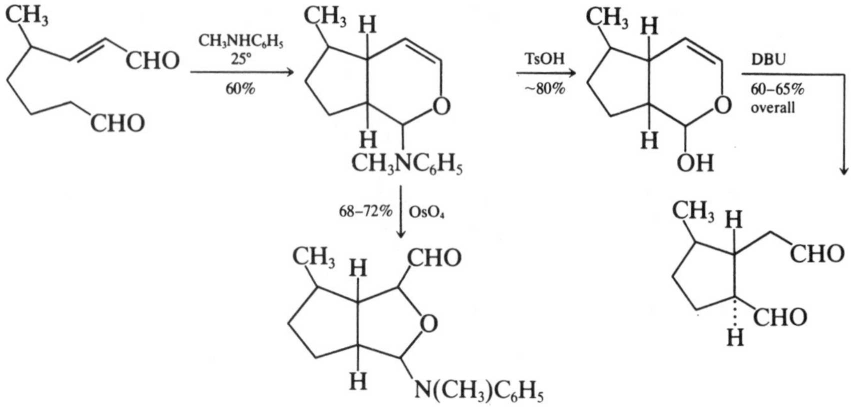

The products are *cis*-fused. They are hydrolyzed by TsOH in aqueous THF to lactols, which can be oxidized by Fetizon's reagent to lactones or converted to dials by DBU. The enol ether group of the adduct can be selectively oxidized by Cl-$C_6H_4CO_3H$ or OsO_4 to provide bicyclic tetrahydrofurans.

[1] S. L. Schreiber, H. V. Meyers, and K. B. Wiberg, *Am. Soc.*, **108**, 8274 (1986).

Methyl (carboxysulfamoyl)triethylammonium hydroxide, inner salt (Burgess reagent).

Dehydration of amides.[1]   Primary amides are converted by this reagent into nitriles at 25° in 82–92% yield. Actually, this reaction occurs more readily than dehydration of a secondary hydroxyl group.

[1] D. A. Claremon and B. T. Phillips, *Tetrahedron Letters*, **29**, 2155 (1988).

Methylene iodide–Zinc–Trimethylaluminum.

Methylenation of α-(N-Boc-amino) aldehydes.[1]   Methylenation of these chiral aldehydes with the Wittig reagent or with CH_2I_2–Zn–$TiCl_4$ (**13**, 114) is accompanied by extensive racemization. However, the neutral reagent **1** obtained from CH_2I_2, Zn, and $Al(CH_3)_3$ converts these aldehydes to the protected allylamines in 40–75% yield and in >99% ee.

Example:

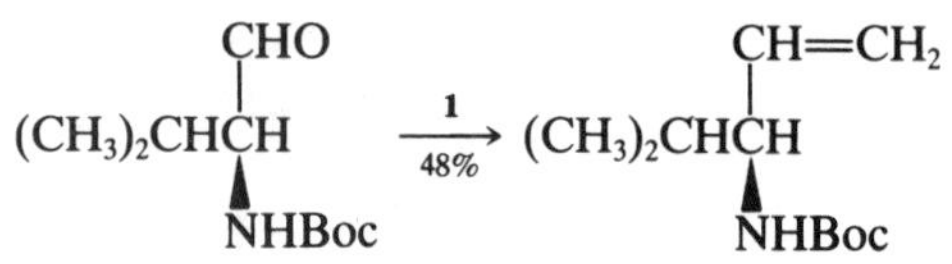

[1] T. Moriwake, S. Hamano, S. Saito, and S. Torii, *Chem. Letters*, 2085 (1987).

N-Methylephedrine.
Chiral **trans-β-*lactams*.**   The silyl ketene acetal (**1**), derived from (1S,2R)-N-methylephedrine, reacts in the presence of TiCl₄ with benzylideneaniline (**2**) to give as the major products *anti-* and *syn-***3** in the ratio ⩾10:1. Cyclization of the mixture gives the *trans*-β-lactam (**4**) in 95% ee.

[1] C. Gennari, I. Venturini, G. Gislon, and G. Schimperna, *Tetrahedron Letters*, **28**, 227 (1987).

2-Methylfuran.
Enediones; tetraols.   Pentitols can be synthesized by the use of 2-methylfuran as the precursor to a chain of five carbon atoms and of 2,3-O-isopropylidene-D-glyceraldehyde (**2**) as the source of chirality. Thus condensation of lithiated 2-methylfuran with **2** furnishes *anti-***3**, which can be converted into the *syn-*isomer by oxidation and hydride reduction. After protection of the hydroxy group as a silyl ether or a benzyloxymethyl (BOM) ether, the furan group is converted into an enedione group by reaction with bromine. The terminal ketone of **4** can be selectively protected as the dimethyl ketal (**5**) and then another chiral center can be introduced by stereoselective reduction of the C₄-carbonyl group. The diastereoselectivity of this reduction can be controlled by choice of the reducing agent and of the protecting group of the C₃-hydroxyl group. Use of a bulky silyl protecting group results consistently in products with the 3,4-*syn* configuration on DIBAH

reduction (equation I). In contrast, reduction with zinc borohydride of the ketone in which the adjacent hydroxyl group is protected as the BOM ether is 20:1 *anti*-selective (equation II). The divergent result is attributed to chelating ability of the BOM group. Thus all four possible configurations of a pentitol group can be derived from the condensation of **1** and **2**.

¹ S. Pikul, J. Raczko, K. Ankner, and J. Jurczak, *Am. Soc.*, **109**, 3981 (1987).

Methyllithium.

(E)-Enol silyl ethers.[1]   A new highly stereoselective route to (E)-enol silyl ethers involves addition of CH_3Li to silyl ketones substituted at the α'-position by a SC_6H_5 group such as **1**. The adduct (**a**) undergoes a Brook rearrangement and

fragmentation to give (E)-**2**. Use of other alkyl-, alkenyl-, or aryllithiums results in somewhat lower stereoselectivities.

The (Z)-silyl enol ether (**4**) of **1** can be obtained with high stereoselectivity by reaction of **3** with $C_6H_5(CH_3)_2SiLi$.

[1] H. J. Reich, R. C. Holtan, and S. L. Borkowsky, *J. Org.*, **52**, 312 (1987).

N-Methyl-N-phenyl(dimethylalano)amide, $(CH_3)_2AlN(CH_3)C_6H_5$ (**1**).

The aluminum amide is prepared by reaction of N-methylaniline with $Al(CH_3)_3$ in CH_2Cl_2 at 25 °.

Selective alkylation of ketones.[1]   This reagent forms a complex so much more rapidly with aldehydes than with ketones that selective alkylation of a keto group in the presence of an aldehyde group with an alkyllithium or Grignard reagent is possible. The opposite chemoselectivity is achieved with the bulky methylaluminum bis(2,6-di-*t*-butyl-4-phenoxide) (MAD, **13**, 203; this volume).

Example:

[1] K. Maruoka, Y. Araki, and H. Yamamoto, *Tetrahedron Letters*, **29**, 3101 (1988).

Methylthiomethyl tolyl sulfone (**13**, 192–193).

syn-β-*Hydroxy-α-amino acids*.[1] α-Hydroxy aldehydes can be converted to one-carbon homologated β-hydroxy-α-amino acids by reaction with the anion (**1**) of methylthiomethyl tolyl sulfone followed by trichloroacetyl isocyanate to provide the trichloroacetylcarbamate **2**. When treated with K_2CO_3 in CH_3OH, **2** cyclizes to the oxazolidinone **3**. Oxidation to a sulfoxide followed by Pummerer rearrangement provides a methyl thioester (**4**), which is hydrolyzed to the β-hydroxy-α-amino acid **5** in 45% overall yield from the aldehyde.

[1] M. Hirama, H. Hioki, and S. Itô, *Tetrahedron Letters*, **29**, 3125 (1988).

N

Nafion-H.

Review. Olah *et al.*[1] have reviewed use of this resin sulfonic acid as a solid catalyst in organic synthesis (181 references).

[1] G. A. Olah, P. S. Iyer, and G. K. S. Prakash, *Synthesis*, 513 (1986).

Nickel, activated (12, 335).

o-Xylylene. This compound, or a related species, can be generated with activated Ni from an α,α'-dihalo *o*-xylene and trapped by electron-deficient alkenes.[1]
Example:

[1] S. Inaba, R. M. Wehmeyer, M. W. Forkner, and R. D. Rieke, *J. Org.*, **53**, 339 (1988).

Niobium(III) chloride–Dimethoxyethane, NbCl$_3$·DME (1).

This soluble Nb(III) reagent is obtained as a brick-red solid by reduction of NbCl$_5$ in DME with Bu$_3$SnH.

2-Amino alcohols.[1] The reagent reacts with a typical imine such as N-benzylbenzenimine to form a product that is hydrolyzed to benzylphenylamine. The precursor is presumably a metallaaziridine, which is cleaved by electrophiles with insertion into the carbon–metal bond. Reaction of the intermediate (**a**) with a carbonyl compound results in a 2-amino alcohol after hydrolysis with moderate *anti*-selectivity.

Example:

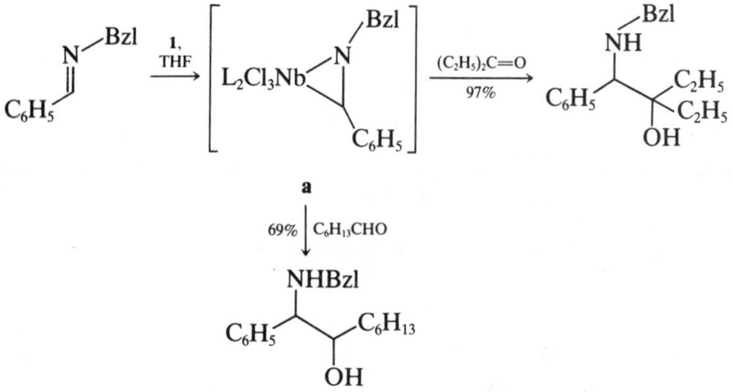

a

69% | $C_6H_{13}CHO$

NHBzl

C_6H_5 ⟍ ⟋ C_6H_{13}
        OH

(*syn/anti* = 4.5 : 1)

[1] E. J. Roskamp and S. F. Pedersen, *Am. Soc.*, **109**, 6551 (1987).

Niobium(IV) chloride, $NbCl_4(THF)_2$. This reagent (**1**) can be obtained[1] by reduction of $NbCl_5$ with aluminum powder in CH_3CN followed by solvation with THF; m.p. 110° dec., yield 50%.

vic-*Diamines*.[2] This low-valent niobium reagent converts the trimethylsilyl-imine of benzaldehyde to a diimido complex (**2**), which is hydrolyzed by aqueous KOH to a *vic*-diamine (equation I). The trimethylsilylimines can be prepared by

(I) 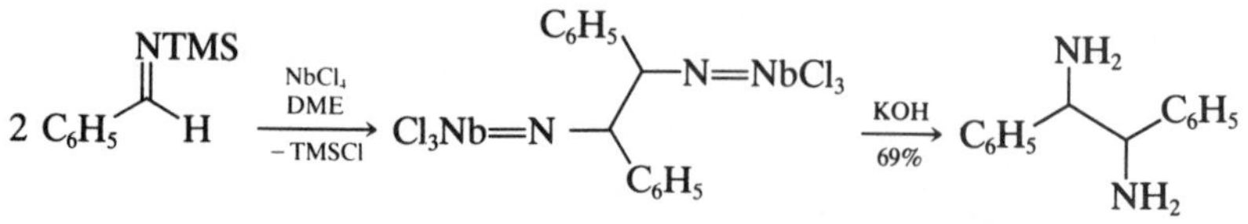 example with **2** and **3** (*dl/meso* = 19:1)

reaction of nonenolizable aldehydes or ketones with $LiN[Si(CH_3)_3]_2$ and then with $ClSi(CH_3)_3$.[3]

The niobium imines can also be prepared by reaction of a niobium halo hydride with a nitrile. Thus a niobium reagent prepared from **1** and Bu_3SnH reacts with nitriles to form a niobium–imido complex, which is hydrolyzed to a *vic*-diamine.

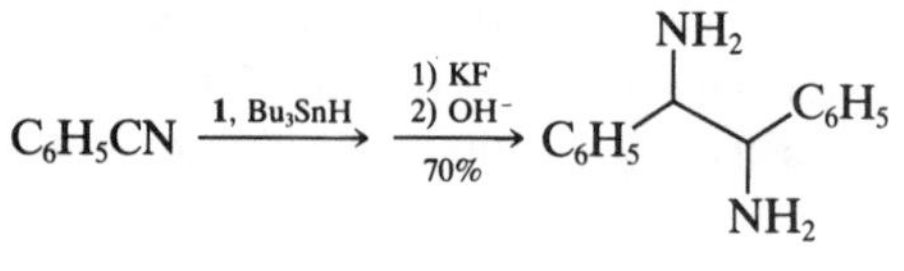

3 (*dl/meso* = 8:1)

[1] L. E. Manzer, *Inorg. Chem.*, **16**, 525 (1977).
[2] E. J. Roskamp and S. F. Pedersen, *Am. Soc.*, **109**, 3152 (1987).
[3] E. W. Colvin and D. G. McGarry, *J.C.S. Chem. Comm.*, 539 (1985).

Nitrosonium tetrafluoroborate, NOBF₄ (1).

O-*Glycosides from* S-*glycosides*.[1] Methylthio or phenylthio glycosides on treatment with **1** in CH_2Cl_2 at 0–25° form a species that reacts with a primary or secondary alcohol to form an O-glycoside. The stereochemistry of this glycosylation can be controlled by a neighboring group. Thus when an acetoxy group is present at C_2, a 1,2-*trans*-O-glycoside is formed, regardless of stereochemistry of the S-glycoside.

—NHNH₂ ⟶ —N₃.[2] This salt converts arylhydrazines at low temperatures into unstable nitroso derivatives that decompose to aryl azides in about 85% yield. Acylhydrazines are also converted by this reaction into acyl azides in 50–75% yield.

[1] V. Pozsgay and H. J. Jennings, *J. Org.*, **52**, 4635 (1987).
[2] *Idem*, *Tetrahedron Letters*, **28**, 5091 (1987).

Norephedrine.

2-*Alkenyloxazolidines*.[1] In the presence of pyridinium tosylate or BF_3 etherate, the N-protected norephedrine (**1**) cyclizes with the α,β-unsaturated acetal **2** to give the 2-alkenyloxazolidine **3** as the major product. Cuprates add to **3** from the *si* face with high selectivity to give adducts (**4**), which are readily converted to (S)-3-alkylsuccinaldehydes in high enantiomeric purity.

Asymmetric reduction of oxime ethers.[2] The complex (**1**) of (−)-norephedrine with BH_3 (2 equiv.) reduces prochiral oxime ethers to optically active amines; the stereoselectivity is controlled mainly by the configuration of the prochiral nitrogen group. Thus the enantioselectivity of reduction of the oxime ether of the alkyl aryl ketone **2** is determined by the geometry of the oxime ether. Even dialkyl ketoxime

2	**3**
anti-	S, 92% ee
syn-	R, 92% ee

ethers are reduced in 79–86% ee, which is generally higher than the enantioselectivity obtained in reduction of dialkyl ketones.

[1] A. Bernardi, S. Cardani, T. Pilate, G. Poli, S. Scolastico, and R. Villa, *J. Org.*, **53**, 1600 (1988).
[2] Y. Sakito, Y. Yoneyoshi, and G. Suzukamo, *Tetrahedron Letters*, **29**, 223 (1988).

O

Organoaluminum compounds.

Conjugate alkylation of nitroalkenes. R_3Al or the etherate reacts rapidly with α,β-unsaturated nitro compounds to form products of 1,4-addition.[1]
Example:

$$CH_3(CH_2)_5\overset{\overset{\textstyle NO_2}{|}}{C}{=}CH_2 + Al(i\text{-}Bu)_3 \xrightarrow{\text{hexane}}$$

$$\left[\begin{array}{c} O{\diagdown}\underset{||}{\underset{\textstyle N}{}}{\diagup}OAl(Bu\text{-}i)_2 \\ CH_3(CH_2)_5\overset{}{C}{-}CH_2Bu\text{-}i \end{array} \right]$$

KMnO$_4$,OH$^-$ | H$_3$O$^+$

86% | | 96%

$$CH_3(CH_2)_5\overset{\overset{\textstyle O}{||}}{C}CH_2Bu\text{-}i \qquad CH_3(CH_2)_5\overset{\overset{\textstyle NO_2}{|}}{CH}{-}CH_2Bu\text{-}i$$

[1] A. Pecunioso and R. Menicagli, *J. Org.*, **53**, 45 (1988).

Organocerium(III) reagents, $RCeX_2$. These reagents are obtained *in situ* by reaction of RLi with CeI_3 or $CeCl_3$.

Addition with carbonyl compounds.[1] These reagents form 1,2-adducts with carbonyl compounds at $-65°$ in high yield without enolization or reduction. 1,2-Adducts are also formed from α,β-enones or -enals. Yields are generally over 90%.

Chiral amines.[2] These reagents also add to imines and this reaction can be used for synthesis of optically active amines. Thus $RCeCl_2$ adds to SAMP-hydrazones (**12**, 30) to form hydrazines in good yield and high diastereoselectivity. These are reduced to optically active amines by hydrogenation catalyzed by Raney nickel. The hydrazines are prone to oxidation, but can be isolated as the stable carbamates.

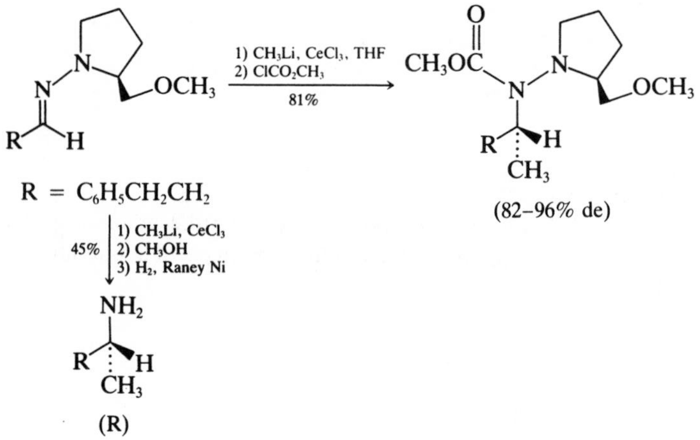

R = C₆H₅CH₂CH₂

R = $C_6H_5CH_2CH_2$

(82–96% de)

(R)

[1] T. Imamoto, T. Kusumoto, Y. Tawarayama, Y. Sugiura, T. Mita, Y. Hatanaka, and M. Yokoyama, *J. Org.*, **49**, 3904 (1984).
[2] S. E. Denmark, T. Weber, and D. W. Piotrowski, *Am. Soc.*, **109**, 2224 (1987).

Organocopper reagents.

Preparation from activated Cu(0).[1]   An activated Cu*, prepared by lithium naphthalenide reduction of CuI·PBu₃ (**12**, 140), reacts with primary alkyl bromides at −50 to −78° to form alkylcopper reagents that undergo 1,4-addition to cyclohexenone in moderate to high yield. This conjugate addition is facilitated by ClSi(CH₃)₃ and a phosphine.

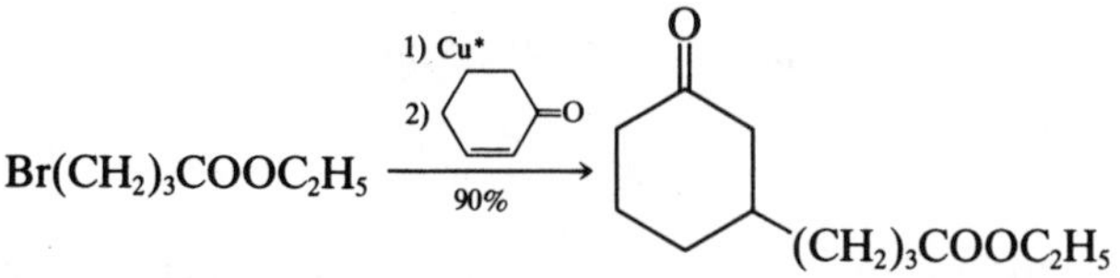

$Br(CH_2)_3COOC_2H_5$

The organocopper reagents prepared in this way cleave epoxides in high yield with substitution of the alkyl group at the less hindered site. This cleavage has been used to effect intramolecular cyclization (second example).

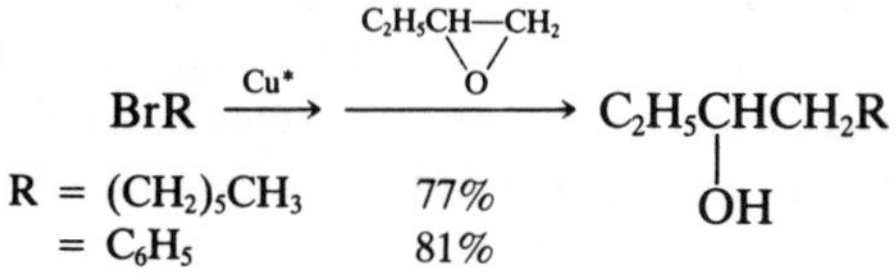

R = (CH₂)₅CH₃   77%
= C₆H₅   81%

$$6:1$$

(*endo*)   (*exo*)

α-Alkoxycopper reagents.[2] These reagents are thermally unstable, but (benzyloxy)methylcopper (**1**) can be obtained by reaction of (benzyloxy)methyllithium with pure, recently prepared CuBr·S(CH₃)₂ (**10**, 104) in a mixture of diisopropyl sulfide and THF at −78°. The reagent undergoes conjugate addition to unhindered α,β-enones in moderate yields. As expected (**11**, 368–369), addition of BF₃ etherate markedly enhances the reactivity and permits additon to even hindered enones (equation I).

(I) $C_6H_5CH_2OCH_2Cu$ + $(CH_3)_2C=CHCOCH_3$ $\xrightarrow[\text{77\%}]{\substack{BF_3 \cdot O(C_2H_5)_2 \\ THF, \, -78°}}$

       1

$$C_6H_5CH_2OCH_2\underset{\underset{CH_3}{|}}{\overset{\overset{CH_3}{|}}{C}}CH_2COCH_3$$

(Benzyloxy)methylcopper is inherently more reactive in conjugate additions than the corresponding homocuprate, $(BzlOCH_2)_2CuLi$, but addition of chlorotrimethylsilane activates the homocuprate markedly. However, even the activated homocuprate does not add to hindered enones at low temperatures, and this reaction also suffers from loss of one equivalent of the reagent.

The same techniques permit synthesis of glucosylcopper reagents derived from α- and β-D-glucopyranosides. These reagents when activated by BF₃ etherate can undergo conjugate addition to unsubstituted enones with retention of configuration of the anomeric carbon, but with no significant diastereoselectivity (equation II). Yields are low, however, in addition to hindered enones.

(II)

(B)

(1:1)

Stereoselective addition of cuprates to aldehydes.[3]   Addition of Bu_2CuLi to the formyl group of methyl γ-oxocarboxylates (**1**) results in *trans*-disubstituted γ-lactones (**2**) with high selectivity. The stereoselectivity is ascribed to chelation,

3 (*trans/cis* =
65–93:35–7)

2 (*trans/cis* =
95–97:5–3)

since the $TiCl_4$-catalyzed addition of allyltrimethylsilane (*cf.*, **7**, 370) to **1** is also *trans*-selective. Stereoselective addition of cuprates to β-alkoxy aldehydes has also been reported (**10**, 193).

(Trimethylstannyl)copper.[4]   This reagent is prepared as the complex **1** as shown. Reaction of 1-trimethylsilyl-1,3-diynes (**2**) with **1** (excess) forms *trans*-bis-(trimethylstannyl)enynes **3**. Methyllithium reacts selectively with the trimethyl-

$$(CH_3)_3SnLi + CuBr \cdot S(CH_3)_2 \xrightarrow{\text{THF}} (CH_3)_3SnCu \cdot S(CH_3)_2LiBr$$

1

2, R = C_6H_{13}

3

stannyl group at C_3 to form a vinyllithium reagent, which can be protonated (**4**) or alkylated. The trimethylstannyl group at C_4 can undergo a subsequent trans-metallation followed by protonation or alkylation (**5**). These two sequences provide stereodefined tri- and tetrasubstituted enynes.

4

5

The (trimethylsilyl)ethynyl group of the products can undergo further transformation to furnish substituted dienes or even alkynes.

Additions to cyclopentenones.[5] Conjugate addition of cuprates to 4-substituted cyclopentenones can show moderate to high *trans*-diastereoselection, which can be attributed to a steric effect. Surprisingly, addition of lithium dimethylcuprate to (R)-5-methoxy-2-cyclopentenone also shows high *trans*-diastereoselectivity (equation I). The stereoselectivity is decreased somewhat by addition of $ClSi(CH_3)_3$.

(I)

(53:1)   (98% ee)

The effect of a C_5-methoxy group can even override the steric effect of a C_4-methyl group, and is attributed to a stereoelectronic effect.

Z-Alkenes and Z,Z-dienes. Lithium dialkylcuprates react with acetylene (2 equiv.) at −50° to form lithium (Z)-alkenylcuprates, which can be trapped by electrophiles to produce (Z)-alkenes.[6] The alkenylcuprates can undergo further

reaction with excess acetylene at 0° to form lithium (1Z,3Z)-dienylcuprates. These provide a useful route to various conjugated dienes.[7]

Reaction with 2,3-epoxy alcohols.[8] Both $(CH_3)_2CuLi$ and $(CH_3)_2CuCNLi_2$ react with *trans*-2,3-epoxy alcohols (with a simple alkyl group at C_4) in ether to give about 1:1 mixtures of 1,2- and 1,3-diols. Addition of TMEDA or an imidazolidinone (DMI, **11**, 202) promotes reaction at C_2 to give 1,3-diols, whereas addition of a Lewis acid promotes reaction at C_3 to give 1,2-diols.

ether, −20°	53:47
THF–TMEDA(4:1)	74:26
THF–DMI(4:1)	84:16
$BF_3 \cdot O(C_2H_5)_2$, THF	21:79

S_N2'-*Addition to vinyloxiranes.*[9] Lithium dimethylcuprate reacts with vinyl-oxiranes by a highly *anti*-selective S_N2' addition to provide homoallylic alcohols. Examples:

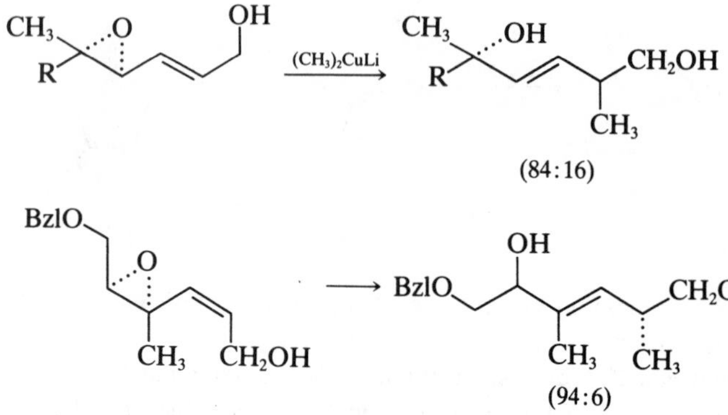

(84:16)

(94:6)

Chiral β-hydroxy esters.[10] Optically pure β,γ-epoxy esters (**2**), obtained as shown from chiral 3-hydroxybutyrolactones (**1**, available from malic acid), react with organocuprates to form optically pure β-hydroxy esters.

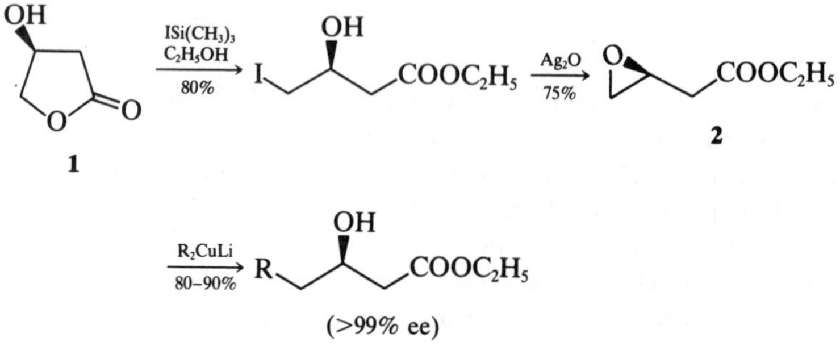

(>99% ee)

$R_2Cu(CN)Li_2$; *reaction with* **vic-**epoxy mesylates.[11] A higher-order cuprate reacts selectively with the epoxide group of the epoxy mesylate **1** to provide **2** with inversion at C_3. Ring closure of **2** furnishes the epoxide **3**, which reacts with a second equivalent of the higher-order cuprate to furnish *meso*-**4**, with inversion at both C_1 and C_3. This two-step reaction provides a route to acyclic alcohols with useful stereocontrol at both adjacent centers.

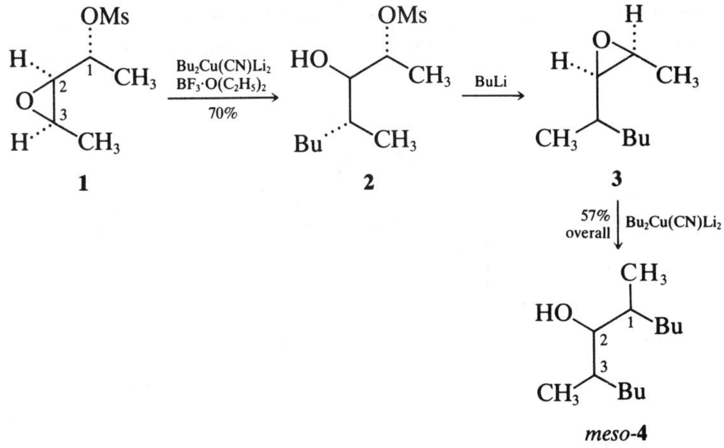

Reactions with cyclopropene.[12] Lithium organocuprates react with the cyclopropenone ketal **1** (**12**, 152–154) to form a copper species (**a**) that behaves as an enolate of a cyclopropanone. Thus it reacts with alkyl halides to form *cis*-2,3-disubstituted derivatives of **1**.

Example:

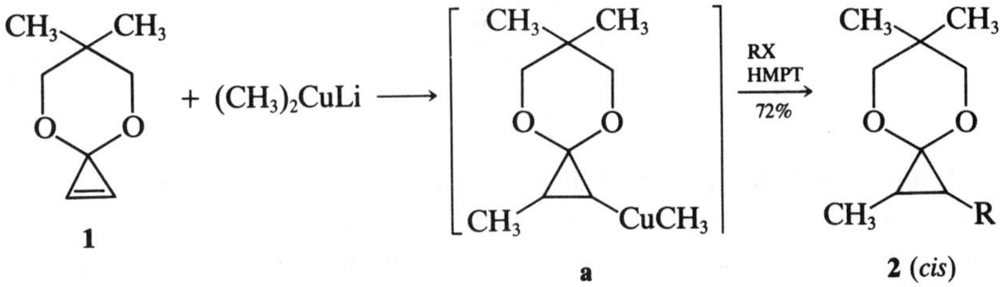

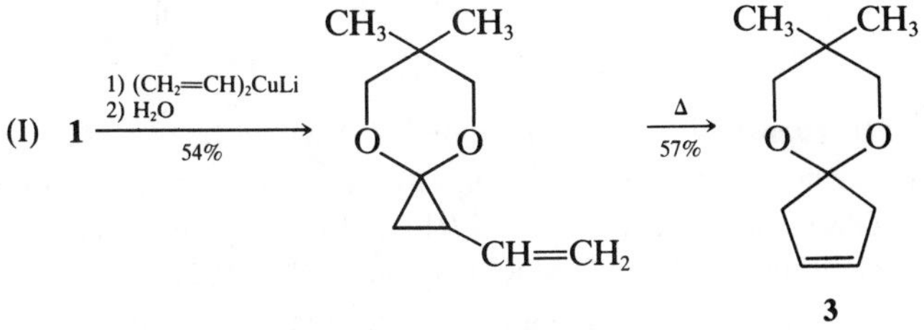

This reaction can be used to effect a [3 + 2] or a [3 + 2 + 2] annelation to a cyclopentenone ketal or a cycloheptadienone ketal, respectively. Thus the adduct of a vinylcuprate with **1** rearranges thermally (240–290°) to 3-cyclopentenone ketals (**3**) in 50–80% yield (equation I). [3 + 2 + 2]Annelation involves preparation of a *cis*-divinylcyclopropane (**4**), which rearranges in the presence of tetrakis(triphenylphosphine)palladium(0) to the 4,5-alkylcycloheptadienone ketal (**5**).

Conjugate addition to unsaturated esters.[13]   The mixed cuprate formed from butyllithium and copper(I) trimethylsilylacetylide, $(CH_3)_3SiC{\equiv}CCu$, is more efficient than $BuCu/BF_3$ or Bu_2CuLi for conjugate addition to α, β-unsaturated esters. Yields of the adduct are markedly improved by addition of $ClSi(CH_3)_3$ (1.3 equiv.) to the reaction.
Example:

Addition of R_2CuLi to bridgehead enones.[14]   Ordinarily organocuprates do not react with a bridgehead halide. However, they can undergo conjugate addition to bridgehead enones generated *in situ* from β-bromo ketones with potassium *t*-butoxide or lithium 2,6-di-*t*-butyl-4-methylphenoxide (**6**,95).

$(CH_3)_2SO$ cuprates.[15]   The carbanion of DMSO is a useful nontransferable ligand for homocuprates. These cuprates are readily prepared by sequential reaction of DMSO with BuLi, CuI, and RLi and undergo the usual reactions with complete retention of the CH_3SOCH_2 group. The $(CH_3)_2SO$ cyanocuprates, $[CH_3SOCH_2Cu(CN)R]Li_2$, are obtained by reaction of dimsyllithium with CuCN

and then RLi. These cuprates show the generally enhanced activity of the higher-order cuprates.

Organobis(cuprates); spiroannelation.[16] 1,4-Dilithiobutane, prepared from 1,4-dichlorobutane and lithium in ether at 0°, on reaction with copper thiophenoxide (2 equiv.) forms a biscuprate, formulated as **1** for convenience. This dimetallic reagent adds to 3-halo-5,5-dimethyl-2-cyclohexenones (**2**) to form the spiro-[4.5]decanone **3** in yields as high as 96%. Cuprates prepared from other Cu(I) sources are less efficient, as is the cuprate prepared from di-Grignard reagents

derived from 1,4-dilithiobutane. This spiroannelation can be extended to synthesis of spiro[4.4]nonanes by addition of **1** to 3-halocyclopentenones. The biscuprate prepared from 1,5-dilithiopentane also undergoes this spiroannelation reaction with β-halo enones. Several unsymmetrical biscuprates are available and also provide spirocycles. However, cuprates in which the metal is attached to an unsaturated center as in **2** or **3** tend to decompose, and yields of products are generally lower.

Examples:

Vinyl cuprates.[17]   Vinyl cuprates can be prepared conveniently by *in situ* trans-metallation of vinylstannanes, available by hydrostannylation of alkynes, with a cuprate such as $(CH_3)_2Cu(CN)Li_2$ (equation I). Reagents prepared in this way effect conjugate addition of the vinyl group to an enone with essentially no transfer of the methyl group.

(I)   R⌒⌒SnBu₃   + (CH₃)₂Cu(CN)Li₂ ⟶

CH₃SnBu₃ + R⌒⌒Cu(CH₃)(CN)Li₂

1

1 (R=H) + [cyclohexenone] $\xrightarrow{95\%}$ [cyclohexanone with CH=CH₂]

2-Thienyl(cyano)copper lithium   [thienyl]–S–Cu(CN)Li (**1**).   The reagent is obtained by reaction of thiophene with BuLi in THF at −78° and then with CuCN at −40°. The reagent is fairly stable and can be stored in THF at −20° for about 2 months. It is inert, but is readily converted by addition of RLi or RMgX into a higher-order mixed cuprate, which is as efficient as the freshly prepared cuprate.[18]
Example:

1 + BuLi ⟶ Bu(2-Th)Cu(CN)Li₂ $\xrightarrow[>90\%]{-78°}$ [cyclohexanone with Bu]

Chiral α-alkyl-δ-alkoxy (E)-β, γ-enoates.[19]   Reaction of $R_2CuLi \cdot BF_3$ with a chiral γ, δ-dialkoxy (E)-α, β-enoate results in diastereoselective α-alkylation via a 1,3-chirality transfer (equation I). Thus the stereochemistry at the α-position of

(I)   [structure: OSiR₃ / CH₃ / OMs, COOCH₃] $\xrightarrow[66\%]{\substack{(CH_3)_2CuLi \cdot BF_3 \\ THF-O(C_2H_5)_2}}$ [structure: OSiR₃ / CH₃ / CH₃, COOCH₃]

(2S,5R > 99% de)

the product is governed by the stereochemistry at the γ-position of the substrate (*anti*-reaction). Somewhat higher yields are obtained by alkylation with $(CH_3)_3Cu(CN)Li_2 \cdot BF_3$ because concomitant γ-alkylation is suppressed.

Addition of RCu to enones; silyl enol ethers.[20] The known ability of $ClSi(CH_3)_3$ to facilitate addition of cuprates to enones has been extended to addition of alkylcoppers to enones. Indeed silyl enol ethers can be obtained readily by addition of RCu to enones in the presence of $ClSi(CH_3)_3$ and TMEDA. A further advantage is that the RCu can be obtained by reaction of RLi with commercial CuI directly.

Examples:

$$CH_2\!\!=\!\!CHCOCH_3 + BuCu \xrightarrow[97-98\%]{} BuCH_2CH\!\!=\!\!C\!\!\begin{array}{l} OSi(CH_3)_3 \\ CH_3 \end{array}$$

$$E/Z = 1:3.2$$

RCu(CN)ZnI.[21] These new copper reagents are prepared by reaction of primary or secondary iodides with zinc that has been activated with 1,2-dibromoethane and chlorotrimethylsilane. The resulting organozinc compounds are then allowed to react with the THF-soluble $CuCN \cdot 2LiCl$ (equation I). Because of the mild conditions, these new reagents can be prepared from iodides containing keto, ester, and nitrile groups.

$$(I) \quad RI \xrightarrow[25-40°]{Zn, THF} RZnI \xrightarrow[0°]{CuCN \cdot 2LiCl} RCu(CN)ZnI$$

These copper reagents do not react with epoxides, but they undergo S_N2' reactions with allylic halides (equation II). These reagents also couple with acyl

$$(II) \quad N\!\!\equiv\!\!CCH_2CH_2CH_2Cu(CN)ZnI + C_6H_5CH\!\!=\!\!CHCH_2Br \xrightarrow[88\%]{0°}$$

$$N\!\!\equiv\!\!C(CH_2)_3CHCH\!\!=\!\!CH_2$$
$$\underset{C_6H_5}{|}$$

chlorides at 0° to give ketones in 80–95% yield, and they react with enones in the presence of $ClSi(CH_3)_3$ to give 1,4-addition products in 62–99% yield.

Example:

$$
\underset{\text{O}}{\bigcirc\!\!\!=} + (CH_3)_3COCCH(CH_2)_3Cu(CN)ZnI \xrightarrow[94\%]{\substack{ClSi(CH_3)_3 \\ \text{ether}}}
$$

$$
\text{(CH}_2)_3\text{CHCOOC(CH}_3)_3 \\
\quad\quad\quad\quad\quad\quad\quad \overset{|}{CH_3}
$$

RCu·BF₃. Yamamoto[22] has reviewed the special properties of RCu combined with BF₃ etherate or AlCl₃ (80 references). In particular, these reagents are generally superior to R₂CuLi for conjugate addition to α,β-enones such as **1**.

$$
\textbf{1} \xrightarrow[47\%]{CH_3Cu \cdot BF_3} \textbf{2}
$$

[1] R. M. Wehmeyer and R. D. Rieke, *J. Org.*, **52**, 5056 (1987); T.-C. Wu, R. M. Wehmeyer, and R. D. Rieke, *ibid.*, **52**, 5057 (1987).

[2] D. K. Hutchinson and P. L. Fuchs, *Am. Soc.*, **109**, 4930 (1987).

[3] T. Kunz and H.-U. Reissig, *Angew. Chem. Int. Ed.*, **27**, 268 (1988).

[4] G. Zweifel and W. Leong, *Am. Soc.*, **109**, 6409 (1987).

[5] A. B. Smith, N. K. Dunlap, and G. A. Sulikowski, *Tetrahedron Letters*, **29**, 439 (1988).

[6] J. F. Normant and A. Alexakis, *Synthesis*, 854 (1981).

[7] M. Furber, R. J. K. Taylor, and S. C. Busford, *J.C.S. Perkin I*, 1809 (1986).

[8] J. M. Chong, D. R. Cyr, and E. K. Mar, *Tetrahedron Letters*, **28**, 5009 (1987).

[9] J. A. Marshall, J. D. Trometer, B. E. Blough, and T. D. Crute, *Tetrahedron Letters*, **29**, 913 (1988).

[10] M. Larchevêque and S. Henrot, *ibid.*, **28**, 1781 (1987).

[11] M. J. Kurth and M. A. Abreo, *ibid.*, **28**, 5631 (1987).

[12] E. Nakamura, M. Isaka, and S. Matsuzawa, *Am. Soc.*, **110**, 1297 (1988).

[13] H. Sakata and I. Kuwajima, *Tetrahedron Letters*, **28**, 5719 (1987).

[14] G. A. Kraus and P. Yi, *Syn. Comm.*, **18**, 473 (1988).

[15] C. R. Johnson and D. S. Dhanoa, *J. Org.*, **52**, 1885 (1987).

[16] P. A. Wender and A. W. White, *Am. Soc.*, **110**, 2218 (1988).

[17] J. R. Behling, K. A. Babiak, J. S. Ng, and A. L. Campbell, R. Moretti, M. Koerner, and B. H. Lipshutz, *Am. Soc.*, **110**, 2641 (1988).

[18] B. H. Lipshutz, M. Koerner, and D. A. Parker, *Tetrahedron Letters*, **28**, 945 (1987).

[19] T. Ibuka, T. Nakao, S. Nishii, and Y. Yamamoto, *Am. Soc.*, **108**, 7420 (1986).

[20] C. R. Johnson and T. J. Marren, *Tetrahedron Letters*, **29**, 27 (1987).

[21] P. Knochel, M. C. P. Yeh, S. C. Berk, and J. Talbert, *J. Org.*, **53**, 2390 (1988).

[22] Y. Yamamoto, *Angew. Chem. Int. Ed.*, **25**, 947 (1986).

Organomanganese reagents.

Selective addition to aldehydes.   Organomanganese halides add to aldehydes selectively in the presence of ketones. This is true for RMnI, RMnBr, prepared from RLi and $MnBr_2$, and for RMnCl, prepared from RLi or RMnX and $MnCl_2$.[1]

Preparation of sec-*or* t-*alcohols.*[2]   A one-pot preparation of either *sec*- or *t*-alcohols involves reaction of RMnI with an acyl chloride to form a ketone complexed with MnCl and stable to further reactions with RMnI. The ketone can be reduced by $LiAlH_4$ or $NaBH_4$ to a *sec*-alcohol or converted into a *t*-alcohol by reaction with an alkyllithium or a Grignard reagent.

RMn(CO)₅.[3]   Alkylmanganese pentacarbonyl complexes are obtained by reaction of alkyl halides with $NaMn(CO)_5$. Under pressures of 2–10 kbar, these complexes undergo an insertion reaction with alkenes to give usually a single manganacycle in which the electron-withdrawing group of the alkene is attached to the same carbon atom as the metal. Demetallation is accomplished by photolysis

which involves loss of carbon monoxide to form a complex that is converted to a ketone on hydrolysis.

The reaction of methylmanganese pentacarbonyl with terminal alkynes proceeds at atmospheric pressure, but the reaction of less reactive manganese complexes with alkynes in general also requires high pressure to obtain manganacycles, which can exhibit the aromatic properties of a metallafuran complex. The insertion reaction of alkynes also usually provides a single adduct. Thus reaction of methylmanganese pentacarbonyl with alkynes results in a single manganacycle **1**

$$CH_3Mn(CO)_5 + C_6H_5C\equiv CH \xrightarrow[67\%]{1\ kbar} \mathbf{1a} \longleftrightarrow \mathbf{1b}$$

Structure **1a**: CH_3–C(=O···Mn(CO)$_4$)–CH=C–C_6H_5

Structure **1b**: CH_3–C(O–Mn(CO)$_4$)=CH–C(C_6H_5)

$\downarrow$ HCl, CH$_3$CN

$$\mathbf{2} \xleftarrow{61\%} \left[\mathbf{a} \right]^+ Cl^-$$

2: CH_3–C(=O)–CH=CH–C_6H_5

a: $\left[CH_3$–C(=O)–CH$_2$–C(···Mn(CO)$_4$)–$C_6H_5 \right]^+$

$\downarrow$ CO

$$\mathbf{3} \xleftarrow{19\%} \left[\mathbf{b} \right]$$

3 (butenolide): ring with CH_3, O, C=O, and C_6H_5

b: CH_3–C(=O)–CH$_2$–C(=C=O)–C_6H_5

formulated as the resonance structures **1a** and **1b**. Treatment of the adduct with a protic acid results in the enone **2** and the butenolide **3** in 61 and 19% yield, respectively. If the demetallation is carried out under carbon monoxide the relative amount of the lactone **3** is increased at the expense of the enone, because of a CO insertion. The adducts can be converted exclusively into butenolides by hydride reduction and CO insertion (equation I).

$$(I)\quad \mathbf{1} \xrightarrow[\text{BuLi}]{\text{DIBAH,}} \left[CH_3\text{–CH(OLi)–CH=C(Mn(CO)_4)–}C_6H_5 \right] \longrightarrow \left[\text{ring with }CH_3, O, C=O, Mn(CO)_3, C_6H_5 \right] \xrightarrow{37\%} \mathbf{3}$$

[1] G. Cahiez and B. Figadere, *Tetrahedron Letters*, **27**, 4445 (1986).
[2] G. Cahiez, J. Rivas–Enterrios, and H. Granger–Veyron, *ibid.*, **27**, 4441 (1986).
[3] P. DeShong, D. R. Sidler, P. J. Rybczynski, G. A. Slough, and A. L. Rheingold, *Am. Soc.*, **110**, 2575 (1988).

Organotin acetylides.

Alkynylation of halo acetals.[1]   In the presence of ZnCl$_2$ (2 equiv.) the organotin acetylide **1** couples with chloromethyl methyl ether or chloromethyl methyl sulfide to form alkynyl ethers or sulfides, respectively (equation I).

(I) CH_3XCH_2Cl + $Bu_3SnC{\equiv}CC_6H_{13}$ $\xrightarrow{\underset{CCl_4}{ZnCl_2,}}$ $CH_3XCH_2C{\equiv}CC_6H_{13}$

1

X = O   62%
X = S   58%

This alkynylation can be used to effect glycosidation of a 1-halo-α-D-glucopyranose (**2**), which undergoes coupling with **1** with retention of configuration to give **3**, which can be converted to the (α-D-glucopyranosyl)methanol (**4**) by Lindlar hydrogenation, ozonation, and reduction.

A similar reaction of **1** with the bromo acetal **5** also proceeds with retention to provide **6**, a useful precursor to unusual amino acids.

[1] D. Zhai, W. Zhai, and R. M. Williams, *Am. Soc.*, **110**, 2501 (1988).

Organotitanium reagents.

Chiral methyltitanium reagents.[1]   The N-sulfonamides (**1**) formed from norephedrine have been used as chiral ligands for a methyltitanium reagent. Thus addition of **1** to $TiCl_4$ and CH_3Li (1:4) and then to isopropanol provides a reagent formulated for simplicity as **2**, which adds to aldehydes to form methylcarbinols. The stereoselectivity is low (30–60%) in the case of aliphatic aldehydes, but is pronounced in reactions with aryl aldehydes. The highest value (90% ee) was observed when Ar of **2** is mesityl and Ar' is *o*-nitrophenyl.

Chiral titanium catalyst for asymmetric Diels–Alder reactions.   A Japanese group[2] recently reported that a chiral titanium reagent (**1**), prepared *in situ* from $TiCl_2(O\text{-}i\text{-Pr})_2$ and the chiral diol **2**, derived from L-tartaric acid, in combination

with 4 Å-sieves serves as an effective catalyst for Diels–Alder reactions of some α,β-unsaturated N-acyloxazolidinones with dienes to provide adducts with useful enantioselectivity. The enantioselectivity is markedly dependent on the solvent, being lowest in benzene and highest in hexylbenzene. In general, useful enantio-selectivity can be obtained with 10 mole % of the chiral catalyst and 1,3,5-trime-thylbenzene (TMB) as solvent.

(92% ee)

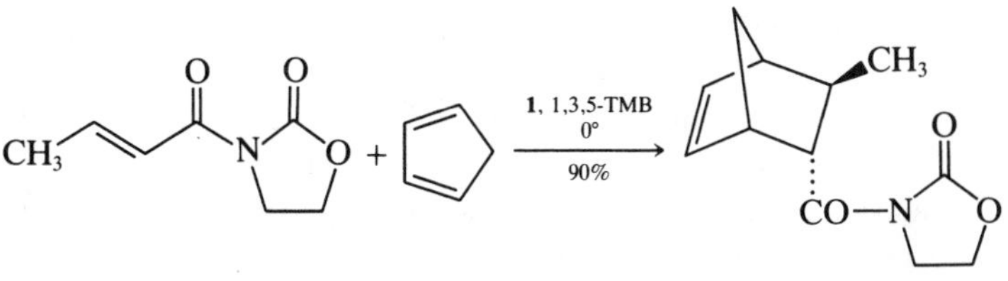

(91% ee)

CH₃Ti(O-i-Pr)₃ (1). Addition of **1** (or CH₃TiCl₃) to γ- and δ-lactols results in 1,4- and 1,5-*syn*-diols with high stereoselectivity. Methylation with CH₃Li, CH₃MgBr, or (CH₃)₃Al shows slight or no asymmetric induction.[3]

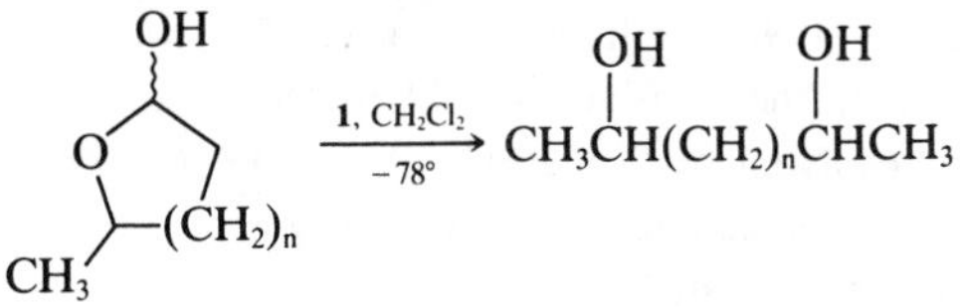

2, n = 1	83%	(*syn/anti* = 12:1)
n = 2	73%	(*syn/anti* = 4.2:1)

[1] M. T. Reetz, T. Kükenhöhner, and P. Weinig, *Tetrahedron Letters*, **27**, 5711 (1986).
[2] K. Narasaka, M. Inoue, and T. Yamada, *Chem. Letters*, 1967 (1986); K. Narasaka, M. Inoue, T. Yamada, J. Sugimori, and N. Iwasawa, *ibid.*, 2409 (1987).
[3] K. Tomooka, T. Okinaga, K. Suzuki, and G. Tsuchihashi, *Tetrahedron Letters*, **28**, 6335 (1987).

Organozinc reagents.
Addition of R₂Zn to aldehydes. In the presence of the β-amino alcohol **1**, (−)-3-exo-(dimethylamino)isoborneol, diethylzinc adds to aromatic aldehydes to

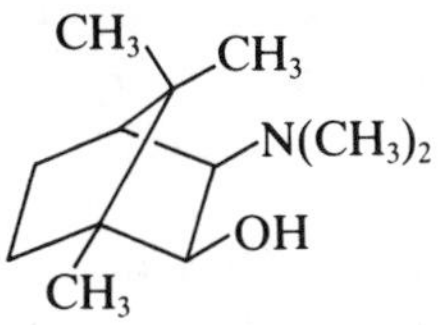

1 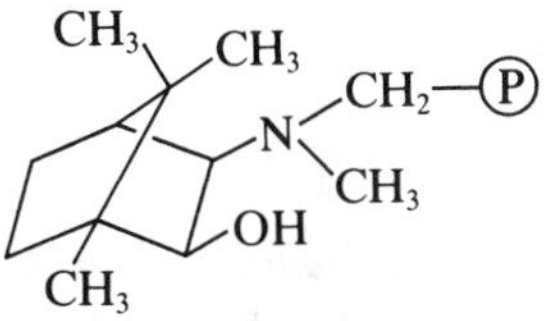   2

form (S)-alcohols in significant enantiomeric excess (90–99% ee), equation I. Some

(I)

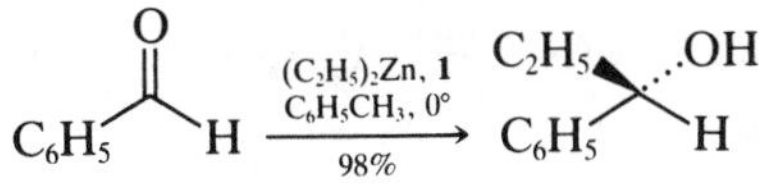

(S), 98–99% ee

α,β-enals and aliphatic aldehyes can also be alkylated with high enantioselectivity (60–96%).[1]

Cornell chemists[2] have prepared a polymeric form (2) of the chiral catalyst, which has the added virtue of recyclability. Both groups observe that reduction rather than alkylation obtains when the ratio of R_2Zn to 1 or 2 is 1:1. The actual catalyst is believed to be a complex of 1 or 2 with 1 equiv. of $(C_2H_5)_2Zn$, which effects transfer of the ethyl group to the aldehyde.

Wynberg[3] has also effected stereoselective addition of $(C_2H_5)_2Zn$ to aryl aldehydes using cinchona alkaloids, particularly quinine and quinidine, which result in (R)- and (S)-alcohols in excess, respectively. The highest enantiomeric excess, 92% ee, was observed with o-ethoxybenzaldehyde catalyzed by quinine.

Alkylzinc iodides.[4]   These reagents are prepared by reaction of alkyl iodides with Zn/Cu in toluene–N,N-dimethylacetamide (DMA). In the presence of 1 equiv. of chlorotrimethylsilane they can add to aldehydes to form alcohols. DMA may be replaced as the cosolvent by N-methylpyrrolidone (NMP), but HMPT retards this reaction. This reaction can be used to obtain γ-, δ-, and ε-hydroxy esters from β-, γ-, and δ-zinc esters (equation I).

(I) $IZn(CH_2)_nCOOC_2H_5$ + ArCHO $\xrightarrow{ClSi(CH_3)_3}$ ArCH(CH_2)_nCOOC_2H_5
                                                                      |
                                                                      OH

n = 2	95%
n = 3	88–95%
n = 4	80%

Another example:

Diastereoselective addition to α-keto esters.[5]   Organozinc reagents, R_2Zn, prepared *in situ* from 2RMgX with ZnX_2, add to the α-keto groups of (–)-menthyl phenylglyoxalate with generally high diastereoselectivity. The choice of solvent

$C_6H_5COCOOMenthyl(-)$ $\xrightarrow[93\%]{R_2Zn}$ $C_6H_5\overset{R}{\underset{OH}{\overset{|}{\underset{|}{C}}}}COOMen$ $\xrightarrow{OH^-}$ $C_6H_5\overset{R}{\underset{OH}{\overset{|}{\underset{|}{C}}}}COOH$

R = Bu

82–84% de   (R)

(THF or ether) or of the halogen has little effect. However, asymmetric induction is low when R is methyl or allyl, and highest when R is *n*-hexyl.

1-*Alkynylzinc bromides*, $BrZnC\equiv CR$.[6] Unlike magnesium and lithium acetylides, these reagents add to an α-alkoxy aldehyde with good to high *syn*-selectivity.

Example:

R' = C_6H_5, *n*-C_6H_{13}

syn/anti = 95:5,   R' = C_6H_5
          = 84:16,   R' = *n*-C_6H_{13}

***ZnCl$_2$·TMEDA.*[7]** Grignard reagents react with ZnCl$_2$·TMEDA to form a species, possibly R$_3$ZnMgX, comparable to R$_3$ZnLi (8,515–516) for conjugate addition to α,β-enones. The method is useful for transfer of primary, secondary, or aryl groups under mild conditions.

[1] M. Kitamura, S. Suga, K. Kawai, and R. Noyori, *Am. Soc.*, **108**, 6071 (1986).
[2] S. Itsuno and J. M. J. Fréchet, *J. Org.*, **52**, 4140 (1987).
[3] A. A. Smaardijk and H. Wynberg, *J. Org.*, **52**, 135 (1987).
[4] Y. Tamaru, T. Nakamara, M. Sakaguchi, H. Ochiai, and Z. Yoshida, *J.C.S. Chem. Comm.*, 610 (1988).
[5] G. Boireau, A. Deberly, and D. Abenhaim, *Tetrahedron Letters*, **29**, 2175 (1988).
[6] K. T. Mead, *ibid.*, **28**, 1019 (1987).
[7] R. A. Kjonaas and E. J. Vawter, *J. Org.*, **51**, 3993 (1986).

Organozirconium compounds.

***Review.*[1]** This review covers preparation and recent synthetically useful reactions of organozirconium compounds (178 references). Many of the recent bond-forming reactions involve zirconocene derivatives.

[1] E. Negishi and T. Takahashi, *Synthesis*, 1 (1988).

Osmium tetroxide.

***Enantioselective osmylation of alkenes.*[1]** Osmium tetroxide forms a 1:1 wine-red complex with the chiral diamine **1**[2] that effects efficient enantioselective dihydroxylation of monosubstituted, *trans*-disubstituted, and trisubstituted alkenes (83–99% ee) at −110° in THF. The enantioface differentiation in all cases corresponds to that observed with *trans*-3-hexene and the complex with (−)-**1**. Essentially complete asymmetric induction is observed with *trans*-1-phenylpropene (99% ee).

$(-)$-1

(90% ee)

OsO₄, $(-)$-1

Directed osmylation.[3]   One step in a synthesis of the carbonucleoside aris-
teromycin (3) required dihydroxylation of the intermediate 1, in which a (nitro-
phenylsulfonyl)methyl group is used as a carboxy equivalent. The desired diol (2)

1, Ad = adenine

2

4

5

is obtained with alkaline permanganate, as shown by conversion to **3**. Surprisingly,
osmylation of **1** results in **4**, formed by attack on the more hindered face of **1**, as
shown by conversion to *lyxo*-aristeromycin (**5**). The unexpected stereocontrol in
osmylation is attributed to coordination of osmium with the nitro group.

Pentitol synthesis.[4]   An asymmetric synthesis of L-arabinitol involves condensation of the (E)-α,β-unsaturated ester (**2**) with the anion of methyl (R)-*p*-tolyl
sulfoxide (**1**). The resulting β-keto sulfoxide (**3**) is reduced stereoselectively by
ZnCl₂/DIBAH (**13**, 115–116) to **4**. Osmylation of **4** with (CH₃)₃NO and a catalytic
amount of OsO₄ (**13**, 224–225) yields essentially a single triol (**5**). Finally, a Pummerer rearrangement of the sulfoxide followed by reduction of an intermediate

affords the tetraacetate **6**. This is converted into the pentaacetate (**7**) of L-arabinitol by debenzylation and acetylation.

Asymmetric catalytic osmylation.[5]   Chiral cinchona bases are known to effect asymmetric dihydroxylation with OsO₄ as a stoichiometric reagent (**10**, 291). Significant but opposite stereoselectivity is shown by esters of dihydroquinine (**1**) and of dihydroquinidine (**2**), even though these bases are diastereomers rather than enantiomers.

These cinchona esters also effect asymmetric dihydroxylation of alkenes in reactions with an amine N-oxide as the stoichiometric oxidant and OsO_4 as the
catalyst. Reactions catalyzed by **1** direct attack to the *re*-face and those catalyzed
by **2** direct attack with almost equal preference for the *si*-face.

This new process has one unexpected benefit: the rates and turnover numbers
are increased substantially with the result that the amount of the toxic and expensive
OsO_4 is considerably reduced (usually 0.002 mole %). The rate acceleration is
attributed to formation of an OsO_4-alkaloid complex, which is more reactive than
free osmium tetroxide. Increasing the concentration of **1** or **2** beyond that of OsO_4
produces only negligible increase in the enantiomeric excess of the diol. In contrast
quinuclidine itself substantially retards the catalytic reaction, probably because it
binds too strongly to osmium tetroxide and inhibits the initial osmylation. Other
chelating tertiary amines as well as pyridine also inhibit the catalytic process.

A further advantage of this ligand-accelerated reaction is that a directing functional group is not essential for enantioselectivity, as in asymmetric epoxidation
and hydrogenation. Even simple alkenes are converted into diols in 20–88% ee

and in high chemical yields. In general, (Z)-disubstituted alkenes are oxidized with lower asymmetric induction than the (E)-isomers, and aromatic alkenes are better substrates than aliphatic alkenes. Thus *trans*-stilbene is converted by the reaction catalyzed by **1** at 0–4° for 7 hours into (+)-*threo*-hydrobenzoin in 89% yield and 88% ee.

The orange-red 1:1 complex of the dihydroquinidine and OsO_4 has been prepared and shown to have the structure shown in Figure I. The OsO_4 is coordinated to the quinuclidine nitrogen by an unusually long bond, and is remote from the quinoline ring. ^{1}H-NMR studies suggest that a slightly different conformation may prevail in solution.[6]

Figure I

[1] K. Tomioka, M. Nakajima, and K. Koga, *Am. Soc.*, **109**, 6213 (1987).
[2] *Idem*, *Tetrahedron Letters*, **28**, 1291 (1987).
[3] B. M. Trost, G.-H. Kuo, and T. Benneche, *Am. Soc.*, **110**, 621 (1988).
[4] G. Solladié, J. Hutt, and C. Fréchou, *Tetrahedron Letters*, **28**, 61 (1987).
[5] E. N. Jacobsen, I. Markó, W. S. Mungall, G. Schröder, and K. B. Sharpless, *Am. Soc.*, **110**, 1968 (1988).
[6] E. N. Jacobsen, S. J. Lippard, I. Marko, C. P. Rao, K. B. Sharpless, and J. S. Svendsen, *Am. Soc.*, in press.

Oxazaborolidines, chiral.

   Enantioselective reduction of ketones.[1]   The ability of diborane in combination with the *vic*-amino alcohol (S)-2-amino-3-methyl-1,1-diphenyl-1-butanol (**12**, 31) to effect enantioselective reduction of alkyl aryl ketones involves formation of an intermediate chiral oxazaborolidine, which can be isolated and used as a catalyst for enantioselective borane reductions (equation I).

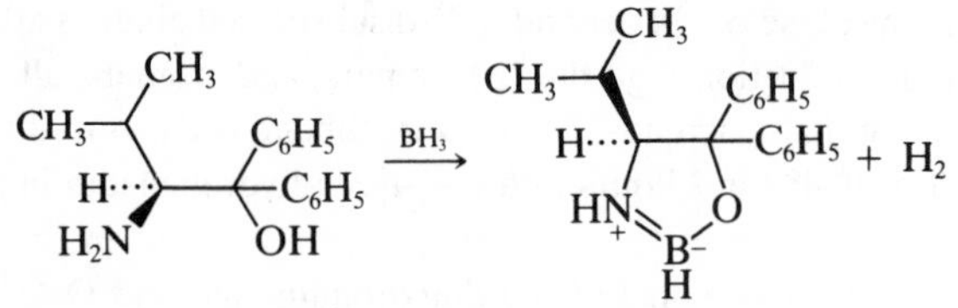

This observation has led to the preparation of more effective bicyclic oxaza-borolidines such as **1**, prepared from (S)-(-)-2-(diphenylhydroxymethyl)pyrrolidine and BH$_3$ (**1a**) or methylboronic acid (**1b**). Both reagents catalyze borane reduction of alkyl aryl ketones to furnish (R)-alcohols in > 95% ee, by face-selective hydride transfer within a complex such as B. Catalyst **1b** is somewhat more effective than

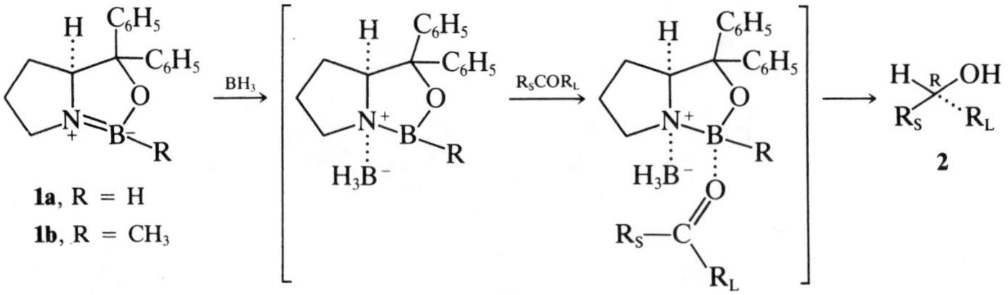

1a, R = H
1b, R = CH$_3$

A   B

1a, and also reduces *t*-butyl methyl ketone to the corresponding (R)-alcohol in 97% ee.

Oxazaborolidine **1b** was used in an enantioselective route to *trans*-2,5-diaryl-tetrahydrofurans.[2] Reduction of the keto ester **3** with BH$_3$ catalyzed by **1b** furnishes

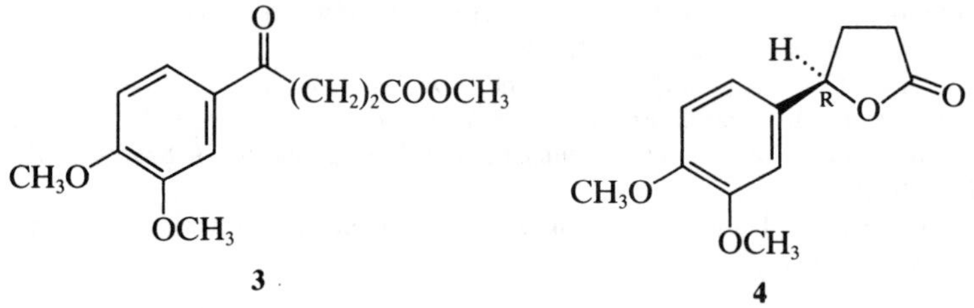

3

4

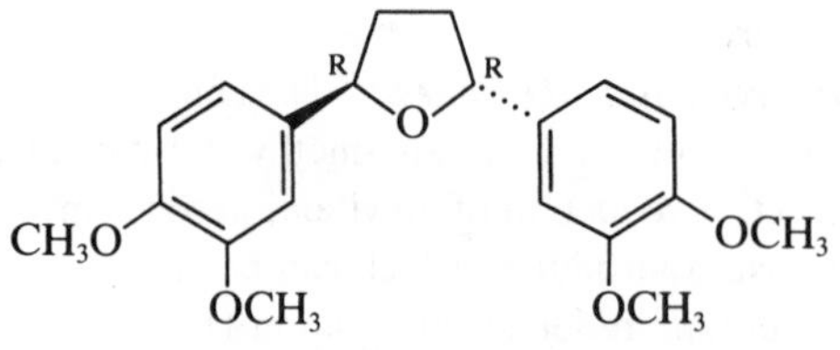

5

the corresponding (R)-alcohol, which cyclizes (NaH) to the (R)-lactone **4**. Reduction (DIBAH) of **4** results in a γ-lactol (1:1 isomers), which after conversion to the corresponding α-bromo ether couples with 3,4-dimethoxyphenylmagnesium bromide to afford *trans*-(2R,5R)-**5** in 70% yield (*trans*/*cis* selectivity = 10:1).

This enantioselective reduction can be used for synthesis of chiral 1-substituted oxiranes.[3] Thus reduction of 2-chloroacetophenone with B_2H_6 catalyzed by **1** (1 mole %) results in (S)-(+)-(chloromethyl)benzenemethanol, which in the presence of base converts to (S)-(−)-phenyloxirane (styrene oxide).

(96.5% ee)

Enantioselective reduction of enones.[4] This methodology can be used to effect enantioselective reduction of enones. Thus the enone **2** is reduced by BH_3 catalyzed

2   **3** (93% ee)

with the (S)-oxazaborolidine (**1b**) to the (R)-alcohol **3** in 93% ee. This alcohol has been used for a total synthesis of natural ginkgolide B (**4**)[5], a potent antagonist of platelet activating factor. This ginkgolide can be converted by reduction into ginkgolide A (**5**)[5], an insect antifeedant.

4, R = OH
5, R = H

[1] E. J. Corey, R. K. Bakshi, and S. Shibata, *Am. Soc.*, **109**, 5551 (1987).
[2] E. J. Corey, R. K. Bakshi, S. Shibata, C.-P. Chen, and V. K. Singh, *ibid.*, **109**, 7925 (1987).
[3] *Idem*, *J. Org.*, **53**, 2861 (1988).

[4] E. J. Corey and A. V. Gavai, *Tetrahedron Letters*, **29**, 3201 (1988).
[5] E. J. Corey and A. K. Ghosh, *ibid.*, **29**, 3205 (1988).

Oxazolidinones, chiral.

α-Azido acids.[1]   Enantiomerically pure α-azido carboxylic acids (**5**) can be obtained by bromination (NBS) of dibutylboron enolates of chiral N-acyloxazolidinones (**2**) followed by displacement with tetramethylguanidinium azide. The products (**4**) can be hydrolyzed to **5** with LiOH in excellent yield without race-

mization. Transesterification to benzyl esters of **5** can be effected with benzyl alcohol and Ti(OBzl)$_4$ in 80–94% yield with only slight racemization.

β-Hydroxy-α-amino acids.[2]   The enolate of the 3-chloroacetyl-2-oxazolidinone **1** reacts with aldehydes to give aldols **2** with high diastereoselectivity. These are useful precursors to *anti*-β-hydroxy-α-amino acids, such as L-*allo*-threonine (**4**).

The synthesis of the rare amino acid 3-hydroxy-4-methylproline (**8**)[3] involves an aldol reaction of the oxazolidinone **5** with methacrolein to provide the α-bromo-β-hydroxy adduct **6**. Azide displacement and removal of the chiral auxiliary gives **7**. On treatment with dicyclohexylborane, **7** undergoes hydroboration–cycloalkylation to provide, after hydrolysis, the methyl ester hydrochloride (**8**) of (2S,3S,4S)-3-hydroxy-4-methylproline in >97% de. This cycloalkylation should be a useful route to cyclic amino acids as well as pyrrolidines.

5   **6** (97% de)

7   2S,3S,4S)-**8**
(>97% de)

Diastereoselective aldol reactions.[4]   The boryl enolates of chiral crotonate imides (**1**) and (**2**) react with aldehydes to form adducts (**3**) and (**4**), respectively, with high diastereoselectivity and complete α-regioselectivity. The method of choice for reductive cleavage of the adducts is formulated for **3**; hydrolysis can also be effected with LiOH and H_2O_2.

1 (X_V—COCH=CHCH₃)   **3** (>98% de)

2 (X_N—COCH=CHCH₃)   **4** (>98% de)

This aldol reaction was employed for an asymmetric synthesis of the azetidinone **9** from the adduct (**5**) of acetaldehyde and **1**.[5] Azetidinone **9** is a versatile precursor to the antibiotic thienamycin **10**. The configurationally stable aldehyde **6**, obtained by ozonolysis of the silyl ether of **5**, undergoes addition with allylzinc chloride to afford **7**, which on transamination is converted to the N-methoxy amide **8**. This product is converted in several steps to the desired **9** in 34% overall yield. An interesting feature of this synthesis is the early incorporation of the hydroxyethyl side chain at C_6, a step that is difficult to effect after formation of the β-lactam ring.

Asymmetric Diels–Alder reactions.[6]   Unlike methyl crotonate, which is a weak dienophile, chiral (E)-crotonyl oxazolidinones when activated by a dialkylaluminum chloride (1 equiv.) are highly reactive and diastereoselective dienophiles. For this purpose, the unsaturated imides formed from oxazolidinones (X_p) derived from (S)-phenylalanol show consistently higher diastereoselectivity than those derived from (S)-valinol or (1S, 2R)-norephedrine. The effect of the phenyl group is attributed in part at least to an electronic interaction of the aromatic ring. The reactions of the unsaturated imide **1** shown in equation (I) are typical of reactions of unsaturated N-acyloxazolidinones with cyclic and acyclic dienes. All the Diels–Alder reactions show almost complete *endo*-selectivity and high diastereoselectivity. Oxazolidinones are useful chiral auxiliaries for intramolecular Diels–Alder

(I)

reaction of trienes; they not only improve *endo*-diastereoselectivity, but they enhance the reactivity.

Hydrolysis of the Diels–Alder products, particularly those formed from hindered imides, is best effected with lithium hydroxide and excess 30% H_2O_2 in aqueous THF at room temperature.

[1] D. A. Evans, J. A. Ellman, and R. L. Dorow, *Tetrahedron Letters*, **28**, 1123 (1987).
[2] D. A. Evans, E. B. Sjogren, A. E. Weber, and R. E. Conn, *ibid.*, **28**, 39 (1987).
[3] D. A. Evans and A. E. Weber, *Am. Soc.*, **109**, 7151 (1987).
[4] D. A. Evans, E. B. Sjogren, J. Bartroli, and R. L. Dow, *Tetrahedron Letters*, **27**, 4957 (1986).
[5] D. A. Evans and E. B. Sjogren, *ibid.*, **27**, 4961 (1986).
[6] D. A. Evans, K. T. Chapman, and J. Bisaha, *Am. Soc.*, **110**, 1238 (1988).

μ-Oxobis[phenyl(trifluoromethanesulfonato)iodine], $O[I(OTf)C_6H_5]_2$ (**1**). The reagent, a mildly air-sensitive, yellow powder, precipitates on reaction of 2 equiv. of $C_6H_5I{=}O$ with triflic anhydride in CH_2Cl_2.

vic-Ditriflates.[1] The reagent reacts with alkenes in CH_2Cl_2 at 25° to form *cis,vic*-ditriflates (50–70% yield) and the co-products C_6H_5I and $C_6H_5I{=}O$. It reacts with butadiene to give an 89:11 mixture of 1,4- and 1,2-adducts.

R. T. Hembre, C. P. Scott, and J. R. Norton, *J. Org.*, **52**, 3650 (1987).

Oxodiperoxymolybdenum(pyridine)(hexamethylphosphoric triamide) (MoOPH).

Oxidative desulfonylation; polyenes. Oxidation of the anion of a primary sulfone with MoOPH results in a β-hydroxy sulfone formed by condensation of

$$(I) \quad 2C_6H_5CH_2SO_2C_6H_5 \xrightarrow[83\%]{\substack{1)\ BuLi \\ 2)\ MoOPH}} \underset{\substack{| \\ OH}}{C_6H_5} \overset{\substack{SO_2C_6H_5 \\ |}}{\underset{}{\diagdown}} C_6H_5$$

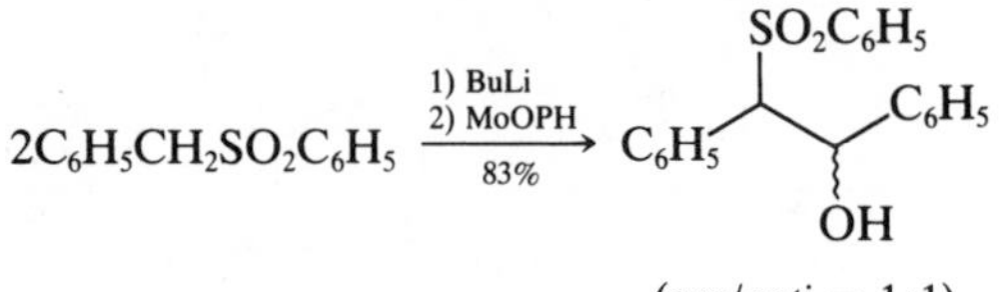

(*syn/anti* = 1:1)

the anion with the aldehyde as formed on oxidation (equation I). This reaction can be used to prepare terpene polyenes containing an (E,Z,Z)-1,3,5-triene system.
Example:

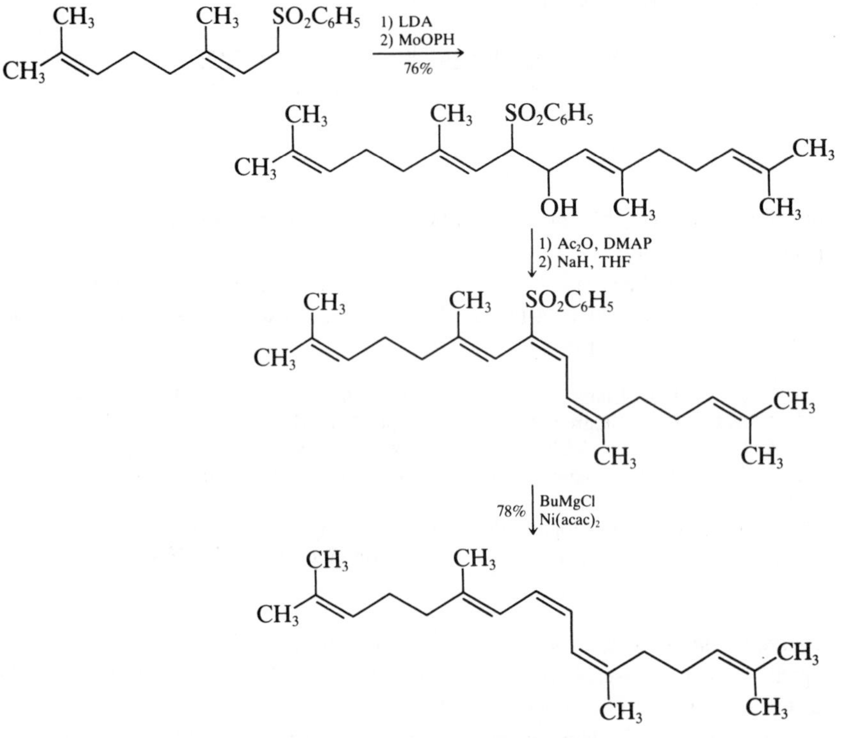

(E,Z,Z, 82%)

α-Diketones.[2]   In the presence of Hg(OAc)$_2$, a variety of internal alkynes are oxidized to α-diketones in 55–90% yield. This system also oxidizes terminal alkynes to α-keto aldehydes.
Example:

$$CH_3(CH_2)_3C{\equiv}CH \xrightarrow[86\%]{\substack{MoO_5 \cdot HMPT, \\ Hg(OAc)_2}} CH_3(CH_2)_3\overset{\overset{\textstyle O}{\|}}{C}{-}CHO$$

[1] M. Capet, T. Cuvigny, C. H. du Penhoat, M. Julia, and G. Loomis, *Tetrahedron Letters*, **28**, 6273 (1987).

[2] F. P. Ballistreri, S. Failla, G. A. Tomaselli, and R. Curci, *Tetrahedron Letters*, **27**, 5139 (1986).

Oxygen, singlet

3-*Alkyl-4-hydroxybutenolides*.[1]   This group is present in some marine sponges and is believed to arise from 3-alkylfurans. This transformation can be realized *in vitro* by singlet oxygen oxidation in the presence of a hindered base such as ethyl-diisopropylamine.

***Hydroxy epoxidation of dienes*.**[2]   Photosensitized oxygenation of dienes when catalyzed by titanium(IV) isopropoxide results in an epoxy alcohol, formed by an oxygen transfer from an allylic hydroperoxide.

Example:

[1] M. R. Kernan and D. J. Faulkner, *J. Org.*, **53**, 2773 (1988).

[2] W. Adam, A. Griesbeck, and E. Staab, *Tetrahedron Letters*, **27**, 2839 (1986); W. Adam and E. Staab, *ibid.*, **29**, 531 (1988).

P

Palladium(II) acetate.

Arylation of cycloalkenes.[1] Aryl halides undergo Heck coupling with cycloalkenes in the presence of a palladium catalyst. The reaction involves addition of an arylpalladium intermediate to the double bond followed by elimination of a palladium hydride.

Example:

$$p\text{-CH}_3\text{CO}_2\text{C}_6\text{H}_4\text{I} + \text{[cyclopentene]} \xrightarrow[94\%]{\underset{\text{Bu}_4\text{NCl, DMF, 20°}}{\text{Pd(OAc)}_2,\ \text{KOAc}}} \text{CH}_3\text{CO}_2-\text{[product]}$$

Allylic acetoxylation.[2] Pd(OAc)$_2$ in HOAc can effect allylic acetoxylation of alkenes, probably via a π-allylpalladium complex, and only a catalytic amount is required in the presence of a cooxidant such as benzoquinone–MnO$_2$. The reaction is not useful in the case of simple alkenes because of lack of discrimination between the two allylic positions, but this acetoxylation can be regioselective in the case of alicyclic alkenes.

Examples:

Oxidative coupling of alkenylstannanes.[3] 1-Alkenylstannanes undergo homocoupling to 1,3-dienes when treated with *t*-butyl hydroperoxide in the presence of catalytic amounts of Pd(OAc)$_2$. Under these conditions 1-alkenylstannanes couple with 2-alkenylstannanes to give 1,4-dienes.

Examples:

$$C_6H_5 \diagup\!\!\!\diagdown Sn(C_2H_5)_3 \xrightarrow[\substack{80\%}]{\substack{t\text{-BuOOH,}\\Pd(OAc)_2}} C_6H_5 \diagup\!\!\!\diagdown\!\!\!\diagup\!\!\!\diagdown C_6H_5$$

68% $\Big|$ CH$_2$=CHCH$_2$Sn(C$_2$H$_5$)$_3$

$$C_6H_5 \diagup\!\!\!\diagdown\!\!\!\diagup\!\!\!\diagdown CH_2$$

(100% E)

[1] R. C. Larock and B. E. Baker, *Tetrahedron Letters*, **29**, 905 (1988).
[2] A. Heumann, B. Akermark, S. Hanson, and T. Rein, *Org. Syn.*, submitted (1988).
[3] S. Kanemoto, S. Matsubara, K. Oshima, K. Utimoto, and H. Nozaki, *Chem. Letters*, 5 (1987).

Palladium(II) acetate–Bis(1,4-diphenylphosphine)butane (dppb).

Dienones. Alkynones, which are readily available by addition of lithium acetylides to N-methyl-N-methoxyamides (**11**, 202), isomerize to dienones when treated with Pd(OAc)$_2$ in combination with a phosphine, particularly dppb. The actual catalyst is probably a hydridopalladium acetate.[1]

Examples:

[1] B. M. Trost and T. Schmidt, *Am. Soc.*, **110**, 2301 (1988).

Palladium(II) acetate–Hydroquinone–(Tetraphenylporphyrin)cobalt.

1,4-Oxygenation of 1,3-dienes.[1] Conjugated dienes can undergo an efficient biomimetic aerobic oxidation in the presence of this triple catalyst. The reaction is believed to involve a π-allylpalladium–benzoquinone complex, and the cobalt catalyst is required for reoxidation of hydroquinone to benzoquinone by O_2. Salcomine (2,360) or cobaloxime (11,135–136) is less efficient for this purpose than cobalt(TPP) (12,138–139). Thus in the presence of 5 mole % each of Pd(OAc)$_2$, hydroquinone, and Co(TPP), 1,3-cyclohexadiene (1) is oxidized in HOAc containing LiOAc to 2 in 89% yield.

1   2 (*trans/cis* = 90:10)

[1] J. E. Bäckvall, A. K. Awasthi, and Z. D. Renko, *Am. Soc.*, **109**, 4750 (1987).

Palladium(II) acetate–Triphenylphosphine.

Cyclization of 1,6-enynes. Pd(OAc)$_2$, particularly when complexed with phosphine ligands, can effect cyclization of 1,6-enynes to a five-membered ring system. This cyclization, **1→2**, provides a key step in a synthesis of the sesquiterpene sterepolide (**3**).[1]

This cyclization is possible even when a quaternary center is formed; in such reactions, N,N'-bis(benzylidene)ethylenediamine, $C_6H_5HC=N(CH_2)_2$ $N=CHC_6H_5$ (**6**), can be more effective than P(C$_6$H$_5$)$_3$ as the ligand.[2]

$$C_6H_5HC{=}N \qquad N{=}CHC_6H_5$$

6

This cyclization can also be used to obtain substituted pyrrolidines, as in the conversion of **7** to **8**.[3]

BzlN⌇⌇CH₂OR $\xrightarrow[60\%]{(R_3P)Pd(OAc)_2}$ BzlN...CH₂OR / CH₂

7   **8**

Cyclization of a 1,6-enyne to a methylenecyclopentane can be effected with a Pd(0) catalyst complexed with a phosphine ligand in combination with a trialkyl-silane as a hydride donor (equation I).[4] Under these conditions a 1,7-enyne can be cyclized to a methylenecyclohexane.

(I) $\xrightarrow[60-88\%]{Pd(0),\ PAr_3,\ R_3SiH}$

E = COOCH₃

Intramolecular arylation of alkenes. The intramolecular palladium-catalyzed condensation of aryl halides with an alkene group (Heck arylation[5]) can actually proceed more readily than the original bimolecular version. These intramolecular cyclizations typically proceed in acetonitrile at room temperature, particularly when catalyzed by Ag₂CO₃, which also inhibits isomerization of the double bond of the product. They can be used to obtain spiro, bridged, and fused systems. Even tetrasubstituted alkenes can participate, with formation of quaternary centers.[6]

Examples:

COOCH₃ $\xrightarrow[86\%]{Pd(OAc)_2,\ P(C_6H_5)_3 \atop Ag_2CO_3}$ COOCH₃   (30:1)

$\xrightarrow{69\%}$ N—CH₃ + *endo*-isomer

3.4:1

This intramolecular Heck cyclization can be extended to diunsaturated aryl iodides, which can undergo two consecutive cyclizations to form polycyclic systems.[7]
Examples:

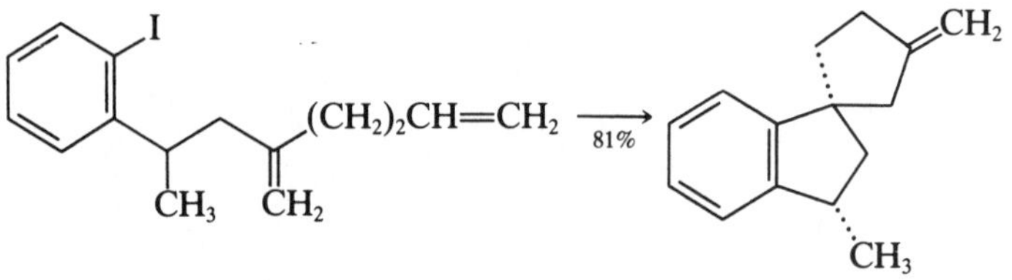

(13:1)

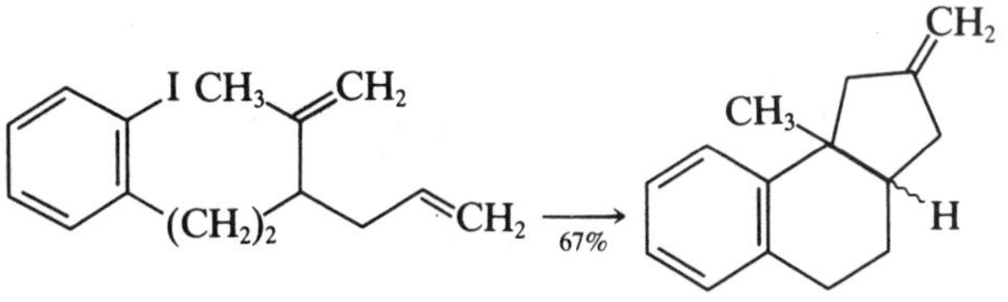

(1.3:1)

[1] B. M. Trost and M. Lautens, *Am. Soc.*, **107**, 1781 (1985); B. M. Trost and J. Y. L. Chung, *ibid.*, **107**, 4586 (1985).
[2] B. M. Trost and D. J. Jebaratnam, *Tetrahedron Letters*, **28**, 1611 (1987).
[3] B. M. Trost and S.-F. Chen, *Am. Soc.*, **108**, 6053 (1986).
[4] B. M. Trost and F. Rise, *ibid.*, **109**, 3161 (1987).
[5] R. F. Heck, *Org. React.*, **27**, 345 (1982).
[6] M. M. Abelman, T. Oh, and L. E. Overman, *J. Org.*, **52**, 4130 (1987).
[7] M. M. Abelman and L. E. Overman, *Am. Soc.*, **110**, 2328 (1988).

Palladium(II) acetate–Tris(2,6-dimethoxyphenyl)phosphine.
    1,3-Enynes.[1] $Pd(OAc)_2$ when complexed with this phosphine (1:1) can effect coupling of two alkynes to provide dimeric 1,3-enynes.

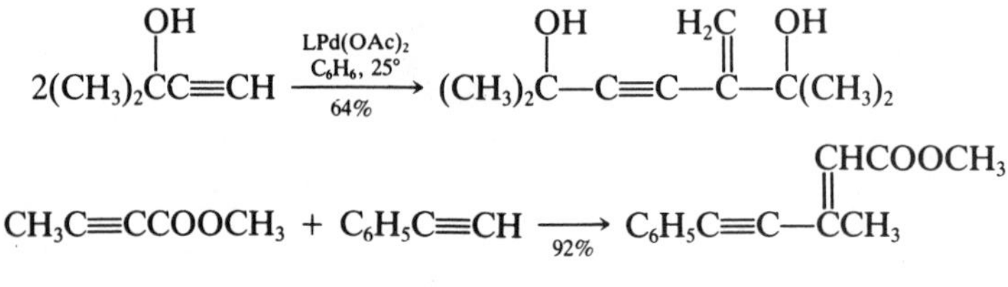

[1] B. M. Trost, C. Chan, and G. Ruhter, *Am. Soc.*, **109**, 3486, (1987).

Palladium(II) chloride–Copper(II) chloride.

Aminocarbonylation of alkenylureas.[1]   In the presence of PdCl$_2$ (with CuCl$_2$ as reoxidant) alkenylureas undergo carbonylation and cyclization to five- and six-membered nitrogen heterocycles.

Examples:

R = CH$_3$ 58%
R = CH$_2$C$_6$H$_5$ 82%

[1] Y. Tamaru, M. Hojo, H. Higashimura, and Z. Yoshida, *Am. Soc.*, **110**, 3994 (1988).

Palladium(II) trifluoroacetate.

Alkylidenecyclopentenediones.[1]   The adduct (**2**) of 1-lithiohexyne to the cyclobutenedione **1** rearranges in the presence of Pd(OCOCF$_3$)$_2$ to alkylidenecyclopentenediones **3**. A similar selective rearrangement obtains in rearrangements to 2-alkylideneindone-1,3-dione monoketals (equation I). In both cases the nonvinylic C–C bond (a) rearranges more rapidly.

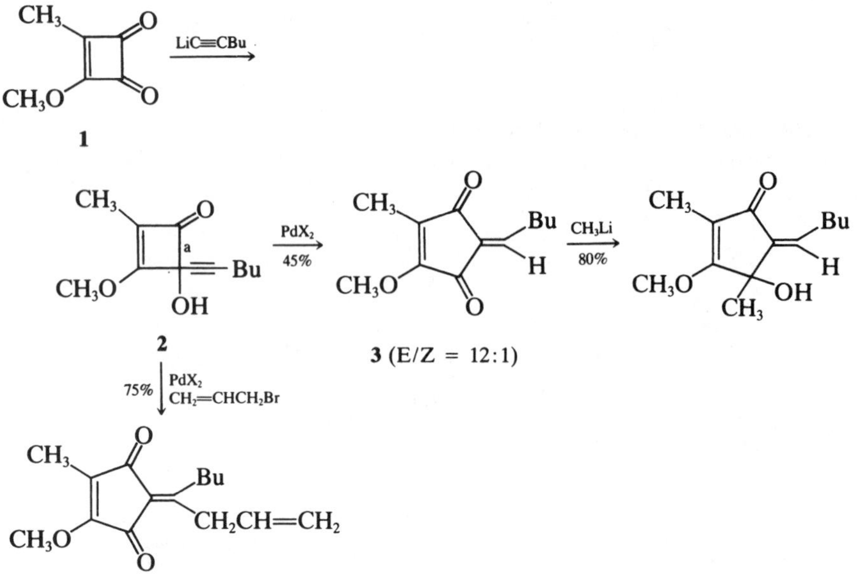

1

2

3 (E/Z = 12:1)

4 (E/Z = >20:1)

(I)

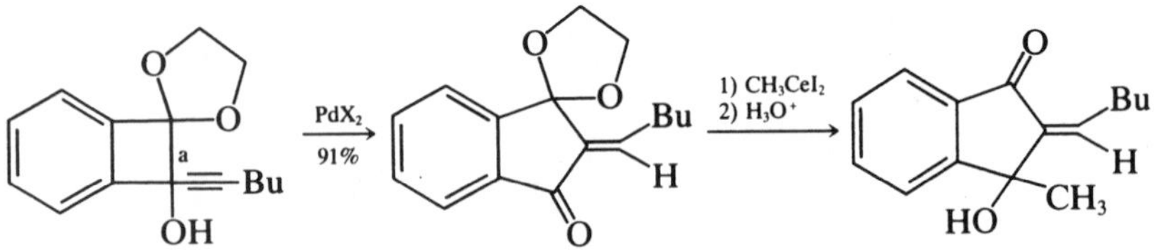

(36:1)

[1] L. S. Liebeskind, D. Mitchell, and B. S. Foster, *Am. Soc.*, **109**, 7908 (1987); L. S. Liebeskind, *Pure Appl. Chem.*, **60**, 27 (1988).

Periodinane (1, 12,378–379).

Oxidation of trifluoromethylcarbinols.[1]   Oxidation of these alcohols to the corresponding ketones with usual oxidants requires rather drastic conditions and an excess of reagent, but can be effected with periodinane (4 equiv.) in CH_2Cl_2 at 25° within 3 hours.

$$RCH_2CHCF_3 \xrightarrow[93-95\%]{1, CH_2Cl_2} RCH_2CCF_3$$

[1] R. J. Linderman and D. M. Graves, *Tetrahedron Letters*, **28**, 4259 (1988).

Peroxomolybdenum–Picolinate N-oxide complex (**1**). The complex is prepared from $Na_2MoO_4 \cdot 2H_2O$, H_2O_2, picolinic acid N-oxide, and Bu_4NOH. It is obtained

1

as a bright yellow hydrate. The complex is soluble in dichloroethane, in which oxidation is faster than in protic media.

This complex oxidizes primary or secondary alcohols in DCE at 50° with little difference in the rate.[1] The aldehydes or ketones are obtained in nearly quantitative yield. In this solvent, epoxidation of double bonds does not compete with alcohol oxidation.

[1] O. Bortolini, S. Campestrini, F. Di Furia, G. Modena, and G. Valle, *J. Org.*, **52**, 5467 (1987).

Phenyl 3-*t*-butylpropiolate, $(CH_3)_3CC\equiv CCOOC_6H_5$ (**1**). The ester is prepared by reaction of lithium *t*-butylacetylide with phenyl chloroformate.

Cyclopentenone annelation.[1] The reaction of the dilithio derivative (**2**) of dimethyl 4-cyclohexene-1,2-dicarboxylate (LDA, THF/HMPT) with **1** at $-45 \rightarrow$ 0° results in the bicyclic enone **3** in 72% yield. The reaction may involve conjugate addition of **2** to **1** to give a ketene (**a**) with loss of lithium phenoxide. Several other

phenyl esters of propiolic acids undergo this reaction, which may be general for vicinal diester dianions.

This bicyclic keto ester (**3**) was used for synthesis of the ginkolide bilobalide **4**.

4

[1] E. J. Corey and W. Su, *Tetrahedron Letters*, **28**, 5241 (1987); *idem*, *Am. Soc.*, **109**, 7534 (1987).

5-Phenyldibenzophosphole, $C_6H_5—P$ $(C_6H_5-DBP, \mathbf{1})$.

The phosphole, m. p. 95–97°, is obtained in 74% yield by reaction of tetraphenylphosphonium bromide with lithium diethylamide.

(E)-*Selective Wittig reagents*.[1]   The reaction of **1** with lithium in THF provides LiDBP, which on reaction with an alkyl halide (2 equiv.) and $NaNH_2$ in THF gives a salt-free ylide such as **2** or **3**, formed by reaction with ethyl iodide or butyl iodide, respectively. These ylides react readily with aldehydes at −78°, but the intermediate oxaphosphetanes are unusually stable and require temperatures of 70–110° for conversion to the phosphine oxide and the alkene, which is obtained in E/Z ratios of 6–124:1. Highest (E)-selectivity is observed with α-branched aldehydes.

Examples:

[1] E. Vedejs and C. Marth, *Tetrahedron Letters*, **28**, 3445 (1987).

1-Phenylethylamine, $C_6H_5\overset{.}{C}H(CH_3)NH_2$ **(1).**

Asymmetric intramolecular Michael cyclization.[1]   Reaction of the ketone **2** with (R)-(+)-**1** (1 equiv.) generates an enamine that adds to the unsaturated ester group to give **3a**. The yield and the ee of **3** are markedly improved when 5 Å

molecular sieves are added. The structure of **3a** was proved by conversion to the lactone **4**, a known precursor to the alkaloid (−)-ajmalicine. Use of (S)-(−)-**1** provides **3b**, a precursor to the bicyclic lactone **5**.

[1] Y. Hirai, T. Terada, and T. Yamazaki, *Am. Soc.*, **110**, 958 (1988).

Phenyliodine(III) bis(trifluoroacetate), $C_6H_5I(OCOCF_3)_2$ **(1).**

p-Benzoquinone monoacetals.[1]   This iodine(III) reagent is effective for oxidation of *p*-alkoxyphenols to *p*-benzoquinone monoacetals in the presence of K_2CO_3 suspended in an alcohol.

Examples:

[1] Y. Tamura, T. Yakura, J. Haruta, and Y. Kita, *J. Org.*, **52**, 3927 (1987).

Phenyliodine(III) diacetate (iodosylbenzene diacetate).

Oxidation of phenols.[1]   The reagent oxidizes 1,2- and 1,4-dihydroxyphenols to the quinones in almost quantitative yield at 25° in methanol. 4-Alkylphenols are oxidized to 4-alkyl-4-methoxycyclohexadienones (mixed quinone ketals) in >90% yield. Monohydric phenols can be oxidized to *p*-quinone diketals on oxidation with 2 equiv. of the reagent in CH_3OH at 25°.

Diimide from hydrazine hydrate.[2]   Diimide can be generated from hydrazine hydrate by oxidation with this hypervalent iodine compound in CH_2Cl_2.

$$NH_2NH_2 \xrightarrow{C_6H_5I(OAc)_2} [HN{=}NH] + C_6H_5I$$

$$80\% \downarrow {C_6H_5C{\equiv}CC_6H_5}$$

$$cis\text{-}C_6H_5CH{=}CHC_6H_5$$

Alkynyl carboxylates, $RC{\equiv}CO_2CR'$.   A wide variety of these previously unknown esters can be prepared in a two-step procedure from **1**, as formulated for the alkynyl benzoates (**2**).[3]

$$C_6H_5I(OCOCH_3)_2 + 2C_6H_5COOH \xrightarrow[CH_3COOH]{\Delta, \ 15 \ mm.} C_6H_5I(OCOC_6H_5)_2 \xrightarrow{LiC{\equiv}CR}$$

$$RC{\equiv}CO_2CC_6H_5 + C_6H_5CO_2Li + C_6H_5I$$

$$\mathbf{2} \ (27{-}57\%)$$

$$\downarrow {H_3O^+}$$

$$RCH_2COOH + C_6H_5COOH$$

Cyclic acetals. In combination with iodine, this reagent resembles lead tetraacetate in the ability to generate alkoxy or aminyl radicals (**12**,243) under thermal or light activation. This radical oxidation can be used to convert β- or γ-hydroxy ethers into dioxolanes or dioxanes.[4]

[1] A. Pelter and S. Elgendy, *Tetrahedron Letters*, **29**, 677 (1988).
[2] R. M. Moriarty, R. K. Vaid, and M. P. Duncan, *Syn. Comm.*, **17**, 703 (1987).
[3] P. J. Stang, M. Boehshar, H. Wingert, and T. Kitamura, *Am. Soc.*, **110**, 3272 (1988).
[4] K. Furuta, T. Nagata, and H. Yamamoto, *Tetrahedron Letters*, **29**, 2215 (1988).

8-Phenylmenthol (1) and -menthone.

L-Amino acids.[1]   Bromination of the ester (**2**) formed from N-Boc-glycine and ($-$)-**1** (R*OH) provides two bromides (**3**), which on reaction with Grignard reagents are converted into a highly reactive iminoacetate (**a**), to which Grignard reagents add with (S)-stereoselectivity to provide **4**. Reductive removal of the chiral auxiliary gives protected L-amino alcohols (**5**), which on oxidation and deprotection yield L-amino acids (**6**) in 82–95% ee. Highest optical yields obtain with *i*-BuMgBr.

Chiral glycolates.   The chiral dioxolanes **1** and **2** are prepared by reaction of 8-phenylmenthone with a protected derivative, $(CH_3)_3SiOCH_2COOSi(CH_3)_3$, of glycolic acid catalyzed by trimethylsilyl triflate. They are obtained in about a 1:1 ratio and are separable by chromatography. Alkylation of the enolates of **1** and **2** proceeds with marked diastereofacial selectivity. After separation of the major isomers, ethanolysis provides (R)- and (S)-α-hydroxy esters with recovery of the chiral auxiliary.[2]

$$\text{(14–58:1)} \qquad\qquad \text{(24–123:1)}$$

Aldol reactions of **1** and **2** can be used to obtain any one of the four possible stereoisomers of α,β-dihydroxy esters.[3] Thus **1** reacts with aldehydes to provide (2S)-aldols, and **2** reacts to provide (2R)-aldols. The *syn/anti* ratio of the aldols can be controlled by the choice of the enolate counterion. Thus lithium or magnesium enolates provide mainly *anti*-aldols, whereas *syn*-aldols predominate with zirconium enolates. Ethanolysis of the purified adducts yields the optically pure α,β-dihydroxy esters without epimerization with recovery of 8-phenylmenthol.

Chiral β-amino esters.[4] The conjugate addition of primary amines to alkyl crotonates proceeds in satisfactory yield when carried out under high pressure. French chemists have recently examined this reaction using crotonates derived from chiral alcohols. Thus addition of diphenylmethylamine to 8-phenylmenthyl crotonate proceeds in 85–90% chemical yield and in 60% de. Optical yields are increased when the phenyl group is substituted in the *para*-position by bulky groups. Essentially complete diastereofacial control is obtained by use of 8-(β-naphthyl)menthyl crotonate because one face of the double bond is well shielded by the naphthyl group.

[1] P. Ermert, J. Meyer, and C. Stucki, *Tetrahedron Letters*, **29**, 1265 (1988).
[2] W. H. Pearson and M.-C. Cheng, *J. Org.*, **51**, 3746 (1986).
[3] *Idem, ibid.*, **52**, 3176 (1987).
[4] J. d'Angelo and J. Maddaluno, *Am. Soc.*, **108**, 8112 (1986).

Phenylselenenyl benzenesulfonate, $C_6H_5SO_2SeC_6H_5$ **(1)**,10,315;11,407–408;12,394.

2-Phenylsulfonyl-1,3-dienes.[1] This reagent undergoes selective 1,2-addition to 1,3-dienes, particularly at low temperatures. Oxidation of the adducts provides 2-phenylsulfonyl-1,3-dienes.

Example:

Addition to 1-alkynes **(10,408).** Terminal alkynes can be converted into acetylenic sulfones[2] or into allenic sulfones[3] by reaction with **1** followed by selenoxide elimination.

[1] J.-E. Bäckvall, C. Nájera, and M. Jus, *Tetrahedron Letters*, **29**, 1445 (1988).
[2] T. G. Back, S. Collins, U. Gokhale, and K.-W. Law, *J. Org.*, **48**, 4776 (1983).
[3] T. G. Back, M. V. Krishna, and K. R. Muralidharan, *Tetrahedron Letters*, **28**, 1737 (1987).

Phenylsilane–Cesium fluoride.

Azomethine ylides. Reduction of oxazolium salts **(1)** with phenylsilane and cesium fluoride provides an unstable 4-oxazoline **(2)**, which can react as an azomethine ylide with a dipolarophile such as DMAD to give a dihydropyrrole **(3)**.[1]

Example:

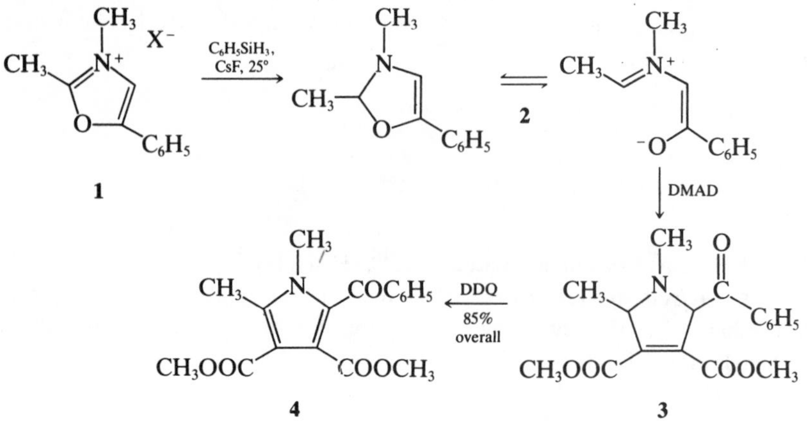

1

2

3

4

[1] E. Vedejs and J. W. Grissom, *Am. Soc.*, **108**, 6433 (1986); **110**, 3238 (1988).

N-(Phenylthio)morpholine, C_6H_5S—N O (1).

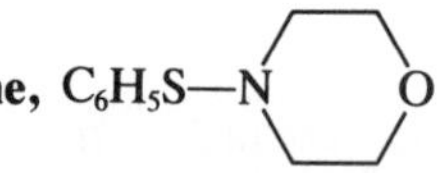

Sulfenoetherification.[1] The reagent in combination with trifluoromethane-sulfonic acid converts suitably unsaturated alcohols into five- to seven-membered cyclic ethers. The cyclization is considered to involve an intermediate episulfonium ion.
Examples:

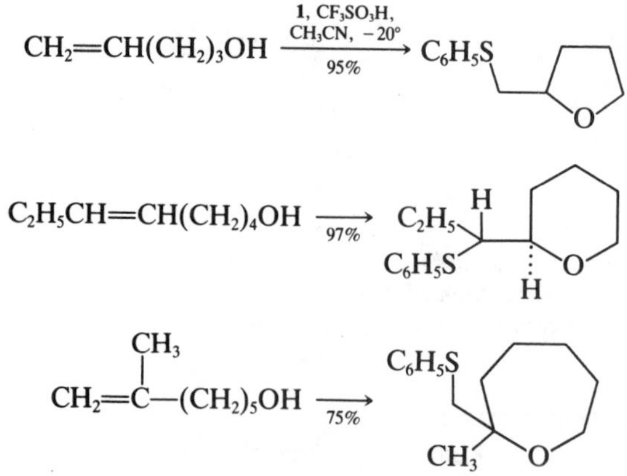

$$CH_2{=}CH(CH_2)_3OH \xrightarrow[95\%]{1,\ CF_3SO_3H,\ CH_3CN,\ -20°} C_6H_5S$$

$$C_2H_5CH{=}CH(CH_2)_4OH \xrightarrow[97\%]{}$$

$$CH_2{=}\overset{CH_3}{\underset{|}{C}}{-}(CH_2)_5OH \xrightarrow[75\%]{}$$

[1] P. Brownbridge, *J.C.S. Chem. Comm.*, 1280 (1987).

Phthaloyl dichloride.

Anthracyclinones.[1]   Phthaloyl dichlorides undergo Friedel Crafts reactions with hydroquinones or the dimethyl ethers to give 1,4-dihydroxyanthraquinones in one step.

Examples:

[1] G. Sartori, G. Casnati, F. Bigi, and P. Robles, *Tetrahedron Letters*, **28**, 1533 (1987).

Pivaldehyde.

Diastereoselective α-alkylation of α-amino acids.[1]   This reaction is based on the observation that 2-(*t*-butyl)-N-benzoyl-1,3-oxazolidines (**1**) are alkylated with high stereoselectivity from the side opposite to that of the *t*-butyl group. These heterocycles are readily obtained by reaction of the sodium salt of an amino acid with pivaldehyde followed by N-benzoylation. Thus the enolate of cis-**1** from (S)-methionine is alkylated by CH_3I to give **2**, which is hydrolyzed to (S)-(+)-2-methylmethionine (**3**).

[1]A. K. Beck and D. Seebach, *Org. Syn.*, submitted (1987).

Potassium 9-alkyl-9-boratabicyclo[3.3.1]nonanes, K 9-R-9-BBNH, $\left[\langle\!\!\!\bigcirc\ \bar{B}HR \right] K^+$.

These ate complexes are prepared by reaction of potassium hydride with a 9-alkyl-9-borabicyclo[3.3.1]nonane (9-R-9-BBN).

Reduction of cyclic ketones.[1]   These complexes selectively reduce cyclic ketones to the less stable alcohol. The most stereoselective reagent is that in which the R group is *t*-butyl; this complex is comparable to lithium trisiamylborohydride in stereoselectivity.

cis, 99.5%

[1] J. S. Cha, M. S. Yoon, Y. S. Kim, and K. W. Lee, *Tetrahedron Letters*, **29**, 1069 (1988).

Potassium *t*-butoxide–Dimethyl sulfoxide.

Three-carbon ring expansion.[1]   The cyclic β-keto esters **1** with a 4-oxopentyl group at the α-position do not undergo the expected aldol condensation on treatment with $KOC(CH_3)_3$ in DMSO, but undergo ring expansion to medium-size ketones (**2**). The reaction may involve aldol and retro-aldol condensation. A similar

n = 5	54%
n = 6	62%
n = 7	78%

reaction of the β-keto ester **3** (n = 5) also results in a ring-expanded ketone (**4**), but homologs (**3**, n = 6 or 7) undergo only the usual aldol condensation.

3, n = 5

4

[1] Z.-F. Xie, H. Suemune, and K. Sakai, *J.C.S. Chem. Comm.*, 612 (1988).

Potassium–18-Crown-6

α-Alkylation of γ-lactones.[1] Potassium dissolves in 18-crown-6 to form K⁻ anions and K⁺ cations complexed by the crown ether. This solution converts γ-lactones in THF into the α-anion, which reacts with alkyl or acyl halides to form α-substituted γ-lactones in >80% yield, without α,α-disubstitution or O-acylation.

[1] Z. Jedliński, M. Kowalczuk, P. Kurcok, M. Grzegorzek, and J. Ermel, *J. Org.*, **52**, 4601 (1987).

Potassium hydride.

Purification. Two laboratories noted that commercial samples of KH[1,2] and NaH[2] are ineffective for conversion of hindered trialkylboranes into the corresponding borohydrides. Both groups find that treatment of the aged metal hydrides with lithium aluminum hydride in THF results in highly active hydrides that react readily even with such hindered trialkylboranes as tris(3-methyl-2-butyl)borane.

[1] J. A. Soderquist and I. Rivena, *Tetrahedron Letters*, **29**, 3195 (1988).
[2] J. L. Hubbard, *ibid.*, **29**, 3197 (1988).

Potassium hydride–Hexamethylphosphoric triamide.

β-Cleavage of homoallylic alcohols.[1] Homoallylic tertiary potassium alkoxides undergo cleavage in HMPT (or DMF) of the allylic C—C bond to give the enolate of a ketone (equation I). A rigid bicyclic system is not essential for the cleavage.

Bis(homoallylic) alkoxides also undergo this β-cleavage. The substrates (1) are available by reaction of a methyl carboxylate with 2 equiv. of allyl- or methallyl-magnesium chloride. The alcohols when heated at 80° in a slurry of KH in HMPT are converted into β,γ- and α,β-unsaturated ketones (2 and 3) in 75–85% yield.

KH/HMPT is also useful for selective β-cleavage of tris(homoallylic) alcohols, as shown by a new synthesis of α-damascone (4). In general, 1,1-dimethyl-2-pro-penyl and benzyl groups are cleaved more readily than a 1-methyl-2-propenyl group, which is cleaved more readily than a 2-propenyl group. A 2-methyl-2-propenyl group is cleaved the most slowly. Thus substitution of the carbon adjacent to the alkoxide favors cleavage. Actually homoallylic potassium alkoxides could undergo oxy-Cope rearrangement or β-cleavage, and generally the former reaction is preferred unless a quaternary center is generated during rearrangement.

[1]R. L. Snowden and K. H. Schulte–Elte, *Helv.*, **64**, 2193 (1981); R. L. Snowden, S. M. Linder, B. L. Muller, and K. H. Schulte–Elte, *ibid.*, **70**, 1858, 1879 (1987).

Potassium permanganate.
—*CHO* → —*COOH*. This oxidation can be conducted rapidly and in high yield with $KMnO_4$ in $(CH_3)_3COH$ and an aqueous solution of NaH_2PO_4 at pH 4.4–7.0. The reaction is slow in the absence of water.
Example:

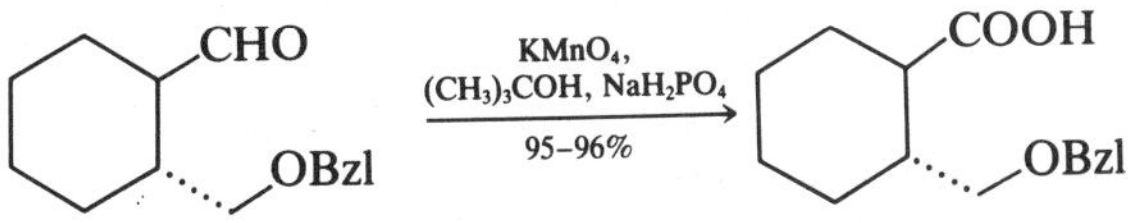

Hydroxyl groups protected as acetonides or as silyl, tetrahydropyranyl, benzyl, or methoxymethyl ethers are stable to these conditions. Yields with $KMnO_4$ are higher than those obtained with $KMnO_4$ and dicyclohexyl-18-crown-6, Bu_4NMnO_4, or $NaMnO_4 \cdot H_2O$.[1]

[1] A. Abiko, J. C. Roberts, T. Takemasa, and S. Masamune, *Tetrahedron Letters*, **27**, 4537 (1986).

Potassium peroxomonosulfate (Oxone).
N-oxides.[1] Oxone does not require catalysis by a ketone for oxidation of nitrogen heterocycles to the N-oxide, but an excess is required for convenient reaction times. Yields can be comparable to those obtained with *m*-chloroperbenzoic acid (80–85%), which is less available and more expensive than Oxone.

[1]T. W. Bell, Y.-M. Cho, A. Firestone, K. Healy, J. Lui, R. Ludwig, and S. D. Rothenberger, *Org. Syn.*, submitted (1988).

Potassium peroxosulfonate–Copper sulfate.
Selective oxidation of dimethylanisoles (**11**,441). An *o*- or a *p*-methyl group in a dimethylanisole that bears an *m*-methyl group is selectively oxidized by $K_2S_2O_8$ catalyzed by $CuSO_4$ to an aldehyde group, which can be oxidized further to a carboxyl group by sodium chlorite.[1]

Example:

[1] F. M. Hauser and S. R. Ellenberger, *Synthesis*, 723 (1987).

2-Pyridinethiol-1-oxide (1).

Decarboxylative halogenation (12,417). The Hunsdiecker reaction is not useful for aromatic acids, but decarboxylative halogenation of these acids can be effected in useful yield by radical bromination or iodination of the thiohydroxamic esters, as reported earlier for aliphatic acids.[1] Thus when the esters **2** are heated at 100° in the presence of AIBN, carbon dioxide is evolved and the resulting radical is trapped by $BrCCl_3$ to provide bromoarenes (**3**). Decarboxylative iodination is effected with iodoform or methylene iodide as the iodine donor.

[1] D. H. R. Barton, B. Lacher, and S. Z. Zard, *Tetrahedron*, **43**, 4321 (1987).

Pyridinium bromide dibromide.

Bromodesilylation; 7-methoxy-1-indanones.[1]   Cyclization contrary to the normal *para*-selectivity of anisole derivatives can be effected by temporary use of an *ortho*-trimethylsilyl group introduced by directed *ortho*-metallation (11,75). Thus the anisole derivative **1** undergoes bromodesilylation and hydrolysis to provide **2**. This product undergoes cyclization to **3** in good yield on conversion to the lithio salt followed by bromine–lithium exchange (8,65–66).

$1 (R^1, R^2 = H, CH_3)$ **2**   **3**

1) DIBAL (81%)
2) CsF, DMF (50–70%)

4

[1]M. P. Sibi, K. Shankaran, B. I. Alo, W. R. Hahn, and V. Snieckus, *Tetrahedron Letters*, **28**, 2933 (1987).

Pyridinium chlorochromate.

$RCH{=}CH_2 \rightarrow RCH_2CHO$.[1] Dialkylchloroboranes, obtained as the major products of hydroboration of 1-alkenes with monochloroborane complexed with dimethyl sulfide, are oxidized by PCC to aldehydes (66–68% yield). Similar oxidation of dialkylchloroboranes derived from cyclic alkenes with PCC gives ketones in 70–85% yield.

Allylic and benzylic oxidation. PCC in refluxing CH_2Cl_2 can oxidize allylic[2] and benzylic[2,3] methylene groups to keto groups in satisfactory yield.

[1] H. C. Brown, S. U. Kulkarni, C. G. Rao, and V. D. Patil, *Tetrahedron*, **42**, 5515 (1986).
[2] E. J. Parish, S. Chitrakorn, and T.-Y. Wei, *Syn. Comm.*, **16**, 1371 (1986).
[3] R. Rathore, N. Saxena, and S. Chandrasekaran, *ibid.*, **16**, 1493 (1986).

Pyridinium dichromate.

$RCHO \rightarrow RCOOCH_3$.[1] Aliphatic aldehydes can be converted directly to methyl esters by PDC (6 equiv.) and methanol in DMF in yields of 60–85%. The method is not useful in the case of aromatic aldehydes or of higher esters.

[1] B. O'Connor and G. Just, *Tetrahedron Letters*, **28**, 3235 (1987).

R

Raney nickel.

Selective oxidation of **sec-alcohols.**[1]   Raney nickel (large excess) in refluxing benzene can oxidize secondary alcohols often in high yield. Addition of a hydrogen acceptor such as 1-octene is not necessary. Primary alcohols are resistant to this oxidation, but can undergo hydrogenolysis. The main difficulty is that a large excess of Raney nickel is required, but the commercial 50% slurry can be used. The main advantage is that the ketones are readily isolated by filtration and removal of solvent.

Hydrogenolysis of **tert-alcohols.**[2]   This deoxygenation can be effected with Raney Ni slurry (Aldrich 50% slurry in water) that has been washed eight times with distilled water and twice with 1-propanol. Thus reaction of a tertiary alcohol with washed Raney Ni for a short time yields a mixture of alkenes that furnish a single alkane on hydrogenation catalyzed by Pd/C.

[1] M. E. Krafft and B. Zorc, *J. Org.*, **51**, 5482 (1986).
[2] M. E. Krafft and W. J. Crooks, III, *J. Org.*, **53**, 432 (1988).

Rhodium(II) acetamide, $Rh_2(NHCOCH_3)_4$ **(1)**. The reagent is obtained by reaction of $Rh_2(OAc)_4$ with molten acetamide. It crystallizes from water as a hydrate.[1]

Stereoselective **trans-cyclopropanation.**[2]   Rhodium(II) carboxylates are generally the preferred catalysts for cyclopropanation of alkenes with diazoacetates (7,313;9,406;10,340) even though they show only low *trans*-selectivity. The *trans*-selectivity can be markedly enhanced by use of rhodium(II) acetamide. Use of rhodium(II) 2,4,6-triarylbenzoates favors *cis*-stereoselectivity.[3]

$Rh_2(OAc)_4$	*trans/cis* = 3.8:1
$Rh_2(NHAc)_4$	*trans/cis* = 10:1

[1] M. Q. Ahsan, I. Bernal, and J. L. Bear, *Inorg. Chem.*, **25**, 260 (1986).
[2] M. P. Doyle, K.-L. Loh, K. M. DeVries, and M. S. Chinn, *Tetrahedron Letters*, **28**, 833 (1987).
[3] H. J. Callot and F. Metz, *Tetrahedron*, **41**, 4495 (1985).

Rhodium(III) chloride.

β-*Silyloxy esters*.[1] Silyl ketene acetals are known to undergo aldol condensation with carbonyl compounds in the presence of $TiCl_4$ (**12**,268) to afford β-silyloxy esters. The same products can be obtained in a one-step reaction of an α,β-unsaturated ester with trimethylsilane and a carbonyl compound in the presence of $RhCl_3 \cdot H_2O$.

Example:

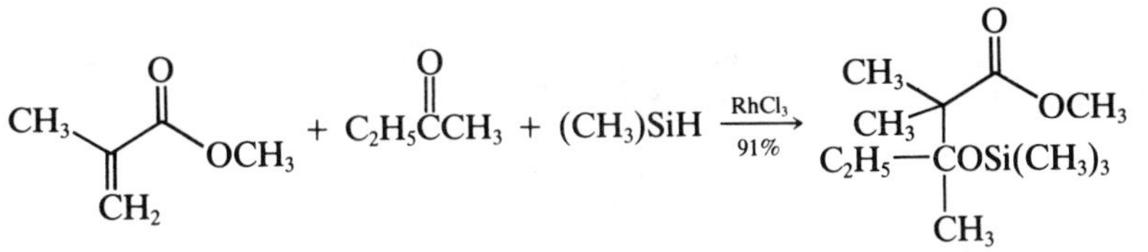

[1] A. Revis and T. K. Hilty, *Tetrahedron Letters*, **28**, 4809 (1987).

Ruthenium(III) chloride.

N-*Heterocyclization*.[1] Anilines react with 1,3-propanediol in diglyme in the presence of $RuCl_3 \cdot 3H_2O$ complexed with Bu_3P to give quinolines in 50–60% yield (isolated). Two isomeric quinolines are formed in a similar reaction with an unsymmetrical 1,3-diol.

$RuCl_3 \cdot 3H_2O$ combined with $P(C_6H_5)_3$ catalyzes a reaction of N-substituted anilines with 2,3-butanediol or 1,2-cyclohexanediol to give 2,3-dimethylindoles or 1,2,3,4-tetrahydrocarbazoles.

Examples:

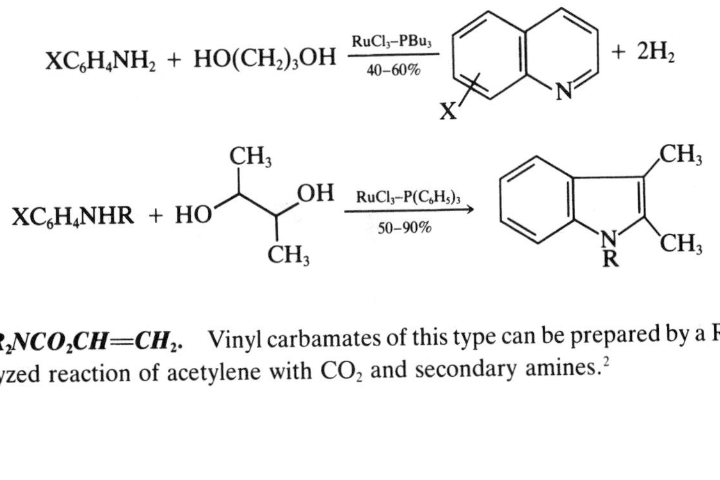

$R_2NCO_2CH{=}CH_2$. Vinyl carbamates of this type can be prepared by a $RuCl_3$-catalyzed reaction of acetylene with CO_2 and secondary amines.[2]

$$\text{[pyrrolidine]}NH + CO_2 + HC\equiv CH \xrightarrow[\text{CH}_3\text{CN, 90}°]{\text{RuCl}_3}$$

$$\text{[pyrrolidine]}NCO_2CH=CH_2 + \text{[pyrrolidine]}NCO_2\overset{\overset{\displaystyle CH_2}{\|}}{C}CH=CH_2$$

(46%)   (2%)

[1] Y. Tsuji, K.-T. Huh, and Y. Watanabe, *J. Org.*, **52**, 1673 (1987).
[2] Y. Sasaki and P. H. Dixneuf, *ibid.*, **52**, 314 (1987).

Ruthenium(IV) oxide.

1,2-Diketones.   The Sharpless procedure for oxidation of alkenes with $NaIO_4$ catalyzed by RuO_2 (**11**,462–463) is equally efficient for oxidation of alkynes to 1,2-diones. In fact, alkenes and alkynes react at a similar rate, but ether, epoxide, and ester groups are stable to the reagent. A 1-silylacetylene is oxidized to an acylsilane. Yields are moderate to high.[1]

The final steps in a total synthesis of (+)-gloeosporone (**3**, a natural germination inhibitor of a fungus) required oxidation of the acetylene group of **1** to a diketone group. The oxidation was carried out in 74% yield by the catalytic RuO_2 procedure of Sharpless. On liberation (pyridine·HF) of the C_7-hydroxyl group, the hydroxy

$$\xrightarrow[\text{74%}]{\substack{\text{NaIO}_4, \text{ RuO}_2 \\ \text{CH}_3\text{CN, CCl}_4, \text{ H}_2\text{O}}}$$

1

2

3

diketone group spontaneously cyclizes to the hemiketal group to provide **3** in 80% yield.[2]

Oxidation of allylic alcohols and enones.[3] These substrates are cleaved to dicarboxylic acids and keto acids by Sharpless catalytic RuO_2 oxidation.

Examples:

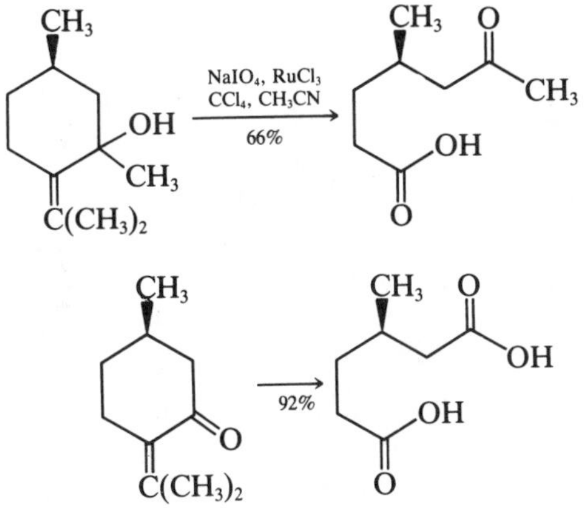

[1] R. Zibuck and D. Seebach, *Helv.*, **71**, 237 (1988).
[2] G. Adam, R. Zibuck, and D. Seebach, *Am. Soc.*, **109**, 6176 (1987).
[3] F. X. Webster and R. M. Silverstein, *J. Org.*, **52**, 689 (1987).

Ruthenium(IV) trifluoroacetate, $Ru(OCOCF_3)_4$ (**1**). The reagent is prepared *in situ* from RuO_2 and CF_3COOH.

Oxidative coupling of aryl tetrahydroisoquinolines.[1] This reagent is superior to thallium(III) trifluoroacetate or vanadium oxyfluoride for nonphenolic oxidative coupling of substrates such as **2** to provide aporphines and homoaporphines (**3**).

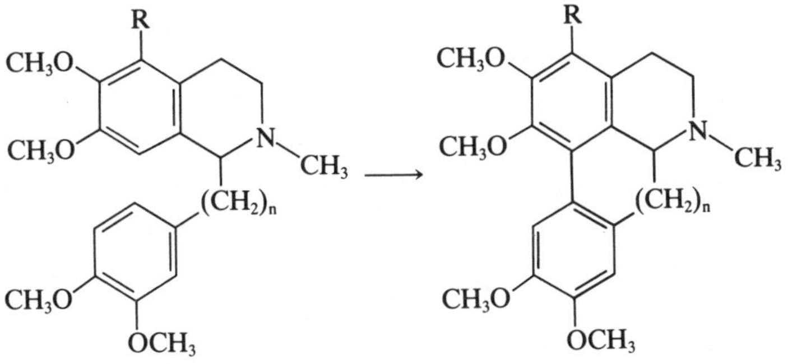

2, n = 1, 2; R = H or OCH₃   **3**

	1	60–76%
	TTFA	47–65%
	VOF_3	40–43%

***Biaryl phenol coupling.*[2]**   The natural dibenzylbutanolide lignans, prestegane A (**1a**) and B (**1b**), are converted to the corresponding bisbenzocyclooctadiene lactones (**2**) by oxidation with RuO_2 in TFA–TFAA in 80–85% yield. The usual reagent for this oxidation, thallium tris(trifluoroacetate), is less efficient (45–50% yields).

1a, $R^1 = CH_3$, $R^2 = H$
1b, $R^1 = R^2 = H$

2

[1] Y. Landais, D. Rambault, and J. P. Robin, *Tetrahedron Letters*, **28**, 543 (1987).
[2] J.-P. Robin and Y. Landais, *J. Org.*, **53**, 224 (1988).

S

Samarium–Methylene iodide.

Cyclopropanation.[1] The carbenoid obtained from the reaction of CH_2I_2 with Sm or Sm/Hg in THF effects cyclopropanation of allylic alcohols, but not of isolated double bonds. The reaction of both cyclic and acyclic allylic alcohols proceeds in generally high yield and often with marked diastereoselectivity.

Examples:

R = Bu	99%	1:1.4
R = *i*-Pr	88%	>200:1

(*exo*)   (*endo*)

[1] G. A. Molander and J. B. Etter, *J. Org.*, **52**, 3942 (1987).

Samarium(III) chloride.

Lewis acid catalysis. Anhydrous lanthanide(III) chlorides, particularly $SmCl_3$, can function as low-cost but efficient nonhomogeneous Lewis acid catalysts for aldol and other reactions. More rapid reactions are observed when the soluble but expensive $Eu(fod)_3$ is used as a lanthanide catalyst.[1]

Example:

$$C_6H_5CHO$$

(CH₃)₃SiCN
SmCl₃, CH₂Cl₂ | 98%

66% | (CH₃)₂C=C⟨OCH₃, OSi(CH₃)₃
SmCl₃, CH₂Cl₂, 20°

C₆H₅—CHCN
      |
      OSi(CH₃)₃

CH₃
|
C₆H₅—CHC—COOCH₃
      | |
      HO CH₃

[1] A. E. Vougioukas and H. B. Kagan, *Tetrahedron Letters*, **28**, 5513 (1987).

Samarium(II) iodide.

Preparation. SmI₂ can also be prepared by reaction of I₂ with samarium powder and THF. A yellow suspension of SmI₃ is formed, which on further reaction with the metal under reflux is converted into SmI₂.[1]

Intramolecular pinacol reduction. SmI₂ can promote intramolecular reductive coupling of keto aldehydes to provide 2,3-dihydroxycyclopentanecarboxylates with marked diastereoselectivity.[2]

Example:

$$\xrightarrow[\text{77%}]{\substack{SmI_2 \\ THF, CH_3OH}}$$

(200:1)

Allenic alcohols.[3] In the presence of SmI₂ and Pd[P(C₆H₅)₃]₄, *sec-* and *tert-*propargylic acetates add to ketones to give allenic alcohols as the only or major product. A mixture of allenic and homopropargylic alcohols is formed from reactions of primary propargylic acetates.

Examples:

$$AcO(CH_2)_9CHC\equiv CH + (CH_3)_2C=O \xrightarrow[\text{46%}]{Pd(0), SmI_2}$$
|
OAc

AcO(CH₂)₉⟩C=C=C⟨H, C(CH₃)₂
H |
                     OH

56%

Barbier-type cyclization of α-(ω-iodoalkyl)-β-keto esters or amides.
Substances of this type (**1a** or **1b**) cyclize in the presence of SmI_2 to *cis*-1,2-disub-
stituted cyclopentanes or cyclohexanes (equation I). Similar diastereoselectivity

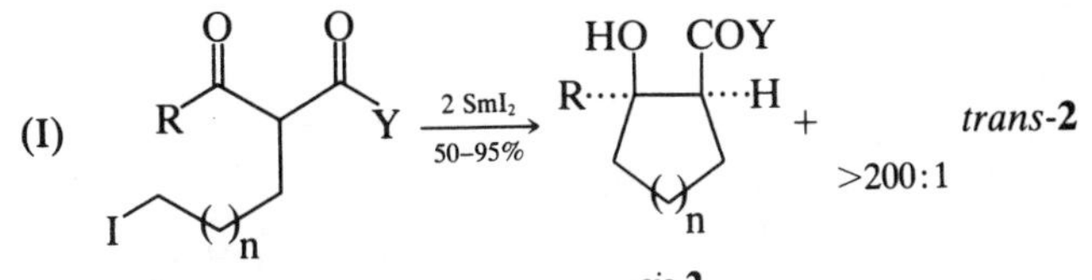

(I)

2 SmI$_2$
50–95%

HO COY

+   *trans*-**2**

>200:1

cis-**2**

1a, n = 1,2, Y = N(CH$_3$)$_2$
1b, n = 1,2, Y = OR'

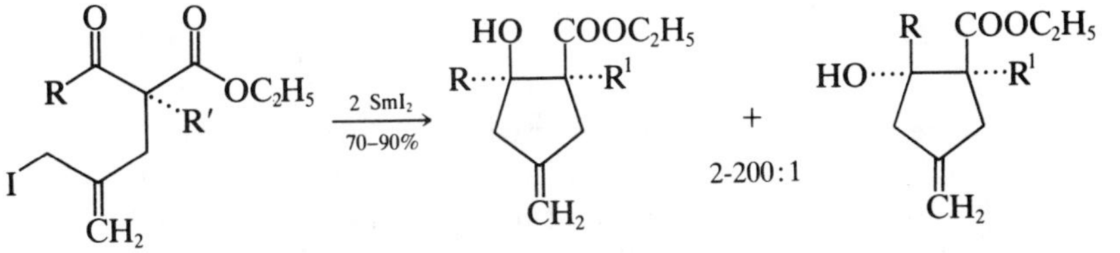

3

2 SmI$_2$
70–90%

HO COOC$_2$H$_5$

+

2-200:1

cis-**4**

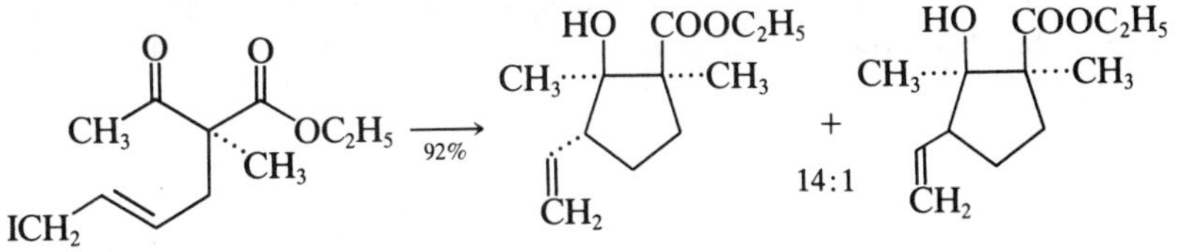

5

92%

HO COOC$_2$H$_5$

+

14:1

obtains in cyclization of substrates containing an allylic halide group, such as **3**
and **5**.[4]

SmI_2 also effects an intramolecular reductive coupling of certain unsaturated β-
keto esters or amides to cyclopentanes with good stereochemical control at three
centers (equation II).[5]

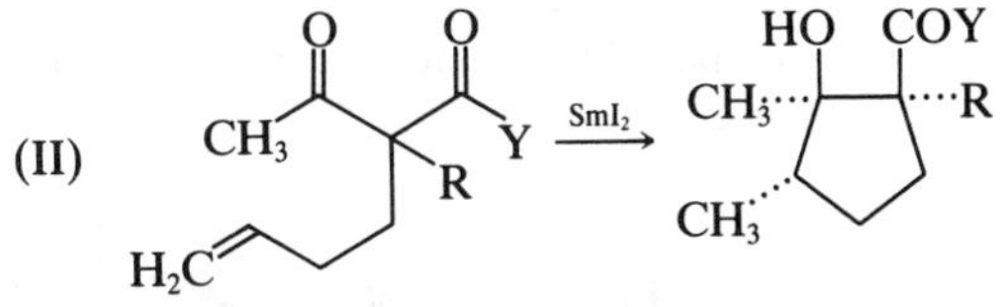

(II)

SmI$_2$

HO COY

Y = OC$_2$H$_5$, R = CH$_3$ 75%   25:1
Y = N(CH$_3$)$_2$, R = H   78%   >200:1

Asymmetric intramolecular Reformatsky reactions.[6]   The bromoacetates (**1**)
of β-hydroxy ketones undergo a Reformatsky-type reaction when treated with SmI_2
to give β-hydroxy-δ-valerolactones (**2**). These products are useful in their own

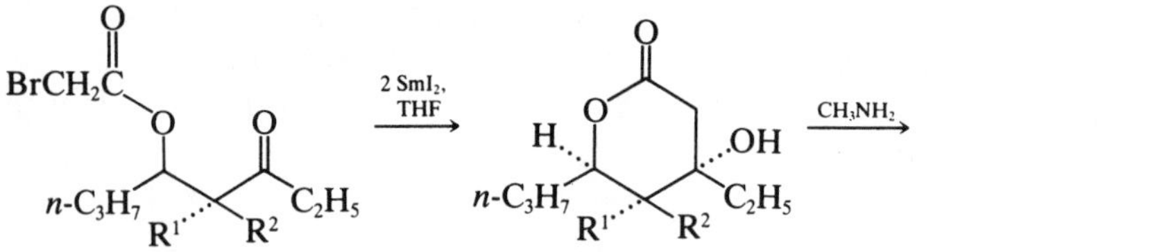

1a, R¹ = H, R² = CH₃   86%   **2**   88%
b, R¹ = CH₃, R² = H   97%   (13:1)   94%

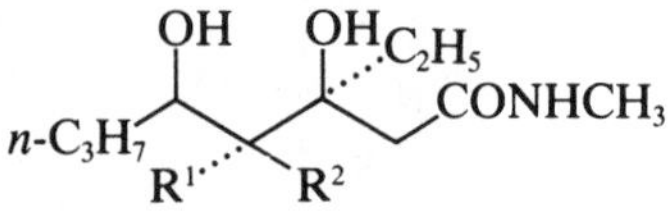

3

right, and are also useful precursors to acyclic *syn*-1,3-diols (**3**). The example shows that the *syn*-1,3-asymmetric induction is observed even with the diastereomeric pairs **1a** and **1b**, although it is diminished in the case of **1b**, in which the α-substituent assumes the axial orientation in the lactone. The same 1,3-asymmetric induction obtains with derivatives of β-hydroxy aldehydes. However, the opposite *anti*-1,3-induction has been observed in one case involving a *t*-butyl β-hydroxyalkyl ketone (**4**).

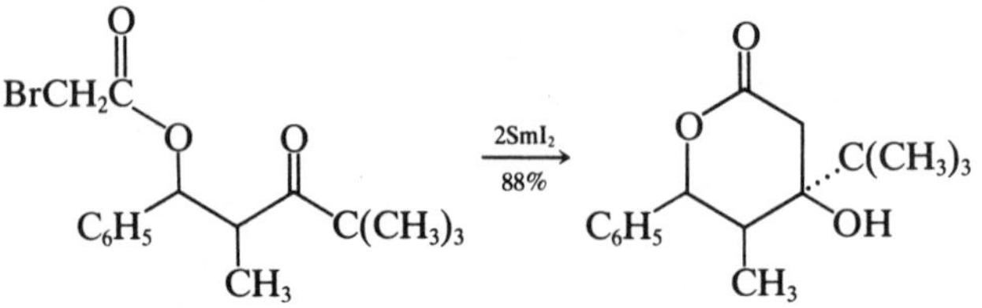

4   **5**

Bicyclic γ-lactones.[7]   These lactones can be obtained by intramolecular reductive cyclization of unsaturated keto esters with SmI₂ in THF–HMPT.
Example:

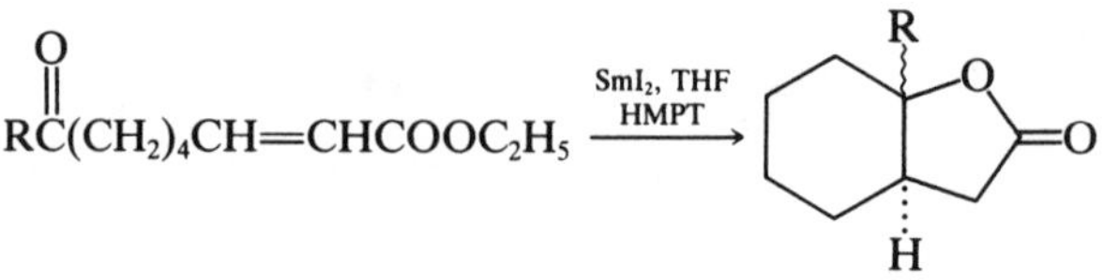

R = CH₃   66%   *trans/cis* = 70:30
R = H   34%   "   = 70:30

Reduction of α,β-epoxy esters to β-hydroxy esters.[8] SmI$_2$ alone reduces these esters to a mixture of α- and β-hydroxy esters. The reaction rate and yield is increased by addition of HMPT. Addition of a chelating agent, TMEDA or N,N-dimethylaminoethanol (DMAE), results in regioselective reduction to β-hydroxy esters (equation I). The system reduces optically active epoxy esters with complete

$$(I) \quad CH_3 \overset{O}{\diagup\!\!\!\diagdown} COOC_2H_5 \quad \xrightarrow[\substack{\text{SmI}_2, \text{HMPT} \\ \text{DMAE, THF} \\ 68\%}]{} \quad CH_3 \overset{OH}{\diagup} COOC_2H_5 \quad + \quad CH_3 \overset{COOC_2H_5}{\diagup\overset{|}{OH}}$$

$$>200:1$$

retention of configuration. It also reduces vinylogous epoxy esters to (E)-δ-hydroxy-β,γ-unsaturated esters (equation II).

$$(II) \quad CH_3 \overset{O}{\diagup\!\!\!\diagdown} \diagup\!\!\!\diagup COOCH_3 \quad \xrightarrow[73\%]{} \quad CH_3 \overset{OH}{\diagup}\diagup\!\!\!\diagup COOCH_3$$

$$(\delta\text{-OH}/\gamma\text{-OH} = >200:1)$$

Magnesium iodide reacts with α,β-epoxy esters to form β-iodo-α-hydroxy esters selectively (200:1). The regioselectivity is attributed to favorable chelation of the iodohydrin. These products are reduced by Bu$_3$SnH to α-hydroxy esters in 75–95% overall yields with retention of the original configuration at the α-position.

Reduction of vinyloxiranes.[9] The substrates are reduced rapidly by SmI$_2$ to (E)-allylic alcohols without effect on keto, ester, or nitrile groups. Chiral substrates are reduced to optically active alcohols with complete retention of stereochemistry. Examples:

(95% ee) → *(95% ee)*

Dihydrofurans.[10] Dihydrofurans can be obtained by reaction of α,α-dibromodeoxybenzoin (**1**) with SmI$_2$ and an alkene. The reaction evidently involves a 1,3-dipolar addition of a ketocarbenoid to an alkene.

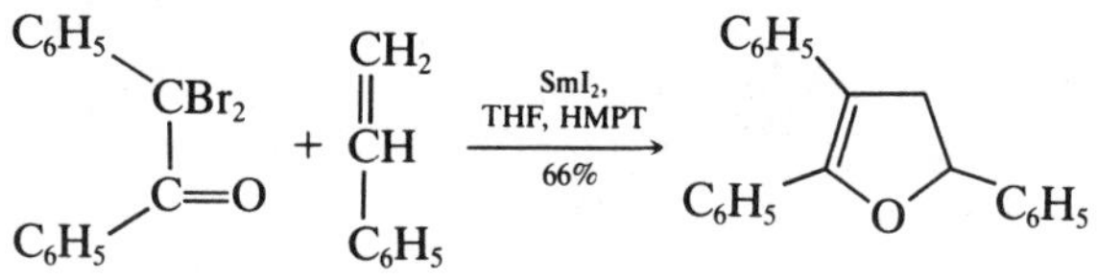

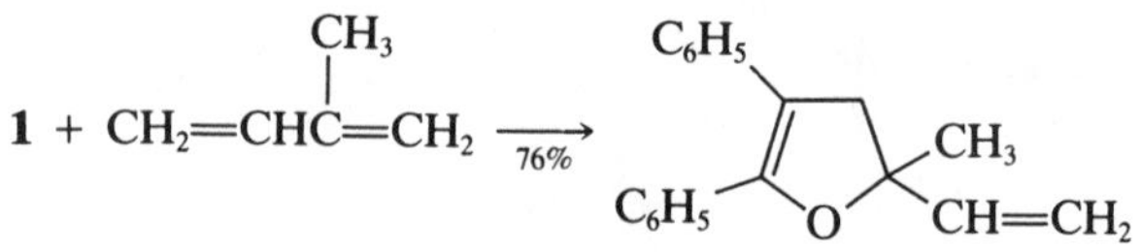

γ-Lactones.[11]   SmI$_2$ promotes reductive cross-coupling of carbonyl compounds with α,β-unsaturated esters resulting in γ-lactones. HMPT catalyzes the rate of this reaction markedly.

Example:

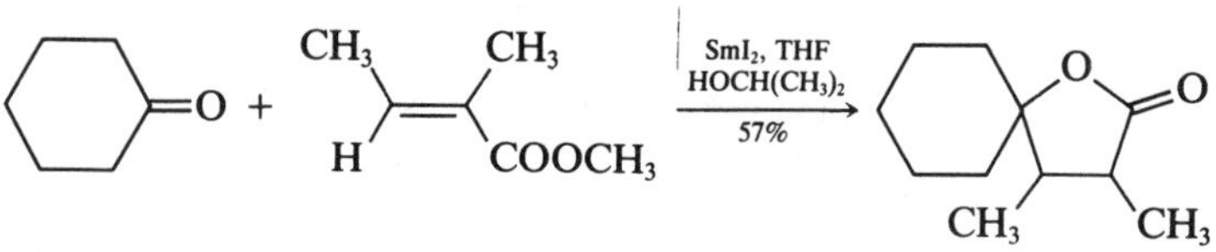

Medium-ring lactones.[12]   Lactones can be obtained by an intramolecular Reformatsky-type reaction of ω-(bromoacetoxy) aldehydes in the presence of SmI$_2$. The β-hydroxy lactones were isolated as the acetates. Yields are high even for 8- to 14-membered lactones.

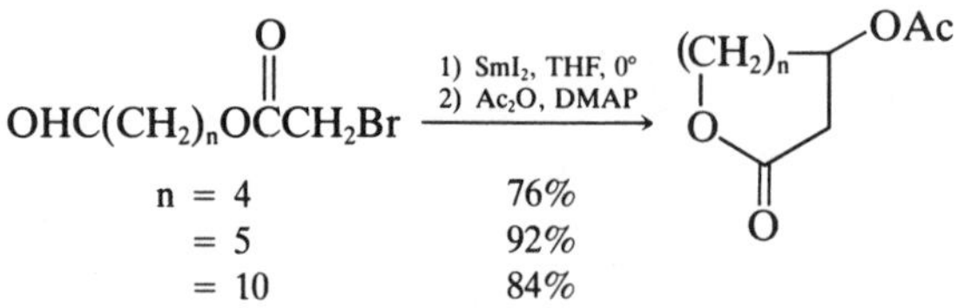

HMPT catalysis.[13]   SmI$_2$ alone can reduce primary alkyl bromides or iodides in high yield, but the reaction is slow even in refluxing THF. In the presence of HMPT (~5 mole %), alkyl, aryl, and vinyl halides, and even alkyl chlorides, are reduced in almost quantitative yield at 25°, often in less than 1 hour.

The combination of SmI$_2$ and HMPT effects a Barbier-type reaction of ω-bromo esters with carbonyl compounds to provide γ- and δ-lactones (equation I).[14]

(I) $C_6H_5(CH_2)_2\overset{\overset{\displaystyle O}{\|}}{C}CH_3$ + $Br(CH_2)_nCOOCH_3$ $\xrightarrow[\text{THF, HMPT}]{\text{SmI}_2,}$

$$\begin{array}{cc} n = 2 & 85\% \\ n = 3 & 55\% \end{array}$$

[1] T. Imamoto and M. Ono, *Chem. Letters*, 501 (1987).
[2] G. A. Molander and C. Kenny, *J. Org.*, **53**, 2132 (1988).
[3] T. Tabuchi, J. Inanaga, and M. Yamaguchi, *Chem. Letters*, 2275 (1987).
[4] G. A. Molander, J. B. Etter, and P. W. Zinke, *Am. Soc.*, **109**, 453 (1987).
[5] G. A. Molander and C. Kenny, *Tetrahedron Letters*, **28**, 4367 (1987).
[6] G. A. Molander and J. B. Etter, *Am. Soc.*, **109**, 6556 (1987).
[7] S. Fukuzawa, M. Iida, A. Nakanishi, T. Fujinami, and S. Sakai, *J.C.S. Chem. Comm.*, 920 (1987).
[8] K. Otsubo, J. Inanaga, and M. Yamaguchi, *Tetrahedron Letters*, **28**, 4435, 4437 (1987).
[9] G. A. Molander, B. E. LaBelle, and G. Hahn, *J. Org.*, **51**, 5259 (1986).
[10] S. Fukuzawa, T. Fujinami, and S. Sakai, *J.C.S. Chem. Comm.*, 919, (1987).
[11] K. Otsubo, J. Inanaga, and M. Yamaguchi, *Tetrahedron Letters*, **27**, 5763 (1986); T. Fujinami, S. Sakai, and A. Nakanishi, *J.C.S. Chem. Comm.*, 624 (1986).
[12] T. Tabuchi, K. Kawamura, J. Inanaga, and M. Yamaguchi, *Tetrahedron Letters*, **27**, 3889 (1986).
[13] J. Inanaga, M. Ishikawa, and M. Yamaguchi, *Chem. Letters*, 1485 (1987).
[14] K. Otsubo, K. Kawamura, J. Inanaga, and M. Yamaguchi, *ibid.*, 1487 (1987).

Samarium(II) iodide–1,3-Dioxolane.

α-Hydroxy aldehydes.[1]   A solution of SmI$_2$ in 1,3-dioxolane reacts with a ketone and iodobenzene at room temperature to form a protected α-hydroxy aldehyde in about 75% yield. The reaction may involve a 1,3-dioxolanyl radical.
Example:

[1] M. Matsukawa, J. Inanaga, and M. Yamaguchi, *Tetrahedron Letters*, **28**, 5877 (1987).

Selenium.

Reaction with Wittig reagents.   Selenium reacts with Wittig reagents to form a selenoaldehyde or -ketone, which reacts further with the Wittig reagent to give a symmetrical *trans*-alkene.[1,2]

Example:

$$(C_6H_5)_3P{=}CHCOOCH_3 + Se \longrightarrow [Se{=}CHCOOCH_3] \xrightarrow{\;1\;}$$

1

$$(C_6H_5)_3P{=}Se + CH_3OOCCH{=}CHCOOCH_3$$
$$(77\%) \qquad\qquad (74\%)$$

Since $(C_6H_5)_3P = Se$ also effects cleavage of $(C_6H_5)_3P = CHR$ to give $RCH = CHR$ and $(C_6H_5)_3P$, selenium can be used in catalytic amounts; this version provides the thermodynamic mixture of the geometric forms of the alkene.

[1] K. Okuma, S. Sakata, Y. Tachibana, T. Honda, and H. Ohta, *Tetrahedron Letters*, **28**, 6649 (1987).
[2] G. Erker, R. Hock, and R. Nolte, *Am. Soc.*, **110**, 624 (1988).

Serine, $HOCH_2CH(NH_2)COOH$.

Amino acid synthesis.[1] Optically pure amino acids can be prepared in two steps from serine, readily available as either the D- or L-enantiomer. Reaction of N-benzylserine (or of N-benzyl-N-Boc-serine) with the preformed Mitsunobu reagent in CH_3CN at $-55°$ provides the protected serine β-lactone (**2**) in almost quantitative yield. The lactone reacts with lithium organocuprates (R_2CuLi) to

2   **3** (99% ee)

provide the desired benzylamino acids (**3**) in moderate chemical and optical yields. The best results are obtained by use of Grignard reagents catalyzed by CuI, which react with almost complete retention of optical purity.

[1] L. D. Arnold, J. C. G. Drover, and J. C. Vederas, *Am. Soc.*, **109**, 4649 (1987); L. D. Arnold, R. G. May, and J. C. Vederas, *ibid.*, **110**, 2237 (1988).

Silver trifluoromethanesulfonate.

C-Glucosidation.[1] Reaction of the protected α-D-glucopyranosyl chloride **1** with silyl enol ethers activated with AgOTf results in α-C-glucosyl esters or ketones.

Example:

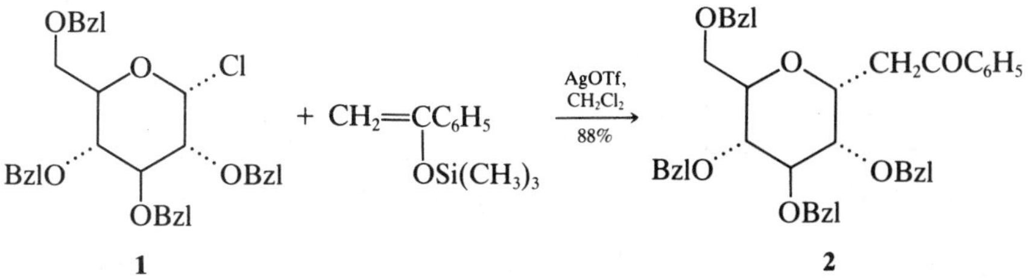

1 + CH$_2$=CC$_6$H$_5$ →   **2**

$\xrightarrow[88\%]{\text{AgOTf,}\atop\text{CH}_2\text{Cl}_2}$

¹ P. Allevi, M. Anastasia, P. Ciuffreda, A. Fiecchi, and A. Scala, *J.C.S. Chem. Comm.*, 101 (1987).

Simmons–Smith reagent.

Angular methylation. In a general synthesis of limonoids, Corey *et al.*¹ prepared the tetracyclic enone **1**, which lacks only the angular methyl group at C$_{13}$ of a limonoid.² Conjugate addition of a methylcopper reagent to **1** fails, probably owing to steric hindrance; however, the Simmons–Smith reagent adds to the corresponding 16α-alcohol to give a single cyclopropane (**2**) in 89% yield. This product is converted into the limonoid **3** by oxidation to the ketone (PDC) followed by cleavage of the cyclopropane ring with Li/NH$_3$. The same reactions when applied to the 16β-alcohol can be used to introduce a β-methyl group at C$_{13}$.

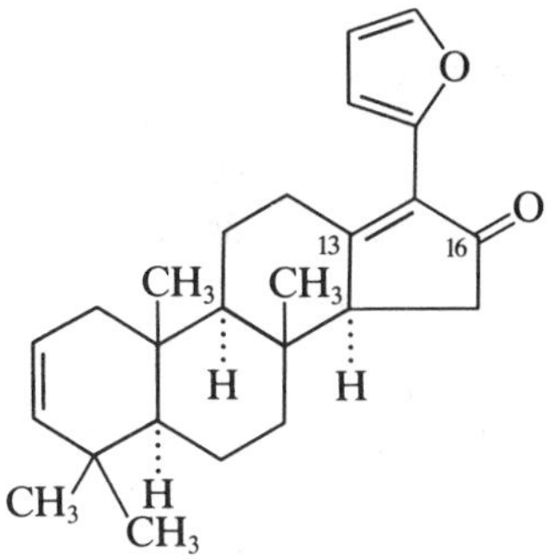

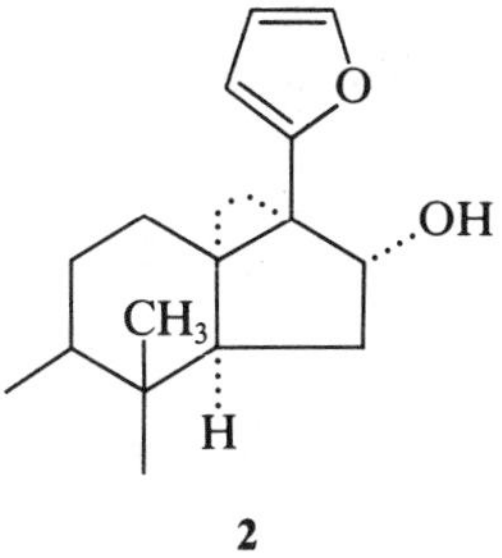

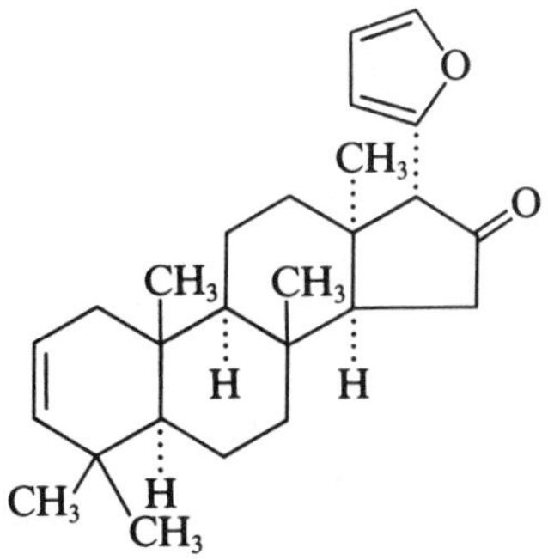

[1] E. J. Corey, J. G. Reid, A. G. Myers, and R. W. Hahl, *Am. Soc.*, **109**, 918 (1987).
[2] D. A. H. Taylor, *Prog. Chem. Nat. Prod.*, **45**, 1 (1984).

Sodium amide.

cis-*Selective Wittig reactions.* Schlosser *et al*[1] have reviewed experimental details for effecting *cis*-selective Wittig reactions. The ylide is best generated with sodium amide (free from NaOH) in an ethereal solvent at 25°. Use of lithium or potassium bis(trimethylsilyl)amide lowers the yield and the *cis*-selectivity. Solvents of choice are THF, ethylene glycol, ether, or *t*-butyl methyl ether. The carbonyl compound should be added to the base slowly and with cooling to maintain the temperature at −75° to −100°. The reaction is then completed at 25°.

Replacement of the triphenyl group by trialkyl groups increases the *trans*-selectivity, but increase in the bulk of the customary phenyl groups can increase *cis*-selectivity. Tris(*o,o*′-difluorophenyl)phosphine (m.p. 125–127°) is recommended as a replacement for triphenylphosphine in *cis*-selective Wittig reactions, particularly of aromatic and α,β-unsaturated aldehydes.

[1] M. Schlosser, B. Schaub, J. de Oliveira–Neto, and S. Jeganathan, *Chimia*, **40**, 244 (1986); B. Schaub, S. Jeganathan, and M. Schlosser, *ibid.*, **40**, 246 (1986).

Sodium benzeneselenolate.

Preparation. This reagent has been prepared by two methods from diphenyl diselenide. The original and most common method involves reduction with $NaBH_4$ in ethanol (**5,**272). A newer method involves reduction with Na in THF (**8,**447). Actually, the first method results in a sodium phenylseleno(triethoxy)borate, Na^+ $[C_6H_5SeB(OC_2H_5)_3]^-$. Although either reagent is satisfactory for some reactions, only the borate complex is useful for reduction of α,β-epoxy ketones to β-hydroxy ketones. However, addition of $B(OC_2H_5)_3$ to uncomplexed $NaSeC_6H_5$ provides a reagent that is equally effective (equation I).[1]

$$
\begin{array}{ll}
+\ NaSeC_6H_5 & 20\% \\
+\ NaSeC_6H_5\ +\ B(OC_2H_5)_3 & 95\% \\
+\ Na^+[C_6H_5SeB(OC_2H_5)_3]^- & 95\%
\end{array}
$$

This reduction of epoxy ketones has been used to prepare a number of santan-olides from the diepoxide (**1**) of α-santonin.[2] Thus reduction of **1** is accompanied by dehydration of the intermediate tertiary alcohol to give dehydroisoerivanin (**2**) in 80% yield.

1   **2**

[1] M. Miyashita, M. Hoshino, and A. Yoshikoshi, *Tetrahedron Letters*, **29**, 347 (1988).
[2] M. Miyashita, T. Suzuki, and A. Yoshikoshi, *ibid.*, **28**, 4293 (1987); *idem*, *Chem. Letters*, 2387 (1987).

Sodium borohydride–Alkoxydialkylboranes.

syn-1,3-*Diols*.[1] β-Hydroxy ketones are reduced by NaBH$_4$ in the presence of $(C_2H_5)_2BOCH_3$[2] to *syn*-1,3-diols. The high diastereoselectivity is ascribed to formation of a boron chelate.

$$syn/anti = 99:1$$

[1] K.-M. Chen, G. E. Hardtmann, K. Prasad, O. Repič, and M. I. Shapiro, *Tetrahedron Letters*, **28**, 155 (1987).
[2] R. Köster, W. Fenzl, and G. Seidel, *Ann.*, 352 (1975).

Sodium borohydride–Aluminum chloride.

Deoxygenation.[1] Diaryl and alkyl aryl ketones are reduced to methylenic hydrocarbons by reaction with NaBH$_4$ and AlCl$_3$ in refluxing THF. Diaryl- and alkylarylcarbinols also undergo this deoxygenation. Yields for both hydrogenolyses are generally above 90%.

[1] A. Ono, N. Suzuki, and J. Kamimura, *Synthesis*, 736 (1987).

Sodium–Ethanol.

Reduction of acyclic α-methyl ketones.[1] Reduction of ketones such as **1** with hydrides such as LiAlH$_4$ results in the *anti* or *syn* isomer as the predominant product. Bouvault–Blanc reduction, Birch reduction, and SmI$_2$ reduction all favor *anti* Cram

LiAlH$_4$	97%	74:26	
Na—C$_2$H$_5$OH	99%	42:58	
Li—NH$_3$	94%	24:76	
SmI$_2$	41%	38:62	

isomers, but at a rather low level. These results suggest that these nonhydride reductions involve protonation of a carbanion intermediate, and that the stereochemical outcome can be used to determine the nature of the reduction.

[1] Y. Yamamoto, K. Matsuoka, and H. Nemoto, *Am. Soc.*, **110**, 4475 (1988).

Sodium borohydride–Methanol.

Selective reduction of esters.[1] Sodium borohydride does not usually reduce esters, but it can reduce normal esters slowly in CH$_3$OH. This reaction has been used to reduce a normal ester selectively in the presence of a vinylogous urethane. Thus the methylene group of the enamine (E)-1, formed from dimethyl acetonedicarboxylate and benzylamine, is alkylated selectively (**2**) and then reduced by NaBH₄ in CH₃OH to **3**. Hydrolysis of the enamine to the ketone (**4**) followed by reduction provides **5**, which is lactonized by a diimide to the *cis-* and *trans-*lactones **6**.

Selective reduction of ketones.[2]   Ketones are reduced selectively in the presence of α,β-enones by NaBH₄ in 50% methanol in CH₂Cl₂ at −78°. In favorable cases the same selectivity can be achieved at 20° by use of only 5% methanol in CH₂Cl₂ and acetic acid as catalyst, but the rate of such reductions is slow.

[1] J. D. Prugh and A. A. Deana, *Tetrahedron Letters*, **29**, 2937 (1988).
[2] D. E. Ward, C. K. Rhee, and W. M. Zoghaib, *ibid.*, **29**, 517 (1988).

Sodium bromite.

Oxidation of alkenes to bromo ketones.[1]   NaBrO₂ reacts with alkenes in acetic acid at 25° to afford α-bromo ketones.

Example:

$$(CH_3)_3CCH{=}CHCH_3 \xrightarrow[83\%]{\substack{NaBrO_2, \\ HOAc}} (CH_3)_3CCCH(Br)CH_3$$

[1] T. Kageyama, Y. Tobito, A. Katoh, Y. Ueno, and M. Okawara, *Chem. Letters*, 1481 (1983); *Org. Syn.*, submitted (1987).

Sodium cyanoborohydride.

cis-3,4-Disubstituted azetidinones.[1]   The key step in a route to these β-lactams involves reduction of the oxime (**1**) of an α-alkyl β-keto ester with NaBH₃CN in an acidic medium to the *syn*-β-hydroxyamino ester **2** with high selectivity. The product, after reduction and hydrolysis to the β-amino acid (**3**), cyclizes to the *cis*-3,4-disubstituted azetidinone **4**.

4

[1] T. Chiba, T. Ishizawa, J. Sakaki, and C. Kaneko, *Chem. Pharm. Bull.*, **35**, 4672 (1987).

Sodium hydride.

Claisen rearrangement.[1]   The allyl ether **1** when heated rearranges and cyclizes slowly and in low yield to the dihydrobenzofuran **2**. However, the dianion (NaH) of **1** rearranges in refluxing DMF mainly to the isomeric dihydrobenzofuran **3**, a precursor to the antibiotic (±)-atrovenetin (**4**).

[1] G. Büchi and J. C. Leung, *J. Org.*, **51**, 4813 (1986).

Sodium hydride–Sodium *t*-amyl oxide–Nickel acetate (NiCRA, **10**, 365)

Activation by ClSi(CH₃)₃.   The rate of hydrogenation of some alkenes catalyzed by this complex reducing agent is increased by $ClSi(CH_3)_3$, although reduction is still sensitive to steric factors. The modified reagent is also useful for hydrogenation of enones to ketones.[1]

Example:

NiCRA   16%
+ ClSi(CH₃)₃   71%

Desulfuration.[2]   This complex as such or in combination with 2,2′-bipyridyl (bpy) or triphenylphosphine (a NiCRAL) can effect desulfuration of heteroarenes, aryl thioethers, dithioketals, sulfoxides, or sulfones in DME or THF at 63° in 1.5–30 hours. NiCRA is sufficient for aryl thioethers, dithioketals, but NiCRALs are more efficient for desulfuration of heteroarenes. Yields can be comparable with those obtained with Raney nickel.

Examples:

$C_6H_5SCH_3 \xrightarrow[98\%]{} C_6H_6$

[1] Y. Fort, R. Vanderesse, and P. Caubere, *Tetrahedron Letters*, **27**, 5487 (1986).
[2] S. Becker, Y. Fort, R. Vanderesse, and P. Caubere, *ibid.*, **29**, 2963 (1988).

Sodium iodide.

β-*Iodovinyl ketones*.[1]   Terminal acetylenic ketones react with NaI in TFA or with ISi(CH₃)₃ in CH₂Cl₂ to form (E)-β-iodovinyl ketones in good yield. Similar reactions with LiBr or BrSi(CH₃)₃ provide (E)-β-bromovinyl ketones. In contrast, the reactions of NaI or LiBr in HOAc provide the (Z)-β-iodovinyl ketones in moderate yield.

These reactions with NaI in TFA and with ISi(CH₃)₃ can be extended to substituted acetylenic ketones, but in this case, the reaction with NaI in HOAc is less stereoselective. An allene intermediate has been identified in the $ISi(CH_3)_3$ reaction.

[1] S. H. Cheon, W. J. Christ, L. D. Hawkins, H. Jin, Y. Kishi, and M. Taniguchi, *Tetrahedron Letters*, **27**, 4759 (1986); M. Taniguchi, S. Kobayashi, M. Nakagawa, T. Hino, and Y. Kishi *ibid.*, **27**, 4763 (1986).

Sodium naphthalenide.

Bridging of macrolides.[1]   A number of natural products are known to consist of oxapolycyclic systems. A potential route to such systems involves bridging of the more readily available macrolides (or the corresponding dithionolides, available by thionation with Lawesson's reagent) by electron-transfer reactions initiated with sodium naphthalenide (Scheme I).

Scheme I

The method can be used to obtain [6,6]-, [6,7]-, [6,8]-, [7,7], [7,8]-, and [8,8]-oxabicyclic systems.

[1] K. C. Nicolaou, C.-K. Hwang, M. E. Duggan, K. B. Reddy, B. E. Marron, and D. G. McGarry, *Am. Soc.*, **108**, 6800 (1986).

Sodium perborate, $NaBO_3$.

Epoxidation of alkenes.[1]   In combination with acetic anhydride, sodium perborate can effect epoxidation of alkenes at room temperature in 16–24 hours. Sulfuric acid can increase the rate, but lower the yields.

Examples:

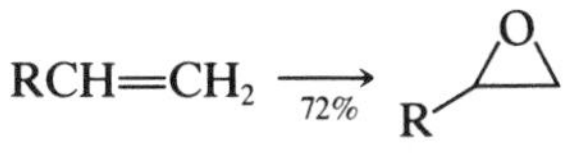
RCH=CH2
72%
O
R

Use of Ac_2O as solvent and H_2SO_4 as catalyst permits a highly exothermic reaction that results in *vic*-acetoxy alcohols.

Example:

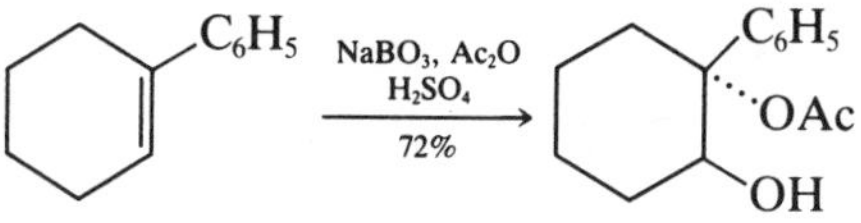
C6H5
NaBO3, Ac2O
H2SO4
72%
C6H5
OAc
OH

[1] G. Xie, L. Xu, J. Hu, S. Ma, W. Hou, F. Tao, *Tetrahedron Letters*, **29**, 2967 (1988).

Sulfur, singlet S_2.

Cyclic disulfides.[1] Singlet sulfur is generated when 2,2′-dithiobenzoylbiphenyl is heated at 80–130° with formation of 9,10-diphenylphenanthrene (equation I). The S_2 can be trapped by a Diels–Alder reaction with 1,3-dienes to form cyclic disulfides in 60–85% yield.

(I)

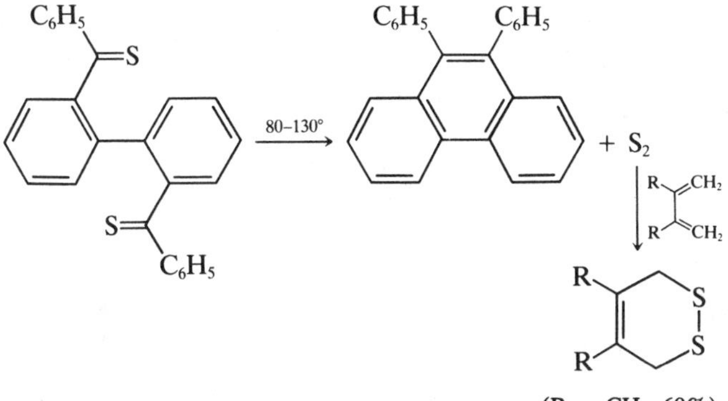
C6H5
S
S
C6H5
80–130°
C6H5
C6H5
+ S2
R
CH2
R
CH2
R
S
S
R

(R = CH_3, 60%)
(R = C_6H_5, 85%)

[1] K. Steliou, P. Salama, D. Brodeur, and Y. Gareau, *Am. Soc.*, **109**, 926 (1987).

Sulfuryl chloride, SO_2Cl_2.

$R^1R^2CHOH \rightarrow R^1R^2CHCl$.[1] This transformation can be effected with retention of configuration in two steps: conversion to the methyl xanthate followed by reaction with sulfuryl chloride (equation I).

$$(I) \quad R^1R^2CHOH \xrightarrow[87-92\%]{\substack{1) \text{ NaH} \\ 2) \text{ CS}_2 \\ 3) \text{ CH}_3\text{I}}} R^1R^2CHO\overset{\displaystyle S}{\overset{\|}{C}}SCH_3 \xrightarrow[57-91\%]{SO_2Cl_2} R^1R^2CHCl + Cl\overset{\displaystyle O}{\overset{\|}{C}}SSCH_3$$

[1] A. P. Kozikowski and J. Lee, *Tetrahedron Letters*, **29**, 3053 (1988).

T

Tetraalkylead–Titanium(IV) chloride.

Tetraalkyllead compounds can be obtained by reaction of Grignard reagents with lead(II) halides or by reaction of 3RLi and RI with PbX_2.[1]

Alkylation of aldehydes.[2] R_4Pb in combination with 1 equiv. of $TiCl_4$ adds selectively to aldehydes in CH_2Cl_2 at temperatures of -78–$0°$. Yield of adducts is greater than 70% in the case of $(C_2H_5)_4Pb$ or Bu_4Pb. Ketones do not react with

$$\text{(I)} \quad C_6H_{11}CHO \xrightarrow[\substack{84-98\%}]{\substack{1)\ TiCl_4,\ CH_2Cl_2,\ -78° \\ 2)\ R_4Pb,\ -78 \to 0°}} C_6H_{11}\underset{\underset{OH}{|}}{C}HR$$

this reagent even at $25°$. Of greater interest this reaction can proceed with high 1,2- and 1,3-asymmetric induction (equations II and III).

$$\text{(II)} \quad CH_3\underset{OBzl}{\diagup}CHO \xrightarrow[81\%]{} CH_3\underset{OBzl}{\diagup}\overset{OH}{\underset{C_2H_5}{\diagup}} + anti\text{-}2$$

syn-2   98:2

$$\text{(III)} \quad CH_3\underset{OBzl}{\diagup}CHO \xrightarrow[66\%]{} CH_3\underset{OBzl\ OH}{\diagup}C_2H_5 + CH_3\underset{OBzl\ OH}{\diagup}C_2H_5$$

78:22

BF_3 or BF_3 etherate can replace $TiCl_4$ as the Lewis acid. But in order to effect clean reactions, the Lewis acid should be added to the aldehyde before addition of the lead reagent. The rate of reaction is highly dependent on the size of the R group. Transfer of an ethyl group is rapid, but a cyclohexyl group is transferred slowly even at $0°$. Similar transfer does not obtain with R_4Sn.

[1] H. Gilman and R. G. Jones, *Am. Soc.*, **72**, 1760 (1950).
[2] Y. Yamamoto and Y. Yamada, *ibid.*, **109**, 4395 (1987).

Tetrabutylammonium fluoride.

Macrodilactonization. The final steps in a synthesis of the 12-membered dilactone intergerrimine 3[1] involve selective cleavage of the (trimethylsilyl)ethyl ester group of **1** with Bu₄NF, which is followed by a spontaneous intramolecular displacement of the mesylate group to form the macrolide **2** in 75% yield. Reaction of **2** with aqueous HF in CH_3CN furnishes the alkaloid **3** in 67% yield.

2, R = SiMe₂-*t*-Bu
3, R = H

This method of cyclization has been used successfully for synthesis of four 11-membered dilactone pyrrolizidine alkaloids,[2] but it does not appear to be useful for cyclization to simple macrolides.

[1] J. D. White and S. Ohira, *J. Org.*, **51**, 5492 (1986).
[2] E. Vedejs, S. Ahmad, S. D. Larsen, and S. Westwood, *ibid.*, **52**, 3937 (1987).

Tetrabutylammonium Oxone. This oxidant can be obtained as a white solid by reaction of commercial Oxone with Bu₄NHSO₄. Unlike Oxone, which is soluble only in aqueous or alcoholic solvents, this salt is soluble in methylene chloride and CH_3CN.

Sulfones.[1] This oxidant is useful for selective oxidation of sulfides to sulfones (**10**,328). Oxidations of acid-sensitive substrates should be buffered with Na_2CO_3.

[1] B. M. Trost and R. Braslau, *J. Org.*, **53**, 532 (1988).

Tetracarbonyldi-μ-chlorodirhodium, $(OC)_2Rh\overset{Cl}{\underset{Cl}{\diagup\diagdown}}Rh(CO)_2$ **(1),12**,112.

α-Pyrones; phenols.[1] α-Pyrones and phenols can be obtained via vinylketenes by carbonylation of cyclopropene derivatives catalyzed by **1**.
Example:

[1] S. H. Cho and L. S. Liebeskind, *J. Org.*, **52**, 2631 (1987).

Tetrakis(triphenylphosphine)palladium(0).

Coupling of aryl triflates with organostannanes (cf., **12**,469–470). This coupling can be effected in the presence of lithium chloride (3 equiv.) and a palladium catalyst, usually $Pd[P(C_6H_5)_3]_4$. Many functional groups are tolerated on either coupling component. Aryl, alkynyl, alkyl, and vinyl groups are transferred to the aromatic ring in high yield, but transfer of allyl groups is less facile and less selective.[1]

Example:

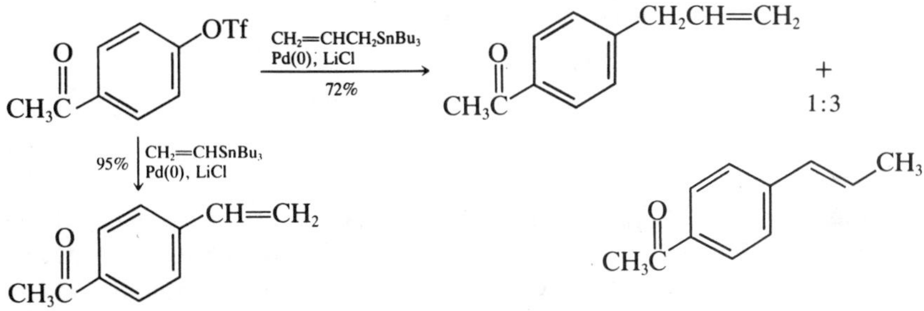

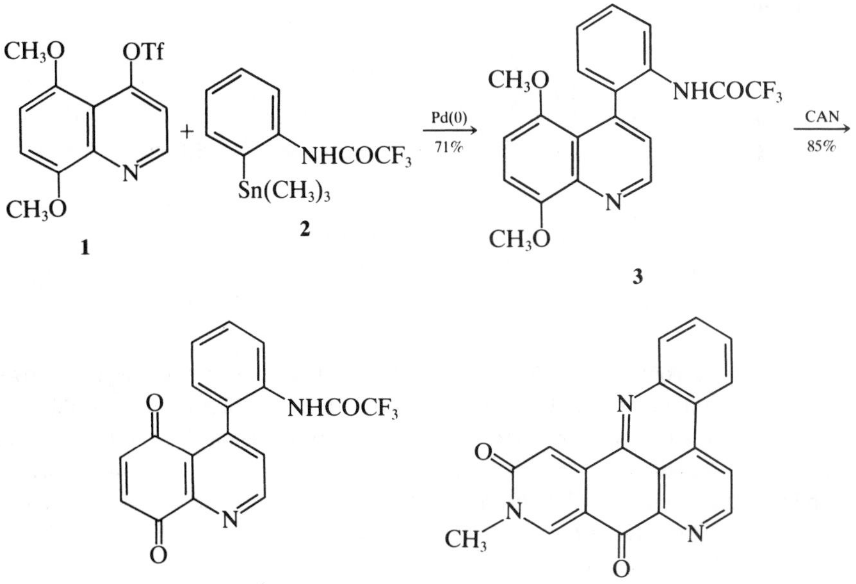

This coupling is a key step in the synthesis of amphimedine (**5**), a cytotoxic alkaloid of a sponge.[2] Thus coupling of the triflate **1** with **2** provides **3**, which is oxidized by CAN to the quinoline-5,8-dione **4**. Conversion of **4** to **5** involves a Diels–Alder addition of an azadiene to the quinone ring and intramolecular cyclization of the adduct.

Macrolides can be obtained in satisfactory yield by a Pd(0)-catalzed intramolecular coupling of an ester with vinyl triflate and vinyltin groups at the two ends (*cf*.,**12**,469–470). This cyclization is insensitive to the ring size; the yield is 56–57% in the cyclization of **1** to form 12- to 15-membered rings.[3]

1 (n = 5–8)   **2**

Coupling of vinylboranes and vinyl halides (**11**,393).[4] This coupling can be effected in the presence of 1–3 mole % of Pd(0) and 2 equiv. of NaOH or NaOCH$_3$, and results in 1,3-dienes formed with retention of the original configuration of both the haloalkene and the vinylborane. This coupling is observed also with aryl, allylic, and benzylic halides. Coupling of vinylboranes with 1-bromoalkynes gives (E)- or (Z)-enynes in good yield with high stereoselectivity.

Examples:

When this coupling is extended to more complex substrates, the rate slows drastically and undesired by-products are formed, even when 1 equiv. of Pd(0) is used.[5] Fortunately the rate is markedly affected by the choice of the base, being highest with thallium hydroxide. Under TlOH conditions the coupling to (E,Z)- and to (Z,Z)-dienes can be effected at 0° often within minutes. Under these improved conditions (Z)-**1** and (Z)-**2** couple to form (Z,Z)-**3** in 73% yield, and (E)-**1** and (Z)-**2** couple to give (E,Z)-**3** in 94% yield.

(Z,Z)-3

Intramolecular Heck reactions.[6]   Heck intramolecular coupling of alkenyl or aryl iodides substituted by 3-cycloalkenyl group is an attractive route to fused, spiro, and bridged polycyclic products. Coupling is achieved with a Pd–phosphine catalyst such as $Pd[P(C_6H_5)_3]_4$ in combination with a base, $N(C_2H_5)_3$ or NaOAc. The coupling tends to produce a mixture of two isomeric alkenes, in which the newly formed bond is allylic or homoallylic to the ring juncture.

Examples:

Similar intramolecular Heck reactions with $Pd(OAc)_2$ and a base (KOAc) or a phosphine have been reported by two other groups.[7,8] These conditions also result in two isomeric polycyclic alkenes.

Hydroxylamination.[9]   In the presence of this Pd(0) catalyst, allyl esters react with hydroxylamines to provide N-allylhydroxylamines, which are reduced by zinc in dilute HCl to secondary allylamines.

Example:

Denitration of allylic nitro groups.[10]   Allylic nitro compounds when complexed with Pd(0) are reductively removed by a variety of hydride donors. The regioselectivity of denitration to give 1- or 2-alkenes can be controlled by the choice of ligand and by the hydride donor. Thus reduction by formates gives 1-alkenes, and use of $NaBH_4$ or $NaBH_3CN$ favors formation of 2-alkenes.

$P(C_6H_5)_3$, HCOONH$_4$ 55%	92%	8%
$P(C_6H_5)_3$, NaBH$_3$CN 98%	32%	68%

Coupling of RCOCl with Bu₄Pb.[11]   In the presence of this Pd(0) complex, acid chlorides couple with Bu₄Pb at 65° to form ketones, RCOBu, in high yield, with utilization of two of the four butyl groups attached to lead.

[1] A. M. Echavarren and J. K. Stille, *Am. Soc.*, **109**, 5478 (1987).
[2] *Idem, ibid.*, **110**, 4051 (1988).
[3] J. K. Stille and M. Tanaka, *ibid.*, **109**, 3785 (1987).
[4] N. Miyaura, K. Yamada, H. Suginome, and A. Suzuki, *ibid.*, **107**, 972 (1985).
[5] J. Uenishi, J.-M. Beau, R. W. Armstrong, and Y. Kishi, *ibid.*, **109**, 4756 (1987).
[6] E. Negishi, Y. Zhang, and B. O'Connor, *Tetrahedron Letters*, **29**, 2915 (1988).
[7] R. C. Larock, H. Song, B. E. Baker, and W. H. Gong, *ibid.*, **29**, 2919 (1988).
[8] M. M. Abelman, T. Oh, and L. E. Overman, *J. Org.*, **52**, 4130 (1987).

[9] S.-I. Murahashi, Y. Imada, Y. Taniguchi, and Y. Kodera, *Tetrahedron Letters*, **29**, 2973 (1988).
[10] N. Ono, I. Hamamoto, A. Kamimura, and A. Kaji, *J. Org.*, **51**, 3734 (1986).
[11] J. Yamada and Y. Yamamoto, *J.C.S. Chem. Comm.*, 1302 (1987).

Tetramethoxycarbonylpalladacyclopentadiene, (1).

The complex is obtained in high yield by reaction of dimethyl acetylenedicarboxylate with bis(dibenzylideneacetone)palladium.[1] The complex is only slightly soluble in the usual solvents.

[2+2+2]*Cycloaddition*.[2]   In combination with a triarylphosphine, **1** catalyzes the cyclization of a 1,6-enyne (**2**) to a 1,3-diene (**3**), and it effects [2+2+2] cycloaddition of **2** with an alkyne to give **4**, which is isomeric with the product (**5**) of Diels–Alder addition of **3** with an alkyne.

[1] K. Moseley and P. M. Maitlis, *J.C.S. Dalton*, 169 (1974).
[2] B. M. Trost and G. J. Tanoury, *Am. Soc.*, **109**, 4753 (1987).

Tetramethylammonium triacetoxyborohydride, $(CH_3)_4NBH(OAc)_3$ (**1**).
   Preparation:

$$(CH_3)_4NOH + NaBH_4 \xrightarrow[81\%]{} (CH_3)_4NBH_4 \xrightarrow{HOAc, C_6H_6} \mathbf{1} \text{ (m.p. 98°)}$$

***anti*-Selective reduction of β-hydroxy ketones.**   This hydride reduces β-hydroxy ketones in $HOAc/CH_3CN$ (1:1) at $-40°$ to 25° to 1,3-diols with high *anti-*

selectivity (*anti*/*syn* = 92–98:8–2). The report includes evidence that the reduction involves ligand exchange to form an alkoxydiacetoxyborohydride (**a**) that reduces the ketone by intramolecular hydride delivery. A substituent (*anti* or *syn*) at the α-position exerts only a slight effect on the stereoselectivity. The reagent does not reduce ketones lacking a β-hydroxyl group.

(*anti*/*syn* = 96:4)

This reagent can also effect sequential diastereoselective reduction of a hydroxy diketo ester to afford an *anti*, *anti*-triol ester as the major product. The only other significant product is the *anti*, *syn*-triol (equation I).

(I)

85:13

(50%)

[1] D. A. Evans, K. T. Chapman, and E. M. Carreira, *Am. Soc.*, **110**, 3560 (1988).

Tetramethyldisilazane, $HN[SiH(CH_3)_2]_2$ (**1**).

   Intramolecular hydrosilation. Tamao et al.[1] have extended their hydrosilylation–oxidation sequence (**12**,243–245) to allyl and homoallyl alcohols as an approach to 1,3-diols. When applied to cyclic homoallylic alcohols, only a *cis*-1,3-diol is obtained, presumably by way of a cyclic intermediate (equation I). Acyclic homoallylic alcohols can be converted by this approach to either *syn*- or *anti*-1,3-

(I) 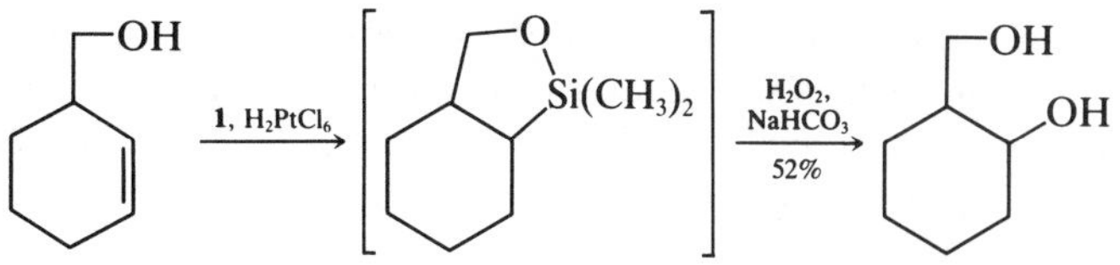 $\xrightarrow{\text{1, H}_2\text{PtCl}_6}$ $\xrightarrow[\text{52\%}]{\text{H}_2\text{O}_2,\ \text{NaHCO}_3}$

(*cis*, 100%)

diols, depending on the geometry and extent of substitution of the double bond. Slight stereocontrol is observed with (E)- or (Z)-disubstituted alkenes, whereas *anti*-control is evident with trisubstituted alkenes. 1,2-*syn*-Selectivity is observed with allylic alcohols, and increases with increasing size of the allylic substituent.

Examples:

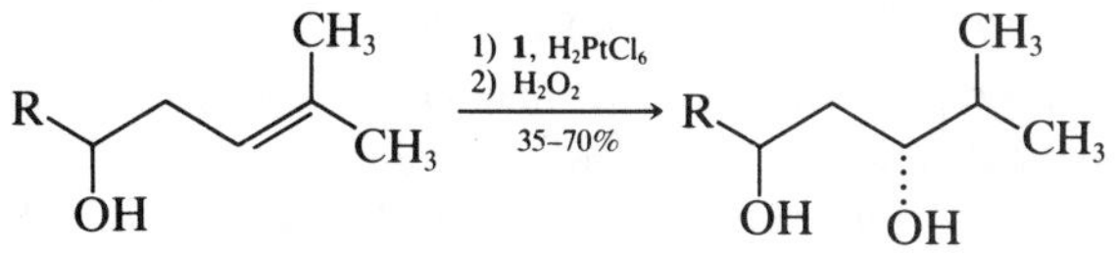

(*anti/syn* = 5.3:1)

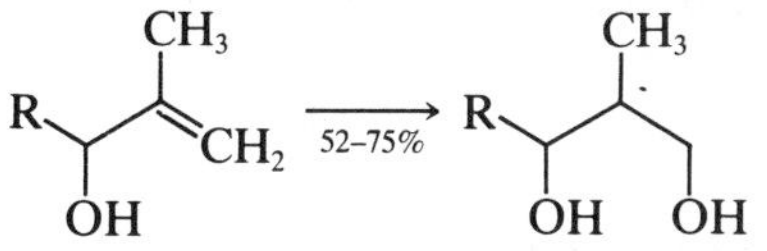

(*syn/anti* = 2.5–100:1)

This intramolecular hydrosilylation can be extended to α-hydroxy enol ethers (2-alkoxy-1-alkene-2-ols) to provide access to 2,3-*syn*-1,2,3-triols.[2] In this case a neutral catalyst, Pt(0)–vinylsiloxane,[3] is preferred over H_2PtCl_6.

Examples:

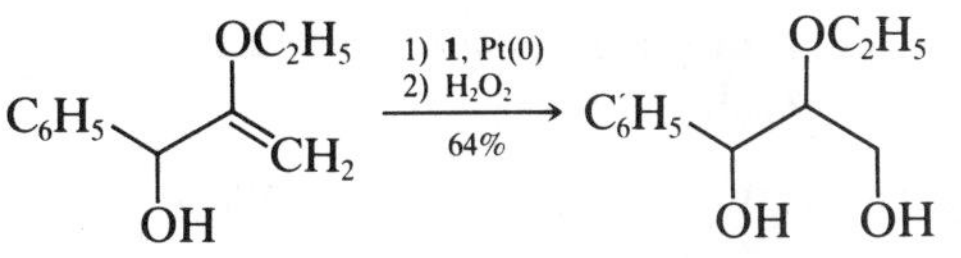

(*syn/anti* = 42–94:1)

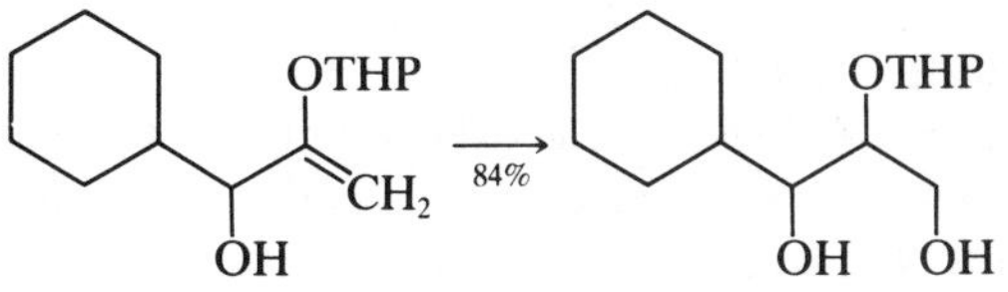

(*syn/anti* = >99:1)

Since the starting materials are available by reaction of aldehydes with lithiated vinyl ethers, this sequence is useful for conversion of aldehydes into these triols. Indeed this sequence can be used to convert an optically active glyceraldehyde into optically pure pentitols with high *syn*-stereocontrol.

[1] K. Tamao, T. Nakajima, R. Sumiya, H. Arai, N. Higuchi, and Y. Ito, *Am. Soc.*, **108**, 6090 (1986).
[2] K. Tamao, Y. Nakagawa, H. Arai, N. Higuchi, and Y. Ito, *ibid.*, **110**, 3712 (1988).
[3] L. N. Lewis and N. Lewis, *ibid.*, **108**, 7228 (1986).

2,2,6,6-Tetramethylpiperidinyl-1-oxyl.

Oxidation of $-CH_2OH \rightarrow -CHO$ (cf., **12**,479–480).   This oxidation can be effected in high yield with sodium hypochlorite (slight excess) in buffered H_2O/CH_2Cl_2 with this nitroxyl radical and KBr as the catalysts.[1] The oxidation is exothermic, and the temperature should be maintained at 0–15° with a salt–ice bath. Saturated primary alcohols are converted to aldehydes in 88–93% yield; yields are lower in the case of unsaturated substrates. Addition of quaternary onium salts permits further oxidation to carboxylic acids.

[1] P. L. Anelli, C. Biffi, F. Montanari, and S. Quici, *J. Org.*, **52**, 2559 (1987); P. L. Anelli, F. Montanari, and S. Quici, *Org. Syn.*, submitted (1988).

Tetrapropylammonium tetraoxoruthenate(VIII), $(Pr_4N)RuO_4$ (1).

The reagent is obtained as a green solid by reaction of RuO_4 with Pr_4NOH in aqueous NaOH. Supplier: Aldrich.

Catalytic oxidant.[1]   In combination with N-methylmorpholine N-oxide (**7**,244) as the stoichiometric oxidant, this ruthenium compound can be used as a catalytic oxidant for oxidation of alcohols to aldehydes or ketones in high yield in CH_2Cl_2 at 25°. Addition of 4Å molecular sieves is generally beneficial. Racemization is not a problem in oxidation of alcohols with an adjacent chiral center. Tetrabutylammonium perruthenate can also be used as a catalytic oxidant, but the preparation is less convenient.

[1] W. P. Griffith, S. V. Ley, G. P. Whitcombe, and A. D. White, *J.C.S. Chem. Comm.*, 1625 (1987).

Thallium(III) nitrate (TTN).

Ring contraction of glycals.[1]   The oxidative ring contraction of cyclic alkenes (**4**,492–493) can be applied to protected glycals. Thus oxidation of the D-glucal **1** with TTN in CH_3CN provides the 2,5-anhydro-D-manose **2**, whose structure was established by conversion to the 2,5-anhydro-D-mannitol derivative (**3**).

Oxidation of 3,4-dihydro-2*H*-pyran (**4**) with TTN in CH₃OH provides 2-(dimethoxymethyl)tetrahydrofuran **5** in essentially quantitative yield.

[1] A. Kaye, S. Neidle, and C. B. Reese, *Tetrahedron Letters*, **29**, 1841 (1988).

Thexylchloroborane–Dimethyl sulfide.

RCOOH →RCHO (**12**,485).[1] This borane (2 equiv.) reduces acids at room temperature to thexylboronic acid and aldehydes, which are best isolated as the sodium bisulfite adduct. Yields of aliphatic aldehydes are in the range 80–95%. Reduction of aromatic acids is slow, and yields are significantly lower.

[1] H. C. Brown, J. S. Cha, N. M. Yoon, and B. Nazer, *J. Org.*, **52**, 5400 (1987).

Tin(II) bromide.

α-Methylene ketones.[1] In the presence of SnBr₂, bromomethyl methyl ether reacts with silyl enol ethers to give α-bromomethyl ketones, which undergo dehydrobromination with DBU to give α-methylene ketones. Higher yields obtain if the overall reaction is conducted in CH₂Cl₂ at 25° without isolation of the intermediate.

Examples:

[1] M. Hayashi and T. Mukaiyama, *Chem. Letters*, 1283 (1987).

Tin(IV) chloride.

Rearrangement of allylic acetals. 3-Acetylpyrrolidines can be obtained in good yield by an acid-catalyzed rearrangement of 5-methyl-5-vinyloxazolidines. The rearrangement involves an aza-Cope rearrangement followed by Mannich cyclization (equation I).[1]

A related Lewis acid-catalyzed rearrangement provides a synthesis of tetrahydrofurans from allylic acetals.[2] The starting material (**1**) is prepared by reaction of an α-hydroxy ketone with vinylmagnesium bromide, followed by acetalization of the resulting diol with an aldehyde. On exposure to $C_2H_5AlCl_2$ or $SnCl_4$ at $-70 \rightarrow -10°$, **1** rearranges to the all-*cis*-product **2**, which rearranges in base to the epimer (**3**). Optically active allylic acetals rearrange with no loss of enantiomeric purity. This rearrangement may involve a hetero-Cope rearrangement followed by an aldol or a Prins cyclization and then a pinacol rearrangement.

1

2

3

This rearrangement of allylic acetals can also be used for furan annelations, in which the formation of the new tetrahydrofuran ring is coupled with ring enlargement of the starting ring.[3] The same *cis*-fused bicyclic tetrahydrofuran is formed from either one of the *cis*- or *trans*-allylic diols used as starting materials.

Cyclization of mixed acetals (13,300).[4]   This reaction is a particularly useful route to eight-membered cyclic ethers (oxocanes) and provides the first practical route to a natural oxocene, (−)-laurenyne (3), from an optically active mixed acetal 1. Thus cyclization of 1 followed by O-desilylation affords 2 as the only cyclic product. Remaining steps to 3 involved C-desilylation, for which only HF/pyridine is useful, introduction of unsaturation into the C_2-side chain, and extension of the C_8-side chain. Exploratory studies showed that unsaturation at the β- or γ-positions to the cite of cyclization of 1 prevent or retard cyclization with a wide variety of Lewis acids. The cyclization is apparently more tolerant of substitution in the terminator position, C_3–C_8, of the oxocene.

$$\text{1} \xrightarrow[\text{37\%}]{\substack{\text{1) SnCl}_4 \\ \text{2) Bu}_4\text{NF}}} \text{2}$$

(−)-3

[1] L. E. Overman, M. Kakimoto, M. E. Okazaki, and G. P. Meier, *Am. Soc.*, **105**, 6622 (1983).

[2] M. H. Hopkins and L. E. Overman, *ibid.*, **109**, 4748 (1987).

[3] P. M. Herrinton, M. H. Hopkins, P. Mishra, M. J. Brown, and L. E. Overman, *J. Org.*, **52**, 3711 (1987).

[4] L. E. Overman and A. S. Thompson, *Am. Soc.*, **110**, 2248 (1988).

Tin(IV) chloride–Zinc chloride.

Propargylic ethers.[1]   These ethers can be prepared by reaction of acetals with 1-trimethylsilylalkynes in the presence of at least 1 equiv. of a Lewis acid such as TiCl$_4$ or SnCl$_4$ (**12**,375). However a combination of SnCl$_4$ and ZnCl$_2$ (10 mole % each) can effect this reaction (equation I). The actual catalyst may be $^+$SnCl$_3$–$^-$ZnCl$_3$.

$$\text{(I)} \quad \text{RCH(OCH}_3)_2 + \text{CH}_3\text{C}\equiv\text{CSi(CH}_3)_3 \xrightarrow[\text{75–80\%}]{\substack{\text{SnCl}_4, \text{ZnCl}_2 \\ \text{CH}_2\text{Cl}_2}} \overset{\displaystyle \overset{\text{OCH}_3}{|}}{\text{RCH}}-\text{C}\equiv\text{CCH}_3$$

[1] M. Hayashi, A. Inubushi, and T. Mukaiyama, *Chem. Letters*, 1975 (1987).

Tin(II) trifluoromethanesulfonate.

Asymmetric sulfenylation.[1]   In the presence of the chiral diamine (S)-1-methyl-2-(piperidinylmethyl)pyrrolidine (**1**), tin(II) enolates of ketones react with

1

phenylthioarenesulfonates to provide β-keto sulfides with high enantioselectivity. The products are useful precursors to chiral epoxides.

Example:

Glycosidation (**13**,302). Glycosidation of 2-acetamido-3,4,6-triacetyl-2-deoxy-α-D-glucopyranosyl chloride (**1**) mediated by Sn(OTf)₂ provides exclusively β-glycosides. The most satisfactory base is 1,1,3,3-tetramethylurea, and CH₂Cl₂ is the preferred solvent.[2]

[1] T. Yura, N. Iwasawa, R. Clark, and T. Mukaiyama, *Chem. Letters*, 1809 (1986).
[2] A. Lubineau, J. Le Gallic, and A. Malleron, *Tetrahedron Letters*, **28**, 5041 (1987).

Titanium(III) chloride–Lithium aluminum hydride.

Reductive elimination of an allylic diol group. A new synthesis of vitamin A involves reduction of the allylic diol **1**, prepared in several steps from β-ionone, with a low valent titanium formed from TiCl₃ and LiAlH₄ in the ratio 2:1. Thus, the allylic diol group of **1** [either (E) or (Z)] is reduced to an (E,E)-1,3-diene group to form the silyl ether (**2**) of vitamin A.[1] When the primary hydroxyl group is protected as an acetate, the reduction gives a mixture of (E)- and (Z)-**2**.

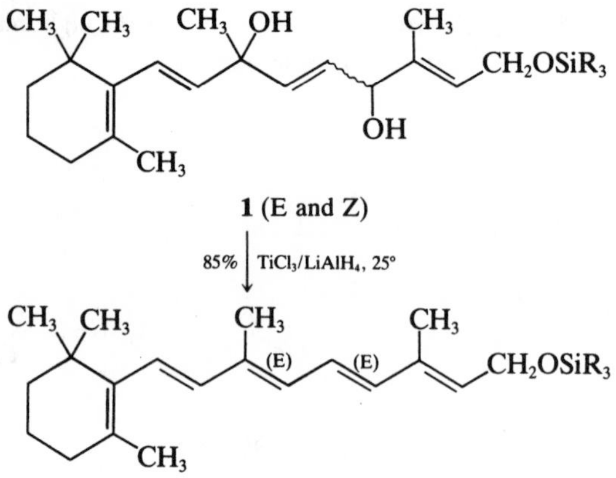

1 (E and Z)

85% | TiCl₃/LiAlH₄, 25°

2

This reaction was also used in a synthesis of 13-*cis*-retinoic acid.[2] Thus reduction of **3** under the same conditions gives the triethylsilyl ether (**4**) of 13-*cis*-retinol, with retention of the geometry of the terminal double bond. This product can be converted to 13-*cis*-retinoic acid by deprotection and oxidation (60% yield).

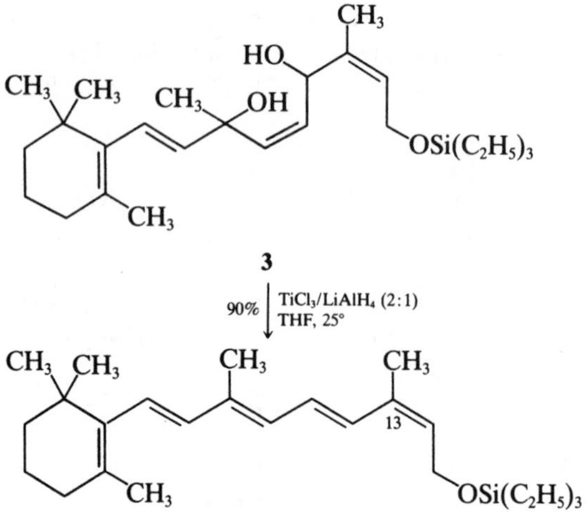

3

90% | TiCl₃/LiAlH₄ (2:1)
THF, 25°

4

[1] G. Solladié and A. Girardin, *Tetrahedron Letters*, **29**, 213 (1988).
[2] G. Solladié, A. Girardin, and P. Matra, *ibid.*, **29**, 209 (1988).

Titanium(III)chloride–Potassium/Graphite, TiCl₃/C₈K (1:2).

McMurry coupling.[1]   Reaction of TiCl₃ with C₈K at 150–160° provides a finely dispersed form of the metal on graphite, which effects coupling of aldehydes or ketones to alkenes in generally high yield. A related reagent prepared by reaction of TiCl$_4$ with C$_8$K at 0° can effect pinacol reduction of aliphatic aldehydes or ketones and effects McMurry coupling of aromatic aldehydes or ketones.

[1] A. Fürstner and H. Weidmann, *Synthesis*, 1071 (1987).

Titanium(III) chloride–Zinc/copper couple.

Dicarbonyl coupling (8,483).   This Ti-catalyzed coupling offers a useful route to cyclic sesquiterpenes such as humulene (**4**).[1] The precursor is obtained by coupling a vinylic zirconium compound (**1**) with the π-allylpalladium complex (**2**) to give, after deprotection, the keto aldehyde **3** in 84% yield. This product couples to humulene as a single isomer in 60% yield.

[1] J. E. McMurry, J. R. Matz, and K. L. Kees, *Tetrahedron*, **43**, 5489 (1987).

Titanium(IV) chloride.

1,2-Rearrangement of epoxy silyl ethers.[1]   When treated with 1 equiv. of TiCl$_4$, α-silyloxy epoxides rearrange to β-hydroxy carbonyl compounds.

Examples:

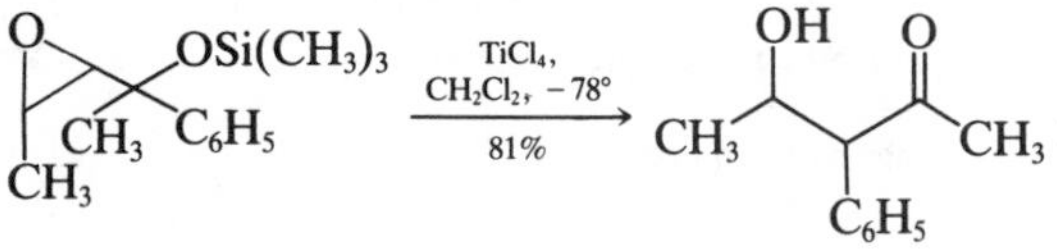

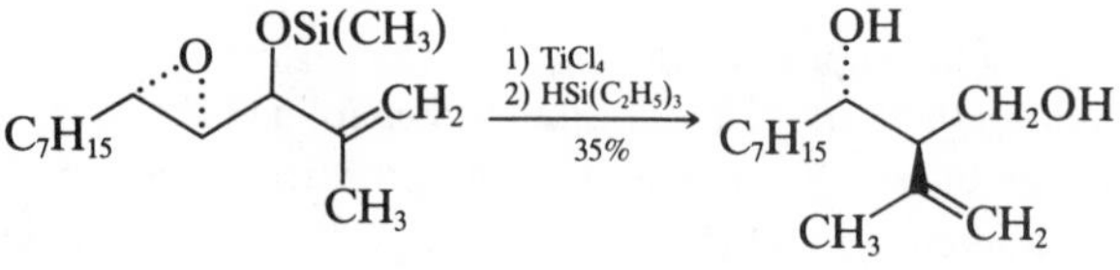

Aryl and alkenyl groups undergo this *anti*-migration more easily than an alkyl group. This rearrangement in combination with Sharpless asymmetric epoxidation affords a stereocontrolled route to aldols and 1,3-diols (second example).

This rearrangement can also be effected with catalytic amounts of either $(CH_3)_3SiI$ or $(CH_3)_3SiOTf$ in the case of α,β-epoxy trimethylsilyl ethers. Yields are dependent on the migratory aptitude of the group that rearranges.[2]

Intramolecular cyclization/enolate trapping of allylsilanes (*cf.*, **12**,496–497).[3] The intermediate enolate formed in the $TiCl_4$-catalyzed cyclization of **1** can be trapped by chloromethyl methyl sulfide to give a decalone derivative with a potential methyl group on the angular position. Actually the reaction results in

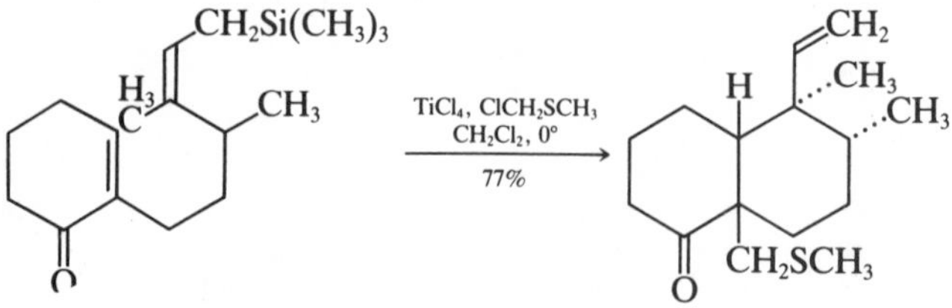

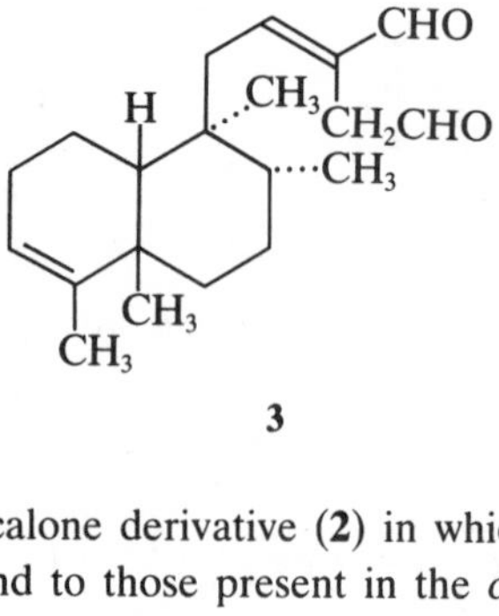

high yield of a single decalone derivative (**2**) in which the four contiguous stereogenic centers correspond to those present in the *cis*-chlorodane diterpene linaridial (**3**).

β-*Chloro carboxylic acids*.[4]   The condensation of ketene silyl acetals with aldehydes mediated by $TiCl_4$ (Mukaiyama reaction, **6**,590) results in β-hydroxy acids. Use of 2 equiv. of $TiCl_4$ results in β-chloro acids in 70–90% yield.
    Example:

$$C_6H_5CHO + CH_3CH=C[OSi(CH_3)_3]_2 \xrightarrow{TiCl_4} \left[\underset{CH_3}{\underset{|}{C_6H_5}} \overset{OSi(CH_3)_3}{\overset{|}{\diagdown}} COOH \xrightarrow{TiCl_4} \right.$$

$$\left. \underset{CH_3}{\underset{|}{C_6H_5}} \overset{OTiCl_3}{\overset{|}{\diagdown}} COOH \right] \xrightarrow{87-90\%} \underset{CH_3}{\underset{|}{C_6H_5}} \overset{Cl}{\overset{|}{\diagdown}} COOH + TiOCl_2$$

[1] K. Maruoka, M. Hasegawa, H. Yamamoto, K. Suzuki, M. Shimazaki, and G. Tsuchihashi *Am. Soc.*, **108**, 3827 (1986).
[2] K. Suzuki, M. Miyazawa, and G. Tsuchihashi, *Tetrahedron Letters*, **28**, 3515 (1987).
[3] T. Tokoroyama, M. Tsukamoto, T. Asada, and H. Iio, *ibid.*, **28**, 6645, (1987).
[4] M. Bellassoud, J.-E. Dubois, and E. Bertounesque, *ibid.*, **29**, 1275 (1988).

Titanium(IV) chloride–Lithium aluminum hydride.
    Reduction of enedicarboxylates.[1]   A low-valent titanium reagent obtained from $TiCl_4$ and $LiAlH_4$ (about 2:1) can reduce enedicarboxylates in the presence of a trace of triethylamine.
    Examples:

$$C_2H_5OOCCH=CHCOOC_2H_5 \xrightarrow[59\%]{\overset{TiCl_4,\ LiAlH_4,\ N(C_2H_5)_3}{THF,\ 100°}} C_2H_5OOCCH_2CH_2COOC_2H_5$$

[1] C. W. Hung and H. N. C. Wong, *Tetrahedron Letters*, **28**, 2393 (1987).

Titanium(IV) isopropoxide.
    Lactamization.[1]   In the presence of this reagent, ω-amino acids cyclize to lactams. Yields are 75–93% for cyclization to 5- and 6-membered lactams.

Examples:

$$H_2N(CH_2)_2\overset{\overset{\displaystyle CH_3}{|}}{C}HCOOH \xrightarrow[81\%]{\substack{Ti(O\text{-}i\text{-}Pr)_4, \\ ClCH_2CH_2Cl, 25°}}$$

$$CH_3NH(CH_2)_3CH_2COOH \xrightarrow[86\%]{}$$

[1] M. Mader and P. Helquist, *Tetrahedron Letters*, **29**, 3049 (1988).

2,4,6-Tri-*t*-butylphenyllithium.

This base (**1**) is prepared by reaction of 2,4,6-tri-*t*-butylbromobenzene with BuLi in THF at 0°.

Kinetic enolates. Alkyllithium reagents have the advantage over lithium amides for deprotonation of ketones in that the co-product is a neutral alkane rather than an amine. This bulky lithium reagent is useful for selective abstraction of the less-hindered α-proton of ketones with generation of the less-stable enolate, as shown previously for a hindered lithium dialkylamide (LOBA,**12**,285). Thus reaction of benzyl methyl ketone (**2**) with **1** and ClSi(CH$_3$)$_3$ at $-50°$ results mainly in the less-stable enolate (**3**), even though the benzylic protons are much more acidic than those of the methyl group, the less hindered ones. Mesityllithium shows

$$C_6H_5CH_2\overset{\overset{\displaystyle O}{||}}{C}CH_3 \xrightarrow[87\%]{\substack{1, \; ClSi(CH_3)_3, \\ THF, \; -50°}} \underset{TMSO}{C_6H_5CH_2}C{=}CH_2 \; + \; \underset{\substack{4 \; (E/Z \; = \; 2:1)}}{\underset{H}{C_6H_5}C{=}C\overset{CH_3}{\underset{OTMS}{}}}$$

$$\underset{\mathbf{2}}{} \qquad\qquad \underset{\mathbf{3}}{} \quad 74:26$$

similar but less pronounced regioselectivity; use of LDA results in formation of **3** and **4** in the ratio 30:70.

[1] M. Yoshifuji, T. Nakamura, and N. Inamoto, *Tetrahedron Letters*, **28**, 6325 (1987).

Tributyltin hydride.

Reviews. The recent use of this reagent in synthesis has been reviewed (187 references).[1] The review includes not only radical dehalogenation but also reductive cleavage of C–O, C–S, C–Se, C–Te, and C–N bonds. The reagent is less toxic than trimethyl- or triethyltin hydride. The review includes some preparative and work-up recommendations.

Curran[2] has reviewed recent applications of the tin hydride method for initiation of radical chain reactions in organic synthesis (191 references). The review covers intermolecular additions of radicals to alkenes (Giese reaction) as well as intramolecular radical cyclizations, including use of vinyl radical cyclization.

Radical cyclization-trapping. Trapping of the initial radical formed in radical cyclizations can proceed with high regio- and stereoselectivity. In such reactions hydrogen atom transfer reactions can be suppressed by use of a catalytic tin system.[3] Thus a system composed of NaBH$_3$CN (2 equiv.), AIBN (0.1 equiv.), and Bu$_3$SnCl (0.1 equiv.) in (CH$_3$)$_3$COH or use of (C$_6$H$_5$)$_3$GeH is usually more effective than a stoichiometric process. This cyclization-trapping can be used for regio- and stereocontrolled formation of two adjacent chiral centers. Thus (+)-prostaglandin F$_{2\alpha}$ (3) has been prepared[4] by addition of two different groups to a derivative (1) of *cis*-2-cyclopentene-1,4-diol. The cyclization-trapping step adds a potential -CH$_2$CHO group to one end of the double bond (used later in a Wittig reaction) as well as the precursor to the unsaturated alcohol chain at the other end of the double bond of 2.

The catalytic method has also been used to effect (hydro)methylation of the double bond of an allylic alcohol with introduction of a methyl group adjacent and *cis* to the hydroxyl group (equation I).[5]

(I)

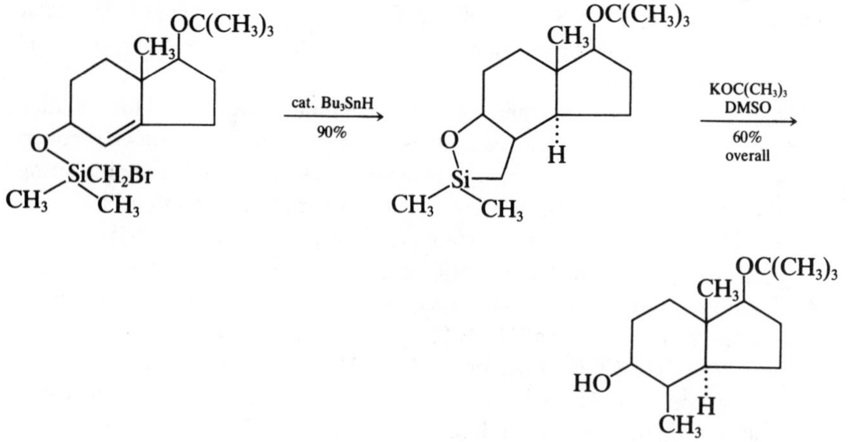

Vinyl radical cyclization of enynes. Two laboratories have reported that tin radicals generated from R_3SnH with AIBN[6] or $B(C_2H_5)_3$,[7] add to the triple bond of an enyne to form vinyltin radicals, which can undergo intramolecular cyclization with the double bond to form five- or six-membered rings substituted by a vinyltin group. The SnR_3 group of the final product can be removed by destannylation with dry SiO_2 in CH_2Cl_2. This enyne cyclization is successful because the addition of a

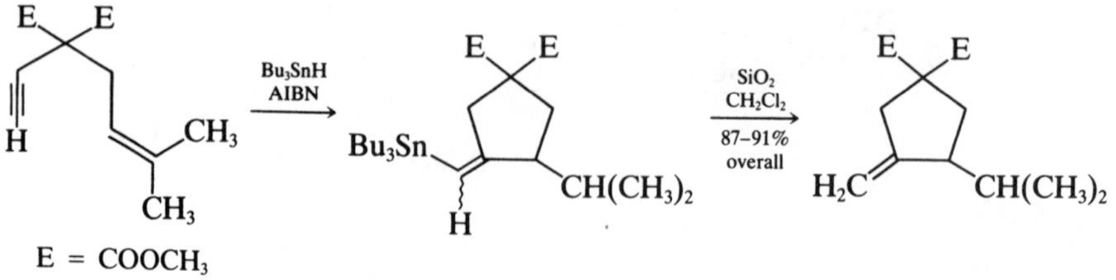

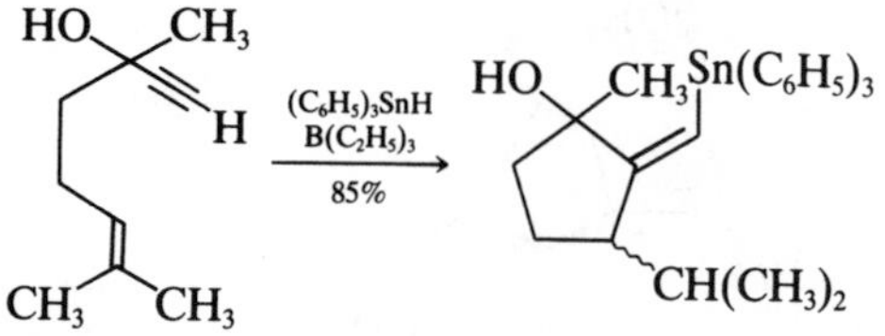

E = COOCH$_3$

trialkylstannyl radical to both double and triple bonds is reversible, but cyclization of vinyl radicals is faster than that of the saturated counterpart obtained by addition to the double bond.

Enynes undergo a similar cyclization to alkylidene thioethers on reaction with thiophenol in the presence of AIBN. This cyclization also involves a vinyl radical (equation I).[8]

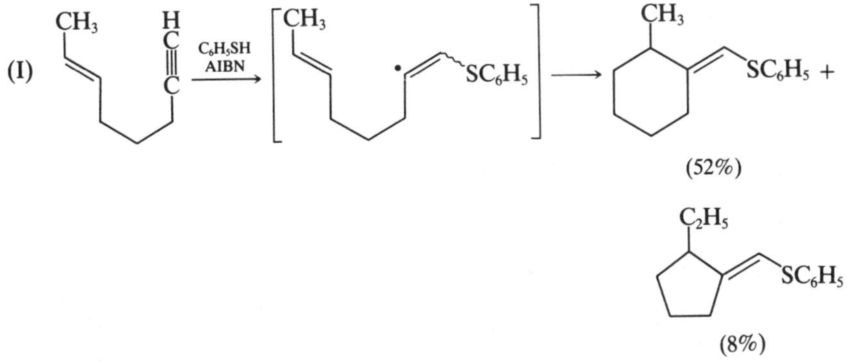

(I)

(52%)

(8%)

Stereocontrolled radical cyclization to furans.[9] Radical cyclization of allyl 2-haloethyl ethers such as **1** show only slight stereoselectivity. However similar cyclization, but involving a dichloromethyl radical, can be highly *cis*-selective, whereas cyclization involving a monochloromethyl radical is *trans*-selective. The

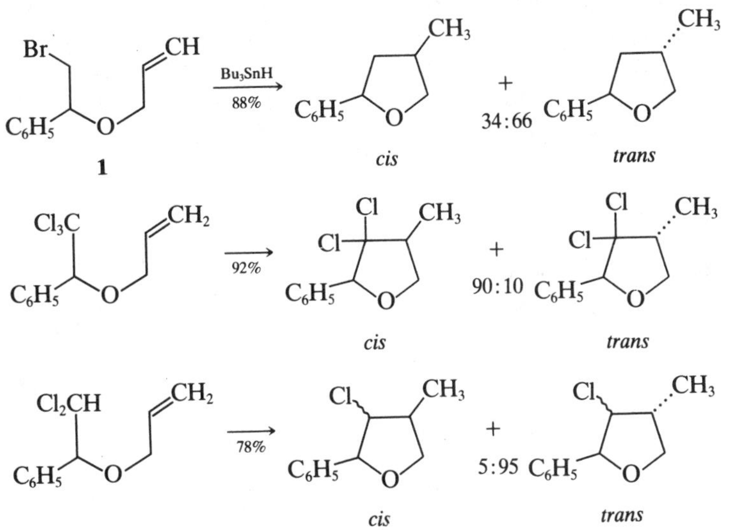

chlorine atoms remaining after cyclization are easily removed by further reaction with Bu_3SnH.

Radical cyclization to macrolides.[10] Cyclization of iodoalkyl acrylates (**1**) by reaction with Bu_3SnH (1 equiv.) in the presence of AIBN is useful for formation of macrolides (**2**) containing 11 or more members. Similar cyclization of iodoalkyl fumarates (**3**) results in two macrolides with the *endo*-product predominating except when n is 16 or higher. Tertiary iodides undergo this free radical cyclization more

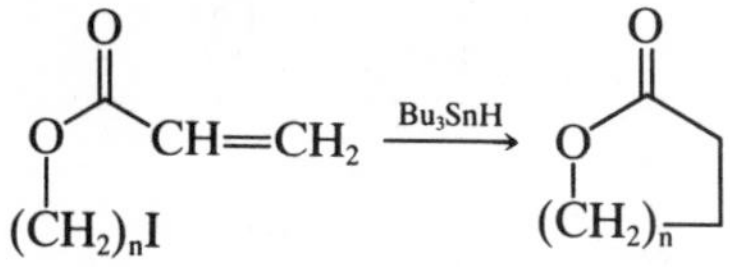

Bu_3SnH

1 **2**

n = 8 27–29%
n = 12 63–67%

50–55%

3 **4** (*endo*) **5** (*exo*)

n = 8 5:1
n = 12 2.5:1

readily than primary iodides, which are more reactive than primary bromides. Dilute conditions are required for these cyclizations, and reduction of the substrates is decreased as the tin hydride concentration is reduced.

Michael-type radical cyclization.[11] A short synthesis of 3-demethoxyerythratidinone (**3**) involves a Michael-type radical addition. Thus **1** on reaction with Bu_3SnH (AIBN) affords **2** as a single isomer in 65% yield. This product is converted by a three-step sequence into **3**.

Bu_3SnH
AIBN
65%
69%
1) CH_3Li
2) C_6H_5SeCl
3) $NaIO_4$

1 **2** **3**

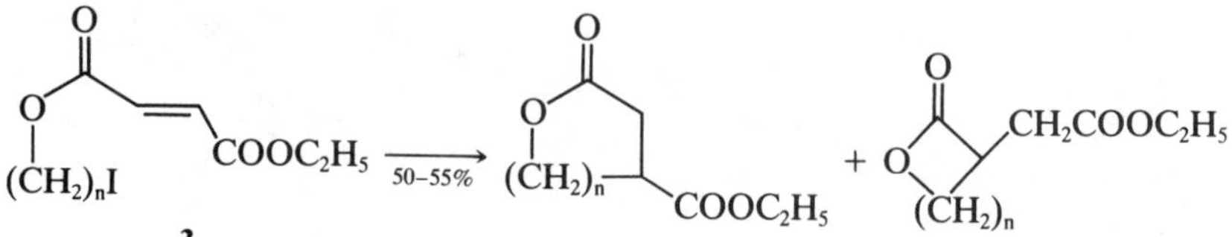

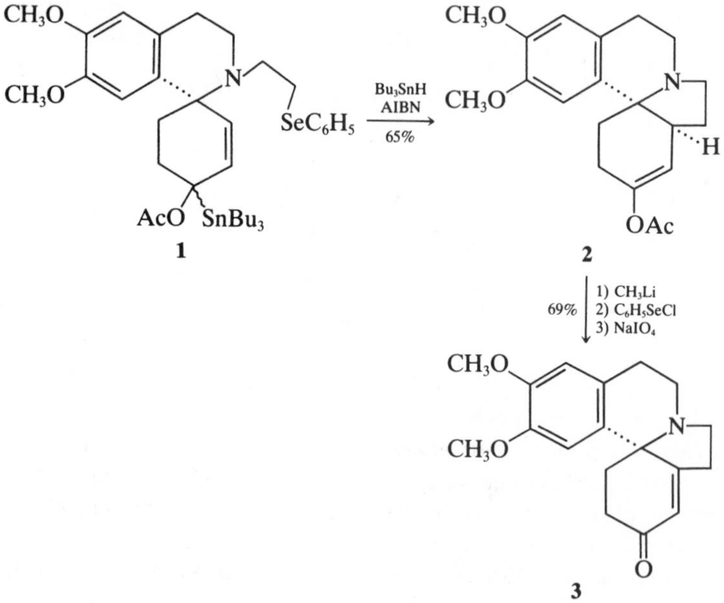

2-Deoxy sugars.[12]   The radical formed on halogen abstraction from acylated glycosyl halides with Bu_3SnH (AIBN) undergoes rearrangement of the acyloxy group from C_2 to C_1 with formation of a 2-deoxy sugar.

Examples:

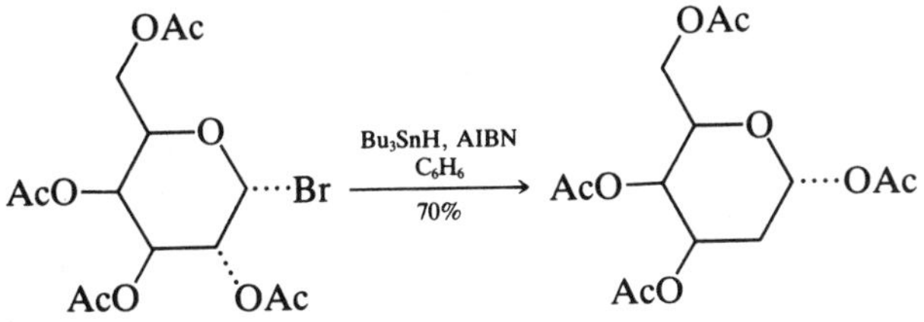

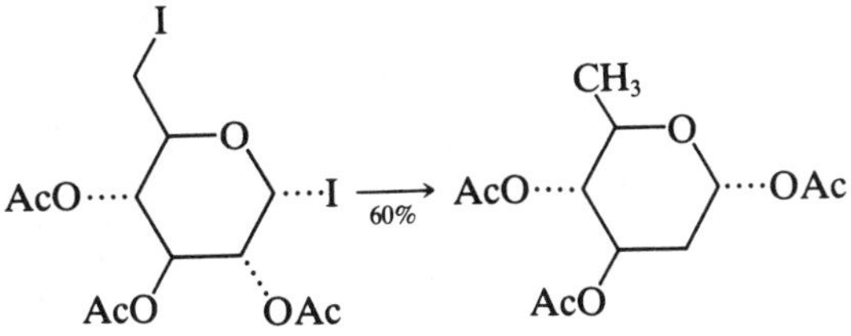

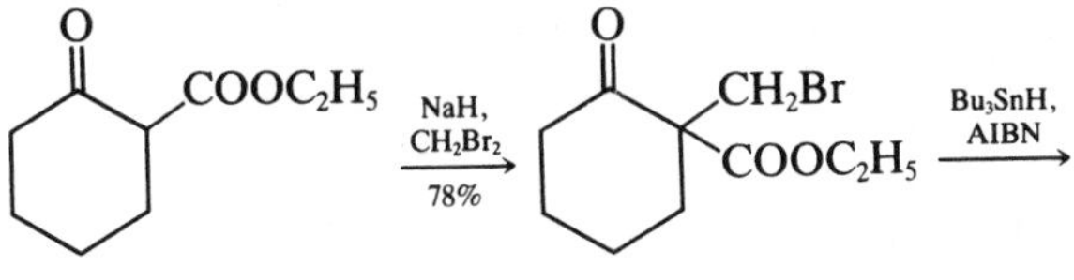

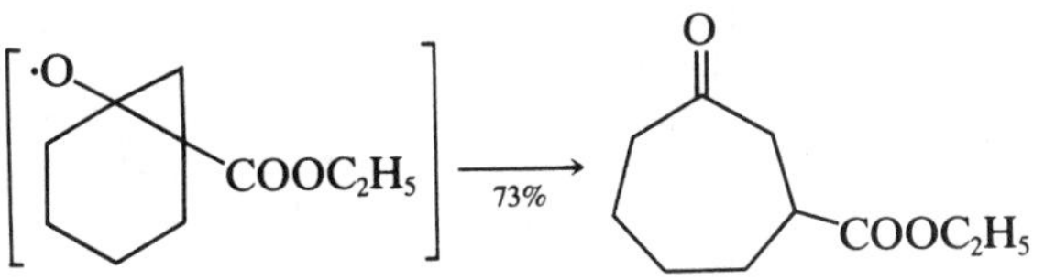

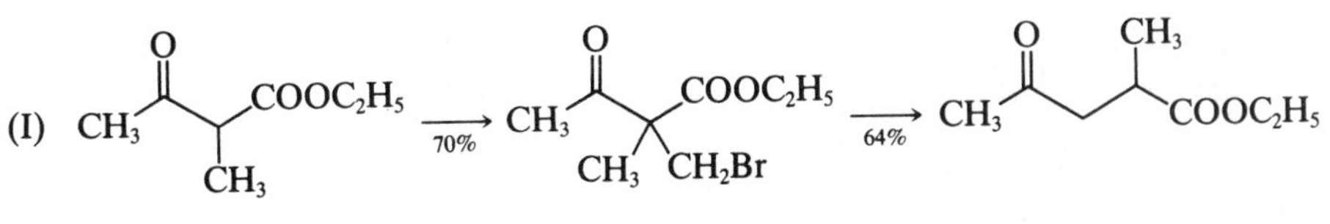

Ring expansion of β-keto esters.   Cyclic β-keto esters (1) can be converted to ring-enlarged γ-keto esters (3) by reaction with methylene bromide to provide the bromomethylated derivative 2. Treatment of 2 with Bu_3SnH (AIBN) generates a primary radical which undergoes ring insertion, possibly via **a**, to provide the γ-keto ester 3. This reaction can be used for homologation of acylic β-keto esters (equation I), but it is particularly useful for ring expansion. Good yields (about 70%) can be obtained in expansion to seven- and eight-membered rings if the rearrangement is carried out under high dilution.[13]

This route cannot be used for a two-carbon ring expansion because of facile direct reduction rather than insertion, but it is useful for three- and four-carbon ring expansion.[14]

[1] W. P. Neumann, *Synthesis*, 665 (1987).
[2] D. P. Curran, *Synthesis*, 417 (1988).
[3] G. Stork and P. M. Sher, *Am. Soc.*, **108**, 303 (1986).
[4] G. Stork, P. M. Sher, and H.-L. Chen, *ibid.*, **108**, 6384 (1986).
[5] G. Stork and M. J. Sofia, *ibid.*, **108**, 6826 (1986).
[6] G. Stork and R. Mook, Jr., *Tetrahedron Letters*, **27**, 4529 (1986); *idem*, *Am. Soc.*, **109**, 2829 (1987); R. Mook, Jr., and P. M. Sher, *Org. Syn.* submitted (1988).
[7] K. Nozaki, K. Oshima, and K. Utimoto, *Am. Soc.*, **109**, 2547 (1987).
[8] C. A. Broka and D. E. C. Reichert, *Tetrahedron Letters*, **28**, 1503 (1987).
[9] Y. Watanabe and T. Endo, *ibid.*, **29**, 321 (1988).
[10] N. A. Porter and V. H.-T. Chang, *Am. Soc.*, **109**, 4976 (1987).
[11] S. J. Danishefsky and J. S. Panek, *ibid.*, **109**, 917 (1987).
[12] B. Giese, K. S. Gröninger, T. Witzel, H.-G. Korth, and R. Sustmann, *Angew. Chem. Int. Ed.*, **26**, 233 (1987); B. Giese and K. S. Gröninger, *Org. Syn.*, submitted (1988).
[13] P. Dowd and S.-C. Choi, *Am. Soc.*, **109**, 3493 (1987); *Org. Syn.*, submitted (1988).
[14] *Idem*, *ibid.*, **109**, 6548 (1987).

Tributyltin hydride–Sodium iodide.

Radical deoxygenation. Tosylates of primary alcohols undergo deoxygenation in 75–100% yield by treatment with NaI and Bu$_3$SnH in DME at 80°. This radical reduction can be applied to tosylates of secondary alcohols, but the rate of reaction is slower and yields are only moderate.[1]

This reaction also converts 1,2-epoxides into secondary alcohols, possibly via an iodohydrin.[2]

Examples:

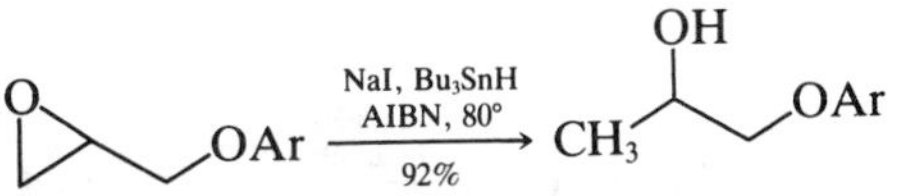

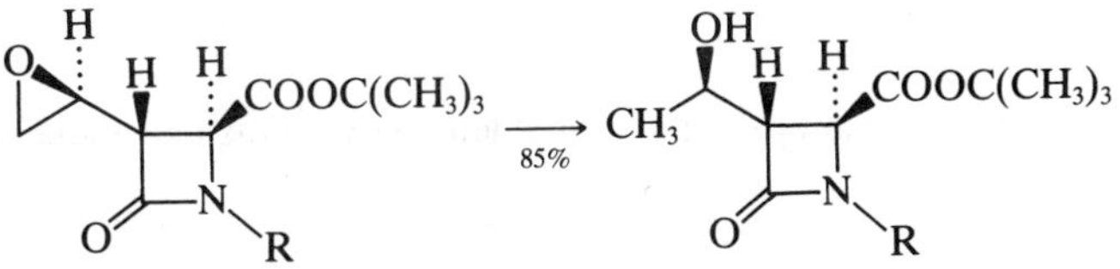

[1] Y. Ueno, C. Tanaka, and M. Okawara, *Chem. Letters*, 795 (1983).
[2] C. Bonini and R. DiFabio, *Tetrahedron Letters*, **29**, 819 (1988).

Tributyltinlithium.

Four-component annelation to alkenolides. Posner *et al.*[1] have reported a one-pot three-step annelation of cycloalkenones to provide, after oxidation, four-atom enlarged macrolides. Thus Michael addition of tributyltinlithium to cyclohexenone (1) and Michael addition of the resulting enolate to ethyl vinyl ketone followed by an aldol reaction results in cyclization to a bicyclic hemiketal (2), which is oxidized by Pb(OAc)₄ to an unsaturated 10-membered lactone (3).

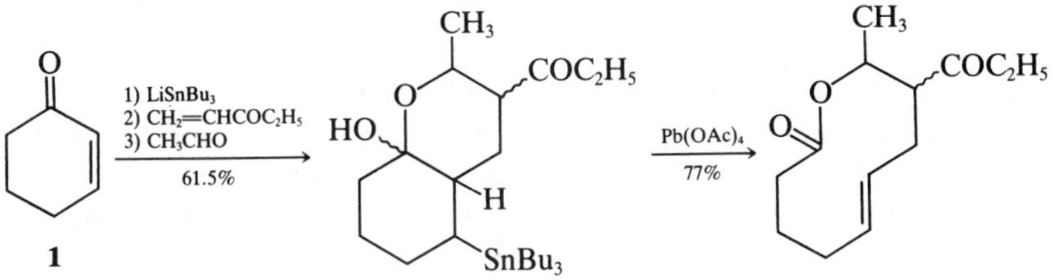

A simpler version of this annelation has been used to obtain phorocantholide (5) via a hemiketal (4).

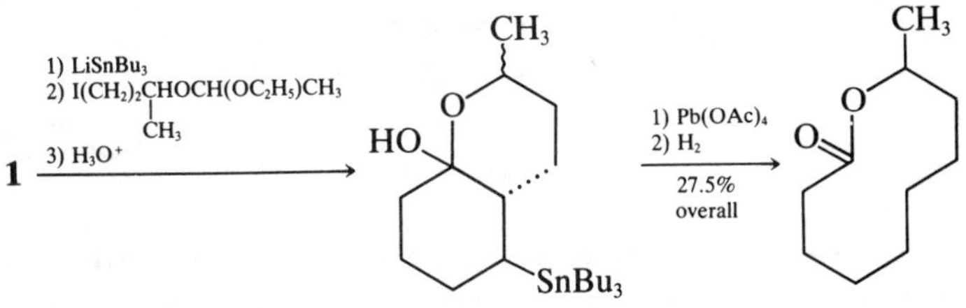

β-*Stannylphosphorous* or β-*silylphosphorous ylides*.[2]   Tributyltinlithium or diphenylmethylsilyllithium react with 1-propenyltriphenylphosphonium bromide to give β-stannyl- or β-silylphosphorous ylides, respectively (equation I).

$$\text{(I)}\quad (C_6H_5)_3\overset{+}{P}\underset{Br^-}{CH}{=}CHCH_3 \xrightarrow[\text{or LiSiR}_3]{\text{LiSnR}_3} (C_6H_5)_3P{=}CH\overset{\overset{\displaystyle CH_3}{|}}{C}HSnR_3(SiR_3)$$

These reagents effect propenylation of aldehydes with high *syn*- and (E)-selectivity.

Example:

$(syn/anti = 15:1,$
$E/Z = 30:1)$

[1] G. H. Posner, E. Asirvatham, K. S. Webb, and S. Jew, *Tetrahedron Letters*, 5071 (1987); G. H. Posner, K. S. Webb, E. Asirvatham, S. Jew, and A. Degl'Innocenti, *Am. Soc.*, **110**, 4754 (1988); K. S. Webb, E. Asirvatham, and G. H. Posner, *Org. Syn.*, submitted (1988).
[2] M. Tsukamoto, H. Iio, and T. Tokoroyama, *Tetrahedron Letters*, **28**, 4561 (1987).

Tricarbonyltris(propionitrile)tungsten, $(C_2H_5CN)_3W(CO)_3$ **(1).**
   Allylic alkylation (*cf.*, **12**, 557).   This W(0) complex in combination with 2,2'-bipyridyl (bpy) catalyzes reactions of nucleophiles with allylic acetates or carbonates, but the chemoselectivity is complementary to that of Pd(0), as shown in equation (I). The W(0)-catalyzed reactions are influenced by inductive and steric

effects, but generally result in cleavage at the more substituted carbon atom. This reaction is useful for monalkylation of allylic diacetates and dicarbonates and for effecting intramolecular cyclization.[1]

[1] B. M. Trost, G. B. Tometzki, and M.-H. Hung, *Am. Soc.*, **109**, 2176 (1987).

Trichloroacetyl chloride Cl_3CCOCl (1).

Acylation of enol ethers.[1]   Reaction of 1 with ethyl vinyl ether in ether provides an intermediate that undergoes dehydrochlorination when heated to provide the trichloromethyl ketone 2, which is converted by base (haloform reaction) to the ester 3 in high yield.

[1] L. F. Tietze, E. Voss, and U. Hartfiel, *Org. Syn.*, submitted (1988); L. F. Tietze, H. Meier, and E. Voss, *Synthesis*, 274 (1988).

Triethyl orthoacrylate, $CH_2{=}CHC(OC_2H_5)_3$ (1), b.p. 162–164°.

The ortho ester is prepared by dehydrobromination of triethyl α-bromoorthopropionate.

Diels–Alder reactions.[1]   In the presence of trimethylsilyl triflate, this orthoester is converted into a 1,1-diethoxyallyl cation, $CH_2{=}CHC^+(OC_2H_5)_2$, which reacts with 1,3-dienes at −78° → 0° to give the corresponding adducts of ethyl acrylate. In the presence of trimethylsilyl triflate, ethyl acrylate can undergo Diels–Alder reactions, but higher temperatures are required and yields are lower.

Example:

[1] P. G. Gassman and S. P. Chavan, *J. Org.*, **53**, 2392 (1988).

Triethylsilane–Trifluoroacetic acid.

Reduction of lactols to ethers.[1]   Five- and six-membered lactols are reduced to cyclic ethers by $(C_2H_5)_3SiH$ and trifluoroacetic acid. The reaction can be highly stereoselective (equation I).

(I)

[1] G. A. Kraus, M. T. Molina, and J. A. Walling, *J.C.S. Chem. Comm.*, 1568 (1986).

Trifluoroacetic acid.

o-Quinodimethanes.[1]   β-Hydroxy trialkyltin compounds are known to undergo proton-induced 1,2-elimination to alkenes (**11**, 424). A related 1,4-elimination can be used to obtain *o*-quinodimethanes. The precursors are obtained by reaction of the dianion of *o*-methylbenzyl alcohols with tributyltin chloride, and the elimination is effected with trifluoroacetic acid at 25° in the presence of electron-deficient alkenes or alkynes as traps.

Examples:

This present method was used successfully for an intramolecular cycloaddition to obtain an octahydrophenanthrene (equation I).

[1] H. Sano, H. Ohtsuka, and T. Migita, *Am. Soc.*, **110**, 2014 (1988).

Trifluoromethanesulfonic acid.

Ionic Diels-Alder reactions (**12**, 531–532).   The allyl cations derived from allyl alcohols or ethers are reactive dienophiles that undergo Diels-Alder reactions at low temperature with high stereoselectivity.[1]

Example:

Similarly, the diethyl acetal of acrolein undergoes facile Diels-Alder reactions in the presence of triflic acid, although yields are only moderate. 2-Vinyl-1,3-dioxolane is recommended as the reagent of choice because of higher yields.[2]

Example:

[1] P. G. Gassman and D. A. Singleton, *J. Org.*, **51**, 3075 (1986).

[2] P. G. Gassman, D. A. Singleton, J. J. Wilwerding, and S. P. Chavan, *Am. Soc.*, **109**, 2182 (1987).

Trifluoromethanesulfonic anhydride.

Cyclobutanones **(11**, 560–561). Ketenimium salts are more reactive than ketenes in [2 + 2] cycloadditions with alkenes to prepare cyclobutanones. The salts are readily available by *in situ* reaction of tertiary amides with triflic anhydride and a base, generally 2,4,6-collidine. The cycloaddition proceeds satisfactorily with alkyl-substituted alkenes and alkynes, but not with enol ethers or enamines.[1]

Example:

Alkenes from **vic-diols**.[2] The 2-dimethylamino-1,3-dioxolanes **1**, obtained by reaction of a 1,2-diol with N,N-dimethylformamide dimethyl acetal in quantitative yield, when treated with ethyldiisopropylamine (4 equiv.) and triflic anhydride (2 equiv.) in dry toluene, decompose to an alkene at room temperature or slightly above. The reaction is stereospecific and is useful for generation of heat-labile alkenes.

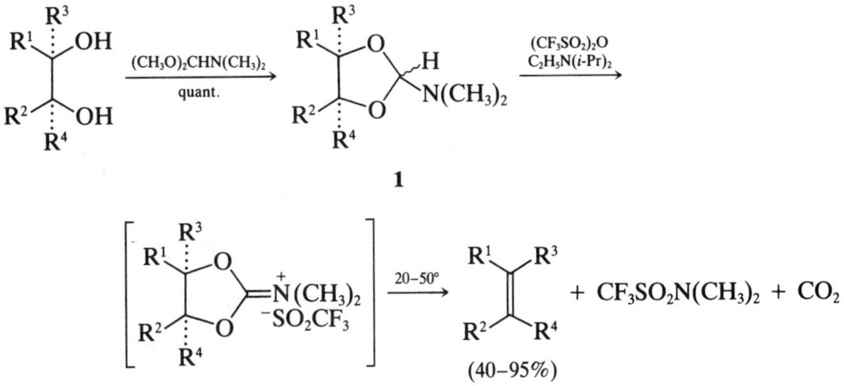

1

a

Cationic cyclization.   A key step in the synthesis of the diterpenes cafestol[3] and atractyligenin[4] involves a novel cation cyclization of bicyclic cyclopropanes to the tetracyclic systems of the diterpenes (equations I and II). Thus treatment of **1** with a slight excess of triflic anhydride and 2,6-lutidine effects cyclization to the rather unstable pentacycle **2** with the kaurene system. The related conversion of **3** to **4** can be effected with triflic anhydride and 2,6-di-*t*-butyl-4-methylpyridine in 1-nitropropane.

(I)

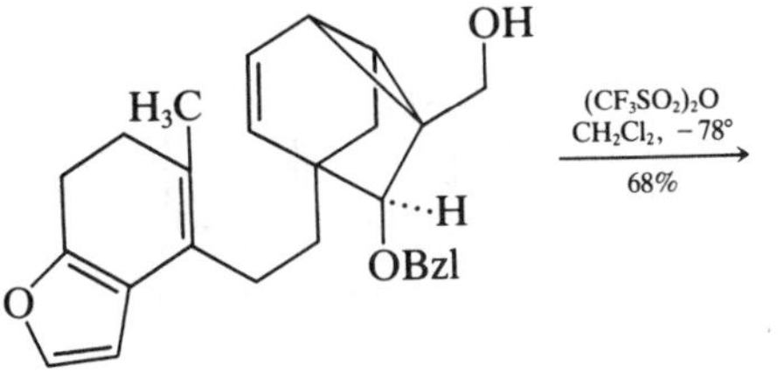

1

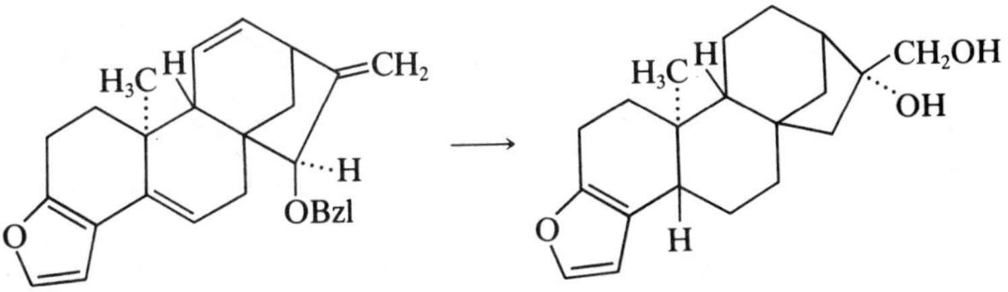

2   Cafestol

(II)

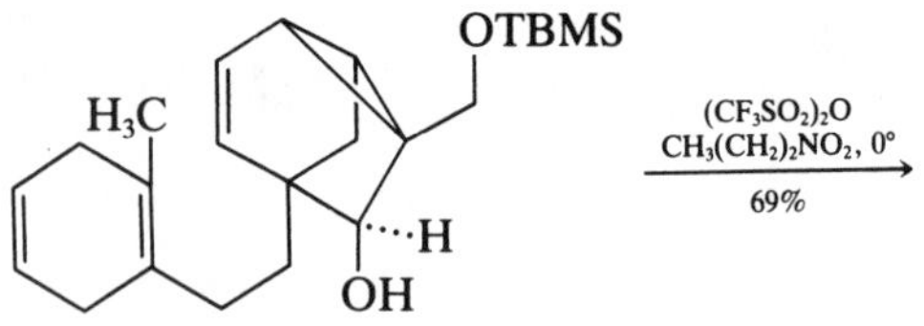

$(CF_3SO_2)_2O$
$CH_3(CH_2)_2NO_2$, 0°
$\longrightarrow$
69%

3

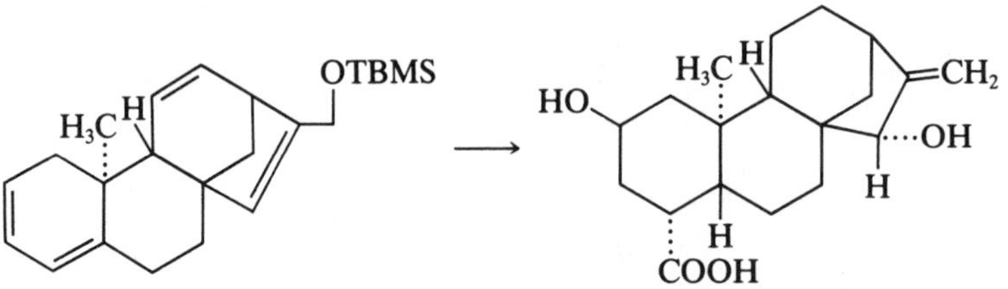

4

Atractyligenin

[1] C. Schmit, J. B. Falmagne, J. Escudero, and H. Vanlierde, and L. Ghosez, *Org. Syn.*, submitted (1987).

[2] J. L. King, B. A. Posner, K. T. Mak, and N. C. Yang, *Tetrahedron Letters*, **28**, 3919 (1987).

[3] E. J. Corey, G. Wess, Y. B. Xiang, and A. K. Singh, *Am. Soc.*, **109**, 4717 (1987).

[4] A. K. Singh, R. K. Bakshi, and E. J. Corey, *ibid.*, **109**, 6187 (1987).

N-(*p*-Trifluoromethylbenzyl)cinchoninium bromide (1), 12, 379–381, 13, 325.

Asymmetric oxygenation of ketones.[1] Oxygenation of the achiral ketone 1 in the presence of this chiral phase-transfer catalyst derived from cinchonine results

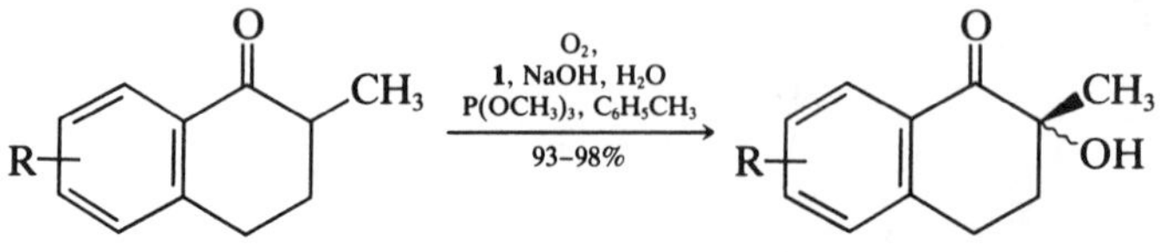

1, R = H, CH$_3$,
OCH$_3$, Cl

2 (70–79% ee)

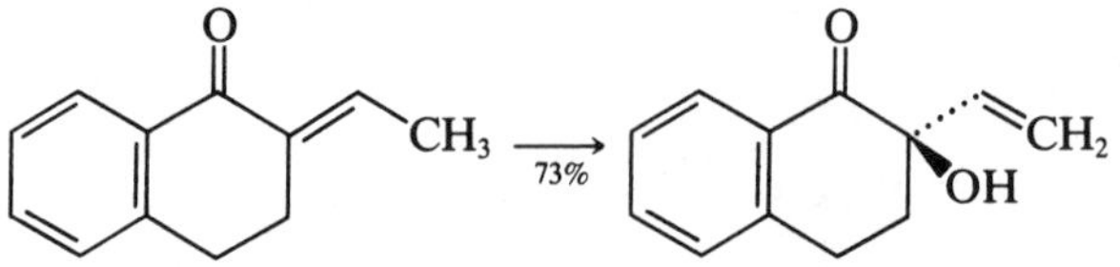

3

4 (R, 55% ee)

in a chiral α-hydroxy ketone in high yield and up to 79% ee. This quaternary ammonium salt is more effective than a number of other chiral phase-transfer catalysts. In géneral, 2-alkyltetralones or -indanones are oxidized in 27–77% ee. The α,β-unsaturated ketone 3 is oxidized to 4 in 55% ee.

[1] M. Masui, A. Ando, and T. Shioiri, *Tetrahedron Letters*, **29**, 2835 (1988).

2,4,6-Triisopropylbenzenesulfonyl azide (trisyl azide), $ArSO_2N_3$ (1), m. p. 41–43°. Preparation from trisyl chloride and NaN_3 (80% yield).[1]

Azidation.[2] Arylsulfonyl azides generally react with enolates to effect net diazo transfer, but this hindered and electron-rich azide can effect azide transfer at the expense of diazo transfer. The nature of the enolate counterion also plays a role, with K being more effective than Na. In addition, acetic acid (or KOAc) is required as the quench for decomposition of the triazine intermediate to the azide with elimination of the arylsulfinic acid, $ArS(O)OH$. By use of these conditions, chiral N-acyloxazolidones such as 2 undergo diastereoselective azidation to give the azides 3 in 75–90% yield and in high optical purity (>91:9). These

carboximides are hydrolyzed by LiOH (2 equiv.) in aqueous THF to (2S)-azido acids (4) with no detectable racemization. However, hydrolysis of 3 [R = $C(CH_3)_3$] is best effected with H_2O_2 (4 equiv.) and LiOH (3 equiv.)[3]

Diazo transfer to these substrates is best effected by reaction of the sodium enolate of 2 with *p*-nitrobenzenesulfonyl azide. A typical diazocarboximide is obtained in 85% yield after a quench with a pH 7 phosphate buffer.

[1] R. E. Harmon, G. Wellman, and S. K. Gupta, *J. Org.*, **38**, 11 (1973).
[2] D. A. Evans and T. C. Britton, *Am. Soc.*, **109**, 6881 (1987).
[3] D. A. Evans, T. C. Britton, and J. A. Ellman, *Tetrahedron Letters*, **28**, 6141 (1987).

2,4,6-Triisopropylbenzenesulfonyl hydrazide.

Vinyllithium cyclizations.[1] The vinyl lithiums formed from trisylhydrazones (9, 486) can participate in intramolecular cyclizations. This anionic cyclization is presently limited to formation of five-membered rings. It has the advantage of greater stereoselectivity than a corresponding radical cyclization.

Examples:

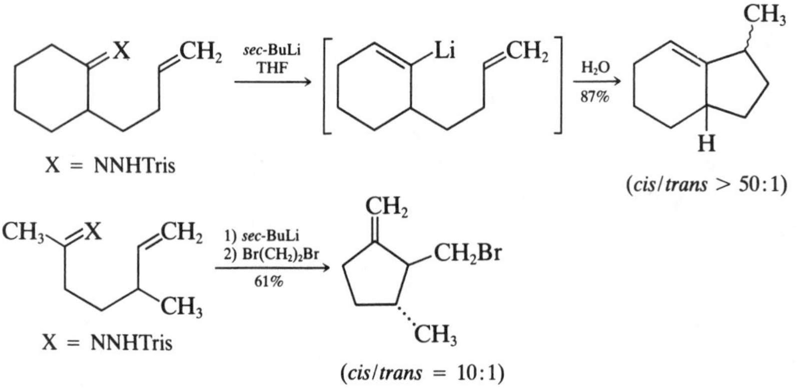

X = NNHTris

(*cis*/*trans* > 50:1)

X = NNHTris

(*cis*/*trans* = 10:1)

[1] A. R. Chamberlin, S. H. Bloom, L. A. Cervini, and C. H. Fotsch, *Am. Soc.*, **110**, 4788 (1988).

Trimethylaluminum–Lithium thiophenoxide.

Michael-aldol cyclization of keto enoates (**10**, 163–164). The oxahydrindene skeleton present in avermectin antibiotics can be generated conveniently by reaction of the keto enoate **1**, prepared from D-ribose, with the ate complex of $Al(CH_3)_3$ and $LiSC_6H_5$. Conjugate addition of the reagent to the enoate group provides an aluminum enolate that undergoes aldol reaction with the ketone group to give **2** in 89% yield.[1] Sulfoxide elimination then provides the oxahydrindene **3**.

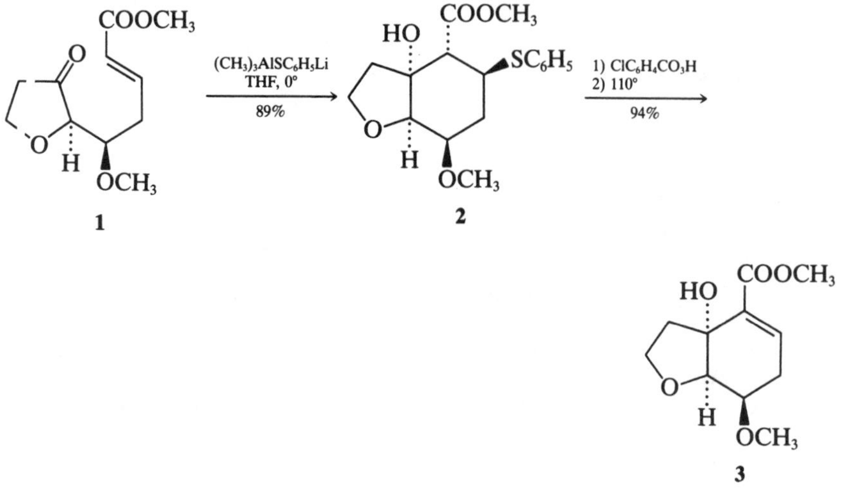

1

2

3

[1] D. M. Armistead and S. J. Danishefsky, *Tetrahedron Letters*, **28**, 4959 (1987).

Trimethylamine oxide.

Anhydrous reagent.[1]   The commercially available material is a dihydrate. Anhydrous reagent can be obtained by distillation of a solution in DMF until the water has been eliminated, followed by slow crystallization of the resultant concentrated solution; m.p. 225–227° dec., 94% yield.

[3 + 2]*Cycloaddition* (**13**, 326).[2] The azomethine ylide (**a**) generated with LDA from trimethylamine oxide adds stereoselectively to the dihydronaphthalenes **1** to provide the benzisoindolines **2**, of use as α-adrenergic agents in therapy.

[1] J. A. Soderquist and C. L. Anderson, *Tetrahedron Letters*, **27**, 3961 (1986).
[2] B. De, J. F. DeBernardis, and R. Prasad, *Syn. Comm.*, **18**, 481 (1988).

Trimethylgermanium chloride.

Aldol condensation with germanium enolates.   The germanium enolate of propiophenone, prepared by reaction with LDA in ether followed by metal exchange with $(CH_3)_3GeCl$, reacts with benzaldehyde to give mainly the *syn*-aldol **2**. However, if the LiCl formed on transmetallation is removed by centrifugation, the salt-free germanium enolate gives mainly the *anti*-aldol **2**. This salt effect is not observed with the germanium enolate prepared with triphenylgermanium chloride.

		syn-2	anti-2
+ BF$_3$·O(C$_2$H$_5$)$_2$	88%	30:70	
+ LiBr	82%	75:25	

[1] Y. Yamamoto and J. Yamada, *J.C.S. Chem. Comm.*, 802 (1988).

Trimethyl orthoformate.

Selective protection of 1,2- or 1,3-diols.[1] Diols react with orthoesters in the presence of an acid catalyst (such as 10-camphorsulfonic acid, CSA) to form an orthoester that can be reduced, without isolation, by DIBAH to a monoacetal

(I)

$$\text{(cyclohexane-1,2-diol)} + (CH_3O)_3CCH_3 \xrightarrow[\text{CH}_2\text{Cl}_2]{\text{CSA,}} \left[\text{orthoester} \right] \xrightarrow[73\%]{\text{DIBAH}}$$

(cyclohexyl structure with OH and OCHOCH$_3$ / CH$_3$)

(equation I). This sequence can provide a method for preferential protection of a secondary over a primary hydroxyl group (equation II).

$$\text{(II)}\quad HOCH_2CHOH \xrightarrow[98\%]{\substack{1)\ (C_2H_5O)_3CH \\ 2)\ DIBAH}} HOCH_2CHOCH_2OC_2H_5$$
$$\qquad\qquad\quad | \qquad\qquad\qquad\qquad\qquad\quad |$$
$$\qquad\qquad\ CH_3 \qquad\qquad\qquad\qquad\qquad CH_3$$

$$(5:1)$$

[1] M. Takasu, Y. Naruse, and H. Yamamoto, *Tetrahedron Letters*, **29**, 1947 (1988).

2-Trimethylsilyloxyfuran, $(CH_3)_3SiO$ (furan structure) (**1**).
Suppliers: Aldrich, Fluka.

Diastereoselective aldol condensations. This furan (**1**) can undergo condensation with aldehydes as a butenolide to form δ-hydroxy-α,β-unsaturated-γ-lactones (**2**). The diastereoselectivity can be controlled by the choice of catalyst. Lewis

$$\text{(I)}\quad \mathbf{1} + RCHO \longrightarrow \text{(anti-2 structure)} + \text{(syn-2 structure)}$$

		anti-**2**	*syn*-**2**
+ TESOTf	85–95%	75–82:25–18	
+ Bu$_4$N⁺F⁻	50–75%	15–35:85–65	

acids, particularly BF_3 etherate or triethylsilyl triflate (TESOTf), favor *anti*-diastereoselectivity; the use of fluoride ion favors *syn*-diastereoselectivity (equation I). These products can be reduced by sodium borohydride in combination with nickel(II) chloride to the corresponding δ-hydroxy-γ-lactones in 90–95% yield.[1]

Butenolides.[2]   When activated by silver trifluoroacetate, this furan is alkylated by primary alkyl iodides or ethyl α-iodoacetate to give 4-alkyl-2-butenolides (2) in 60–80% yield.

[1] C. W. Jefford, D. Jaggi, and J. Boukouvalas, *Tetrahedron Letters*, **28**, 4037 (1987).
[2] C. W. Jefford, A. W. Sledeski, and J. Boukouvalas, *J.C.S. Chem. Comm.*, 364 (1988).

β-Trimethylsilylethyl chloroformate, $(CH_3)_3SiCH_2CH_2OCOCl$ (1). The reagent is prepared by reaction of β-trimethylsilylethanol and phosgene.

N-Debenzylation of t-*amines.*[1]   The reagent converts N-benzyl tertiary amines into the corresponding β-(trimethylsilyl)ethoxycarbonyl derivatives, which are cleaved by fluoride ion (**8**, 470–471) to *sec*-amines.

Example:

The reaction is not useful for N-demethylation because of slow rate and low yield. In fact, selective N-debenzylation is possible with this reagent.

[1] A. L. Campbell, D. R. Pilipauskas, I. K. Khanna, and R. A. Rhodes, *Tetrahedron Letters*, **28**, 2331 (1987).

Trimethylsilylmethanol, $(CH_3)_3SiCH_2OH$ (1). Suppliers: Petrarch Systems, Aldrich.

Hydroxymethylation.[1]   Trimethylsilylmethanol is converted by reaction with butyllithium and CO_2 and then with *sec*-butyllithium into the lithiated lithium carbonate **2**, which converts esters, acid chlorides, and nitriles into hydroxymethyl carbonyl compounds.

$$\textbf{1} \xrightarrow[\text{2) CO}_2]{\text{1) BuLi, THF}} (CH_3)_3SiCH_2OCO_2Li \xrightarrow[-25°]{sec\text{-BuLi,}} (CH_3)_3SiCHOCO_2Li$$
$$\underset{Li}{|}$$
$$\textbf{2}$$

$$CH_3 - \langle \rangle - COOC_2H_5 \xrightarrow{\textbf{2}}$$

$$\left[\begin{array}{c} CH_3SiCHOCO_2Li \\ | \\ \underset{Ar}{\diagup} \overset{\displaystyle C}{\diagdown} O \end{array} \right] \xrightarrow[68\%]{H^+} p\text{-}CH_3C_6H_4\overset{\displaystyle O}{\overset{\|}{C}}CH_2OH$$

[1] A. R. Katritzky and S. Sengupta, *Tetrahedron Letters*, **28**, 1847 (1987).

2-Trimethylsilylmethyl-3-acetoxy-1-propene (**1**), **11**, 578.

Cycloaddition to aldehydes. In the presence of a Pd(0) catalyst, **1** adds to electron-poor alkenes to give methylenecyclopentanes. It also adds to aldehydes to form methylenetetrahydrofurans when tributyltin acetate is used as a cocatalyst.[1]

Example:

[1] B. M. Trost and S. A. King, *Tetrahedron Letters*, **27**, 5971 (1986).

Trimethylsilylmethyl trifluoromethanesulfonate, $(CH_3)_3SiCH_2OTf$ (**1**).

Azomethine ylides.[1] The reaction of **1** with the oxime of an aldehyde results in an iminium salt **2**. Desilylation of **2** (CsF) gives rise to an azomethine ylide (**a**) that undergoes 1,3-dipolar cycloaddition with electron-deficient alkenes (equation I).

The salt **2** reacts with dimethyl acetylenedicarboxylate (DMAD) in the presence of CsF (or NaH) to form an isoxazolidine (**4**), which is formally derived from N-methyl-C-phenylnitrone. This reaction is believed to involve removal of the proton from the OH group of **2** to give **b**, which cycloadds with DMAD to provide **c**, the immediate precursor to **4**.

[1] A. Padwa, W. Dent, and P. E. Yeske, *J. Org.*, **52**, 3944 (1987).

Trimethylsilyl trifluoromethanesulfonate (TMSOTf, **1**).

β-Lactams.[1] Silyl ketene acetals condense with imines in the presence of this catalyst to afford only or mainly *anti*-β-amino esters, which can be cyclized to *trans*-β-lactams.

Example:

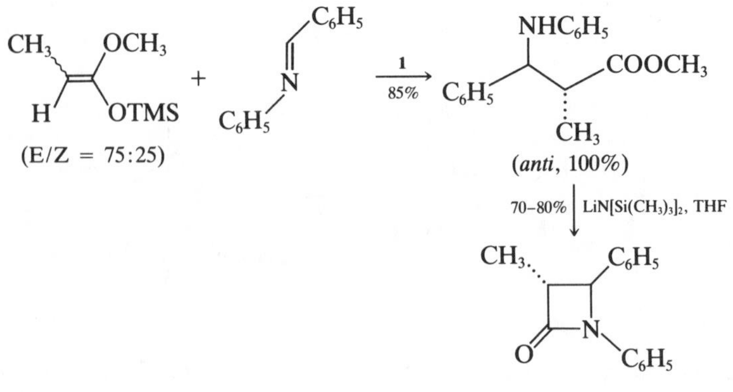

The silyl ketene acetal **2**, prepared from ethyl 3-hydroxybutanoate, undergoes a similar condensation with benzylideneaniline but with *syn*-selectivity. The product cyclizes to a *cis*-β-lactam (**4**).

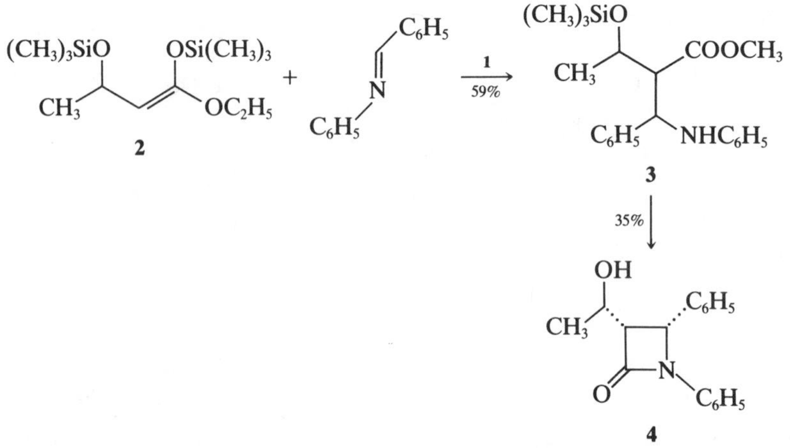

1,3-Dioxolanation.[2]   Use of 1,2-bis(trimethylsilyloxy)ethane catalyzed by $(CH_3)_3SiOTf$ for 1,3-dioxolanation was first reported by Noyori (**10**, 439). The method has been shown by Hwu to be useful for selective 1,3-dioxolanation of carbonyl groups in which one such group is less sterically hindered than another and for selective protection of a keto group in the presence of an enal group. This method is also useful for dioxolanation of acid-sensitive α,β-enals. Yields are typically 75–95%.

[1] G. Guanti, E. Narisano, and L. Banfi, *Tetrahedron Letters*, **28**, 4331, 4335 (1987).

[2] J. R. Hwu and J. M. Wetzel, *J. Org.*, **50**, 3946 (1985); J. R. Hwu, L.-C. Leu, J. A. Robl, D. A. Anderson, and J. M. Wetzel, *ibid.*, **52**, 188 (1987).

Trimethyltinlithium, $(CH_3)_3SnLi$ (**1**). This reagent can be obtained by reaction of $(CH_3)_3SnCl$ in THF with lithium at $0°$ in an ultrasonic vessel.

β-*Stannyl ketones.*[1]   The reagent undergoes conjugate addition to α,β-enals or -enones to give β-stannyl aldehydes or ketones in 50–90% yield. When treated with $TiCl_4$, these adducts undergo an intramolecular reaction leading to cyclopropanols, which can be isolated in fair to high yield in certain cases (equation I). The adducts of cyclohexenones also undergo this intramolecular cyclization, but the

(I)

cyclopropanols (**a**) are unstable and undergo cleavage either to cyclohexanones or to substituted cyclopentanones (equations II and III). The mode of cleavage of the intermediate is determined by the substitution pattern of the enone.

(II)

(39%)   (6%)

(III)

52%

Reaction with unsaturated selenides.[2]   Organotin compounds are useful in synthesis because they undergo Sn/Li exchange and also various condensations with electrophiles. A new route to organotrimethyltin compounds involves reaction of $(CH_3)_3SnLi$ (**1**) with various selenides.

Examples:

$$C_6H_5SeCH_2CH=CH_2 \xrightarrow[90\%]{\substack{1)\ LDA \\ 2)\ C_6H_5CH_2Br}}$$

$$\underset{\underset{CH_2C_6H_5}{|}}{C_6H_5Se} \diagup\!\!\!\diagdown^{CH_2} \xrightarrow[67\%]{\substack{1.\ THF \\ -78°}}$$

$$C_6H_5CH_2CH=CHCH_2Sn(CH_3)_3$$

$$E/Z\ =\ 14:86$$

$$C_6H_5SeCH_2C_6H_5 \xrightarrow[93\%]{\substack{1)\ LDA \\ 2)\ (E)\text{-}BrCH_2CH=CHC_6H_5}}$$

$$\underset{C_6H_5}{\overset{C_6H_5Se}{\diagdown\!\!\diagup}}CHCH_2CH\overset{(E)}{=\!=}CHC_6H_5$$

$$\Big\downarrow{\scriptstyle 90\%}\ \mathbf{1}$$

$$\underset{C_6H_5}{\overset{(CH_3)_3Sn}{\diagdown\!\!\diagup}}CHCH_2CH\overset{(E)}{=\!=}CHC_6H_5$$

This displacement of an allylic C_6H_5Se group by a $(CH_3)_3Sn$ group is observed also with allenyl, propargyl, and benzyl selenides, and in all cases results in the less substituted of the two possible trimethylstannyl compounds.

[1] T. Sato, M. Watanabe, T. Watanabe, Y. Onoda, and E. Murayama, *J. Org.*, **53**, 1894 (1988).
[2] H. J. Reich and J. W. Ringer, *ibid.*, **53**, 455 (1988).

Triphenylphosphine–Diethyl azodicarboxylate.

Mitsunobu glycosidation.[1]   A total synthesis of the antitumor agent phyllanthoside (**1**) requires coupling of a derivative (**2**) of the disaccharide (α/β = 2:1) with a precursor (**3**) to the aglycone phyllanthocin. Simple condensation of **2** and the acid chloride of **3** results mainly in the undesired α-glycoside (*ca.*, 5:1, 92% yield). However, Mitsonobu coupling of **2** and **3** (2 equiv.) with $P(C_6H_5)_3$ and diisopropyl azodicarboxylate (1.5 equiv. each) in THF gives a separable mixture of the β- and α-glycosides (**4**) in the ratio 2:1 in 55% yield. The carbonyl group of β-**4** is then reduced ($NaBH_4$) and the resulting axial hydroxyl group acylated with cinnamoyl chloride. Finally, deprotection of the triethylsilyl groups (HOAc) gives phyllanthoside (**1**).

2   **3**

β-4

Phyllanthoside (1)

[1] A. B. Smith, III, and R. A. Rivero, *Am. Soc.*, **109**, 1272 (1987).

Triphenylphosphine oxide–Trifluoromethanesulfonic anhydride.

Dehydration (cf., 6, 648).[1]   A reagent (1), prepared *in situ* from $(C_6H_5)_3PO$ and Tf_2O in the molar ratio 2:1, effects dehydration, usually at 25°, of amides or oximes to nitriles in >90% yield. It also effects condensation of acids and amines to form amides. The reaction of an aryl carboxylic acid with an *o*-phenylenediamine promoted by **1** provides 2-arylbenzimidazoles in >80% yield (equation I). If the

by-product of these reactions, triphenylphosphine oxide, interferes with isolation of the product, dehydration can be effected with a related reagent prepared from $(C_6H_5)_2PON(CH_2CH_2)_2NCH_3$ and triflic anhydride, which furnishes a water-soluble phosphinamide triflate as the by-product.

[1] J. B. Hendrickson and M. S. Hussoin, *J. Org.*, **52**, 4137 (1987).

6-(Triphenylphosphoranylidene)-2,4-hexadienoic acid, methyl ester (1).

Polyene synthesis.[1]   The Vedejs trienic phosphorane **1** (**6**, 14–15) is useful for synthesis of bifunctional all-*trans* polyenes, since the mixture of (E)- and (Z)-

isomers formed in the Wittig reaction can be isomerized by iodine. Reactions are also cleaner when a radical scavenger is present. The reagent is well adapted to reiterative reactions.

Example:

[1] S. Hanessian and M. Botta, *Tetrahedron Letters*, **28**, 1151 (1987).

Tris(4-bromophenyl)aminium hexachlorostibnate, $(p\text{-BrC}_6\text{H}_4)_3\overset{+}{\text{N}}\text{SbCl}_6^-$ (**1**).

Diels–Alder catalysis.[1] This radical cation can increase the *endo*-selectivity of Diels–Alder reactions when the dienophile is a styrene or electron-rich alkene. This *endo*-selectivity obtains even in intramolecular Diels–Alder reactions. Thus the triene **2**, a mixture of (Z)- and (E)-isomers, cyclizes in the presence of **1** to 0° to the hydroindanes **3** and **4** in the ratio 97:3. Similar cyclization of (E)-**2** results in **3** and **4** in the ratio 98:2; therefore, the catalyst can effect isomerization of (Z)-**2** to (E)-**2**. Even higher stereoselectivity is observed when the styrene group of **2** is replaced by a vinyl sulfide group (SC$_6$H$_5$ in place of C$_6$H$_4$OCH$_3$).

[1] B. Harirchian and N. L. Bauld, *Tetrahedron Letters*, **28**, 927 (1987).

Tris(dibenzylidenacetone)dipalladium·chloroform, Pd$_2$(dba)$_3$·CHCl$_3$ (1).

Reductive cleavage of enedicarbonates.[1] The dicarbonates of enediols in the presence of this Pd(0) complex and triisopropyl phosphite undergo reductive elimination to afford 1,3-dienes.

$$R(CH_2)_7 \overset{\text{Pd(0), P(OR)}_3}{\underset{\text{THF, 25°}}{\longrightarrow}}$$

$$R(CH_2)_7 \diagup\diagdown CH_2 + O{=}P(OR)_3$$
$$(60-79\%)$$

This elimination fails with diacetates of enediols and involves an overall *cis*-elimination in cyclic systems.

Terminal allenes.[2] A synthesis of 1,2-dienes (**3**) from an aldehyde or a ketone involves addition of ethynylmagnesium bromide followed by reaction of the adduct with methyl chloroformate. The product, a 3-methoxycarbonyloxy-1-alkyne (**2**), can be reduced to an allene by transfer hydrogenolysis with ammonium formate catalyzed by a zero-valent palladium complex of **1** and a trialkylphosphine. The choice of solvent is also important. Best results are obtained with THF at 20–30° or with DMF at 70°.

$$\underset{R^2}{\overset{R^1}{\diagdown}}C{=}O + HC{\equiv}CMgBr + ClCOOCH_3 \overset{\text{THF}}{\underset{50-95\%}{\longrightarrow}} \underset{R^2}{\overset{R^1}{\diagdown}}\overset{OCOOCH_3}{\underset{H}{}}$$

2

$$\mathbf{2} \overset{\text{1, Bu}_3\text{P}}{\underset{\underset{65-95\%}{\text{HCOONH}_4, \text{ DMF}}}{\longrightarrow}} \underset{R^2}{\overset{R^1}{\diagdown}}C{=}C{=}CH_2$$

3

[1] B. M. Trost and G. B. Tometzki, *J. Org.*, **53**, 915 (1988).
[2] J. Tsuji, T. Sugiura, and I. Minami, *Synthesis*, 603 (1987).

Tris(diethylamino)sulfonium   difluorotrimethylsilicate, $[(C_2H_5)_2N]_3S(CH_3)_3\bar{S}iF_2$
(TASF, 1).

ArX ⟶ ArCH₃.   In the presence of allylpalladium chloride dimer, $(C_3H_5PdCl)_2$, TASF converts aryl halides, particularly the iodides, into methylarenes in moderate to good yield. The actual reagent is believed to be an ArPd(II)-CH₃ species.[1]

Example:

$$CH_3OCO-\langle\!\!\!\bigcirc\!\!\!\rangle-I + (CH_3)_3SiF_2^- \xrightarrow[84\%]{Pd(II)} CH_3OCOC_6H_4CH_3\text{-}p$$

Oxygenation of a silane.   In the course of an anthracyclinone synthesis, Vedejs and Pribish[2] found that a benzylic trimethylsilane can be converted into a hydroxyl group by reaction with TASF in the presence of oxygen and trimethyl phosphite (**2 → 3**, 95% yield). The conversion is considered to involve formation of a benzylic

anion, which is oxygenated to a hydroperoxide anion, and eventually reduced by the phosphite. No other source of fluoride ion can replace TASF. This oxygenation can extend to some simple enol silanes, but most enol silanes are simply cleaved to the parent carbonyl compound. Even $C_6H_5CH_2Si(CH_3)_3$ can be converted into $C_6H_5CH_2OH$, but in only 20% yield.

Vinylsilanes.[3]   TASF is superior to Bu₄NF for cleavage of $(CH_3)_3Si—Si(CH_3)_3$ to a silyl anion species that reacts with vinyl halides in the presence of a Pd(0) catalyst to form vinylsilanes. The reaction proceeds at 25° in THF/HMPT with retention of the geometry of the double bond.

$$C_6H_5\diagdown\diagup\diagdown I \ + \ (CH_3)_3Si—Si(CH_3)_3 \ \xrightarrow[82\%]{\substack{TASF, \\ Pd(0)}} \ C_6H_5\diagdown\diagup\diagdown Si(CH_3)_3$$

[1] Y. Hatanaka and T. Hiyama, *Tetrahedron Letters*, **29**, 97 (1988).
[2] E. V. Vedejs and J. R. Pribish, *J. Org.*, **53**, 1593 (1988).
[3] Y. Hatanaka and T. Hiyama, *Tetrahedron Letters*, **28**, 4715 (1987).

Tris[3-(heptafluoropropylhydroxymethylene)-*d*-camphorato]europium(III), (+)-Eu(hfc)₃, **12**, 561.

L-Glucose.[1] Further study of this chiral catalyst (**1**) indicates that the (+)-isomer (from D-camphor) consistently favors formation of the L-dihydropyrone from condensation of aldehydes with activated dienes, even though the enantioselectivity is only modest. The effect of a chiral alkoxy group was then examined. The best result was obtained using *l*-8-phenmenthol as the chiral auxiliary. Thus the diene **2** reacts with benzaldehyde in the presence of (+)-Eu(hfc)₃ to give the L-dihydropyrone **3** in 75% yield after acid hydrolysis. The product was converted in several steps to peracetyl-L-glucose (**4**). The high enantioselectivity in this case

is unexpected since the diene component is D-selective, but may result from a "specific interactivity" between the catalyst and the chiral auxiliary.

[1] M. Bednarski and S. Danishefsky, *Am. Soc.*, **108**, 7060 (1986).

Tris[2-(2-methoxyethoxy)ethyl]amine (1, TDA-1), 13, 336–337.

Catalyzed Wittig reactions. Wittig reactions of cyclopropylidenetriphenylphosphorane with carbonyl compounds proceed in low yield under standard conditions. Reactions conducted at 62° in THF with **1** (10 mole %) as the phase-transfer catalyst result in alkylidenecyclopropanes in 60–95% yield.[1]

[1] J. A. Stafford and J. E. McMurry, *Tetrahedron Letters*, **29**, 2531 (1988).

Tris(phenylthio)methyllithium, $(C_6H_5S)_3CLi$ (**1**). Tris(phenylthio)methane is prepared from thiophenol and trimethyl orthoformate.

Cyclopropanation of enones. The adduct (**a**) of **1** to cyclohexenone can be metallated at $-45°$ by *sec*-BuLi to give a dianion **b**, which when quenched at $0°$

gives **2** by a 1,2-hydride transfer. When **b** is quenched at $-45°$, the cyclopropyl ketone **3** is formed. Cohen *et al.*[1] present evidence that the conversion of **b** to **3** involves a lithium bicyclo[1.1.0]butanolate, which can react with electrophiles at either carbon or oxygen.

vic-Disubstitution.[2] Chiral butenolides can serve as useful precursors to carboxylic acids with vicinal, stereochemically controlled secondary and tertiary centers. Thus tris(methylthio)methyllithium (**7**, 412) reacts with the chiral butenolide

2, obtainable from L-glutamic acid, to give only one adduct (**3**). The trimethylthio group effectively controls the direction of alkylation or hydroxylation (MoOPH, **11**, 382) of the enolate of **3**, but can be converted by desulfuration with Raney nickel into a methyl group. A variety of substitution patterns are available by sequential reactions (Scheme I).

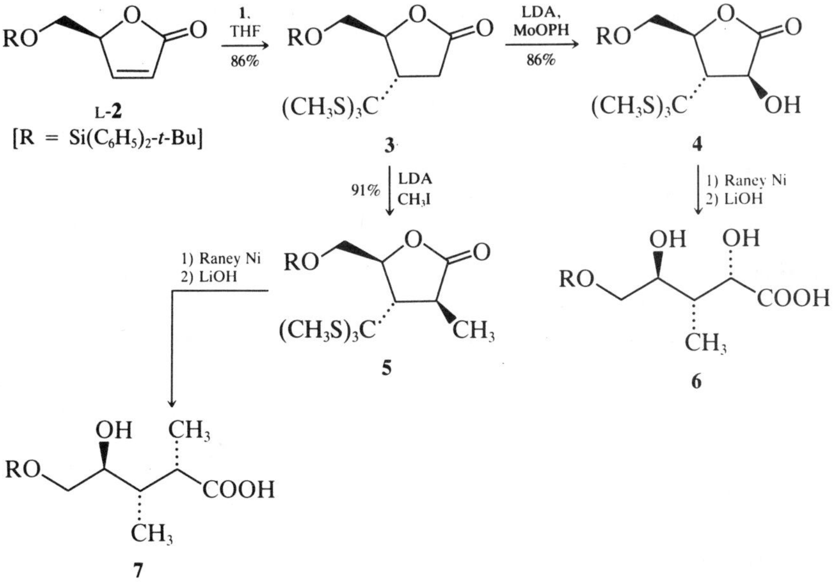

Scheme I

[1] K. Ramig, M. Bhupathy, and T. Cohen, *Am. Soc.*, **110**, 2678 (1988).
[2] S. Hanessian and P. J. Murray, *J. Org.*, **52**, 1170 (1987).

(S)-γ-Trityloxymethyl-γ-butyrolactam, 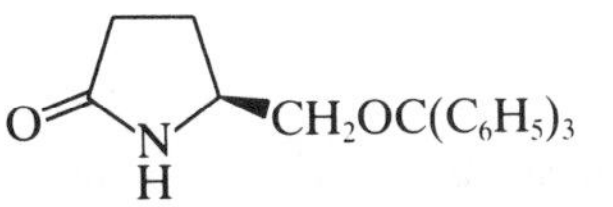 (**1**), m.p. 166°, α_D + 13.7°. The lactam is prepared in a few steps from L-glutamic acid.

Asymmetric conjugate additions to imides.   α,β-Unsaturated imides (**2**) derived from **1** undergo highly diastereoselective Michael reactions with Grignard reagents when catalyzed by CuBr·S(CH_3)_2[1] or with the silyl cuprate $(C_6H_5Me_2Si)_2$-CuLi_2CN[2] to provide optically active β,β-disubstituted carboxylic acids (**4**) after hydrolysis. The bulky trityloxymethyl group effectively shields one face of the double bond from nucleophilic attack.

2

3 (81–91% de)   **4**

[1] K. Tomioka, T. Suenaga, and K. Koga, *Tetrahedron Letters*, **27**, 369 (1986).
[2] I. Fleming and N. D. Kindon, *J.C.S. Chem. Comm.*, 1177 (1987).

Trityl perchlorate, $TrClO_4$.

Michael additions to quinones. In the presence of $TrClO_4$, enol silyl ethers undergo 1,4-addition to benzoquinone to give adducts that cyclize to benzofurans.[1] A similar reaction with diimidoquinones produces indole derivatives.

Example:

γ-Acyl-δ-lactones. These lactones can be obtained with high stereoselectivity by a tandem Michael–aldol reaction between α,β-enones, ketene silyl acetals, and aldehydes catalyzed by trityl perchlorate. A single product is formed from reactions of a mono-β-substituted enone.[2]

Example:

RCHO $\longrightarrow$ RCOOH.[3]   In the presence of $TrClO_4$, *t*-butyl trimethylsilyl peroxide reacts with aldehydes to form acetal-type peroxides, which are converted to
carboxylic acids when heated with piperidine in water.
Example:

$$RCHO + (CH_3)_3COOSi(CH_3)_3 \xrightarrow[70-95\%]{\substack{TrClO_4, \\ CH_2Cl_2}} RCH[OOC(CH_3)_3]_2$$

$$50-90\% \Big\downarrow \substack{Base, H_2O, \\ 90°}$$

$$RCOOH$$

β-*Alkoxy ketones*.[4]   These ketones can be prepared by an aldol-type reaction
of enol ethers with acetals catalyzed by a trityl salt. Methoxymethyl (MOM) enol
ethers are more reactive than methyl enol ethers.

(*syn/anti* = 75:25)

[1] T. Mukaiyama, Y. Sagawa, and S. Kobayashi, *Chem. Letters*, 2169 (1987).
[2] S. Kobayashi and T. Mukaiyama, *ibid.*, 1805 (1986).
[3] T. Mukaiyama, N. Miyoshi, J. Kato, and M. Ohshima, *ibid.*, 1385 (1986).
[4] M. Murakami, H. Minamikawa, and T. Mukaiyama, *ibid.*, 1051 (1987).

Tungsten carbonyl, $W(CO)_6$.
   ***Desulfuration–dimerization of dithioketals*.**[1]   Dithioketals when refluxed with
$W(CO)_6$ in chlorobenzene are converted into the corresponding dimeric alkenes.

$$(C_6H_5)_2C=C(SC_6H_5) \xrightarrow[75\%]{} (C_6H_5)_2C=C=C=C(C_6H_5)_2$$

(E/Z = 1:1)

[1] L. L. Yeung, Y. C. Yip, and T.-Y. Luh, *J.C.S. Chem. Comm.*, 981 (1987).

Tungsten(VI) chloride–Butyllithium.

Deoxygenation of epoxides (**4**, 569–570).   Only a low-valent tungsten reagent of Umbreit and Sharpless[1] effects deoxygenation of the mycotoxin anguidine (**1**) to the alkene. The most effective reagent is prepared from WCl_6 and BuLi in the ratio 1:3. When used in excess (4 equiv. of WCl_6) in THF at 50°, it can effect the desired deoxygenation in 97% yield.[2]

1

[1] M. A. Umbreit and K. B. Sharpless, *Org. Synth.*, **60**, 29 (1982).
[2] W. R. Roush and S. Russo–Rodriguez, *J. Org.*, **52**, 598 (1987).

V

(S)-Valine *t*-butyl ester.

Michael reaction of enamines of α-alkyl-β-keto esters. The chiral lithioen-amine (**1**), prepared from (S)-valine *t*-butyl ester, does not react with methyl vinyl ketone or ethyl acrylate unless these Michael acceptors are activated by ClSi(CH₃)₃

or a Lewis acid. BF₃ etherate is the most effective Lewis acid, but the highest enantioselectivity (87% ee) is obtained by reactions of **1** with methyl vinyl ketone and ClSi(CH₃)₃ (5 equiv.) in THF at − 100°. The enantioselectivity can be reversed by carrying out the reaction in toluene and HMPT (1 equiv.), which serves as a ligand to the lithium cation. The role of the chlorosilane is not known, but probably does not involve a trimethylsilyl azaenolate.

¹ K. Tomioka, W. Seo, K. Ando, and K. Koga, *Tetrahedron Letters*, **28**, 6637 (1987).

Y

Ytterbium.

Pinacol coupling. Yb(0) metal can effect pinacol reduction of aromatic, but not aliphatic ketones.[1] However, it can effect cross-coupling of benzophenone with aliphatic ketones, often in high yield (equation I).

$$(I) \quad (C_6H_5)_2C{=}O \xrightarrow[\text{2) } n\text{-}C_6H_{13}COCH_3]{\text{1) Yb}} \underset{\underset{\underset{CH_3}{|}}{(90\%)}}{(C_6H_5)_2\overset{\overset{\displaystyle HO}{|}}{C}-\overset{\overset{\displaystyle OH}{|}}{C}-C_6H_{13}\text{-}n} \;+\; \underset{(8\%)}{(C_6H_5)_2CHOH}$$

[1] Z. Hou, K. Takamine, Y. Fujiwara, and H. Taniguchi, *Chem. Letters*, 2061 (1987).

Z

Zinc.

Activation. Erdik[1] has reviewed the methods used since 1970 for activation of zinc and of organozinc reagents. Although chemical activation is still useful, ultrasound activation is being used increasingly. Thus sonic activation allows use of ordinary zinc for cyclopropanation of alkenes with CH_2I_2 in 67–97% yield and for Reformatsky-type reactions at room temperatures.

Dechlorination of 4,4-dichlorocyclobutenones.[2] These products of [2 + 2] cycloaddition of dichloroketene with alkynes (**9**, 153) can be reduced satisfactorily and without isomerization by zinc dust in ethanol containing 5 equiv. each of acetic acid and a tertiary amine (preferably TMEDA).

Example:

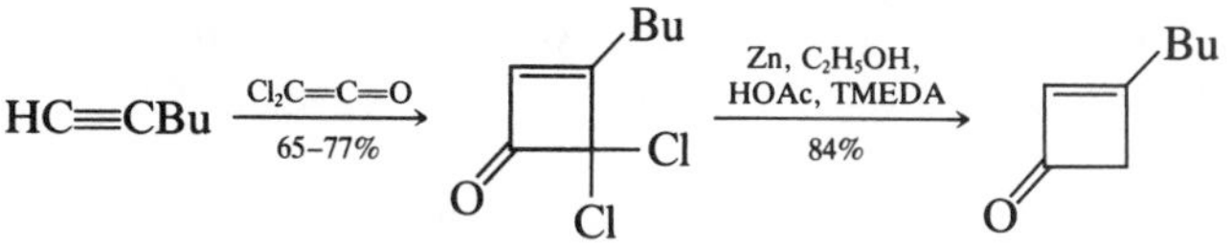

The same conditions effect dechlorination of 4-alkyl-4-chlorocyclobutenones, ' but with considerable isomerization to the more stable cyclobutenone.

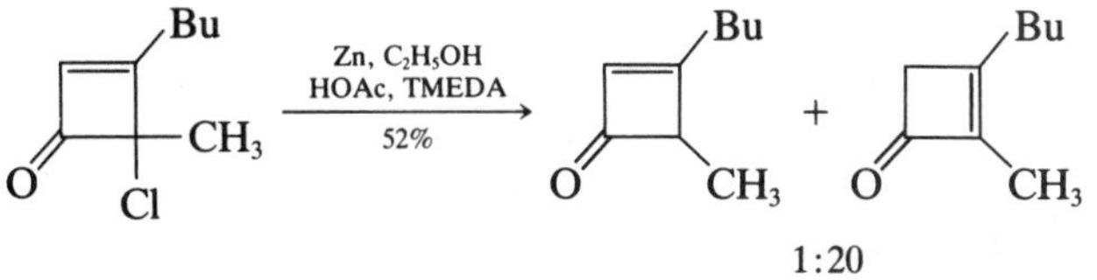

1:20

Stereoselective Reformatsky reaction. The Reformatsky reaction of the chiral 2-azetidinone **1** with 3-(2-bromopropionyl)-2-oxazolidone (**2a**) gives essentially a 1:1 mixture of the diastereomers **3aβ** and **3aα**. However, introduction of two methyl groups at C_4 in **2** markedly improves the β-diastereoselectivity, as does an increase in the temperature from 0 to 67° (reflux, THF). The highest diastereoselectivity (95:5) is observed with the derivative of 4,4-dibutyl-5,5-pentamethylene-2-oxazolidone. The 3β-diastereomer is a useful intermediate to 1β-methylcarbapenems.[3]

The Reformatsky reagent from **2a** shows only slight diastereoselectivity in reactions with aldehydes but reactions with **2b** give 2,3-*syn*-aldols as the major prod-

2a, R = H	82%	
2b, R = CH₃	94%	

$$3\,\beta \qquad 48{:}52$$
$$79{:}21$$

ucts (equation I). This reaction is probably not a practical route to optically active aldols, since a 1:1 mixture of optically active *syn-* and *anti-*aldols is obtained when the oxazolidone group is substituted by an (S)-isopropyl group.[4]

(I)   RCHO + **2b** $\xrightarrow{\text{Zn, THF}}$

*syn-***4**   *anti-***4**

R = C₆H₅, −78° 99%   96:4

[1] E. Erdik, *Tetrahedron*, **43**, 2203 (1987).
[2] R. L. Danheiser and S. Savariar, *Tetrahedron Letters*, **28**, 3299 (1987).
[3] Y. Ito and S. Terashima, *ibid.*, **28**, 6625 (1987).
[4] *Idem, ibid.*, **28**, 6629 (1987).

Zinc–Copper(II) acetate–Silver(I) nitrate.

Reduction of enynes to (Z)-alkenes.[1] Lindlar's catalyst is not useful as a hydrogenation catalyst for reduction of trienynes or of dienediynes. The best results can be obtained in CH₃OH with zinc activated by successive treatment with Cu(OAc)₂ (10%) and AgNO₃ (10%). This reduction results in conversion of the triple bond to a (Z)-double bond. The system does not reduce simple, nonactivated alkynes, and α-branched enynes are reduced slowly. The reduction is effected at 25° with (Z)-enynes, but temperatures of 45° are necessary for the (E)-isomers. Yields of pure tetraenes are 25–65%.

[1] W. Boland, N. Schroer, C. Sieler, and M. Feigel, *Helv.*, **70**, 1025 (1987).

Zinc–Copper(I) chloride.

α,α-Difluoro-β-hydroxy ketones.[1]   The aldol reaction of chlorodifluoromethyl ketones (1) with aldehydes when carried out in the presence of zinc (excess) activated by a trace of CuCl and 4Å molecular sieves results in α,α-difluoro-β-hydroxy ketones in generally good yield.

$$CH_3(CH_2)_5\overset{\displaystyle O}{\overset{\|}{C}}CF_2Cl \; + \; RCHO \xrightarrow[60-100\%]{\underset{THF}{Zn-CuCl,}} \; R\cdots\overset{OH}{\underset{F\;\;F}{\overset{|}{C}}}\cdots\overset{O}{\overset{\|}{C}}(CH_2)_5CH_3$$

1

$$\downarrow \begin{array}{c} ZnCl_2, TMEDA \\ DIBAH \end{array}$$

$$R\cdots\overset{OH}{\underset{F\;\;F}{}}\cdots\overset{OH}{}(CH_2)_5CH_3$$

(90–95% *syn*)

Aldol condensation of chlorodifluoromethyl ketones with ketones can be effected in moderate yield by use of diethylaluminum chloride (1.1 equiv.) and zinc activated with silver (I) acetate.

[1] M. Kuroboshi and T. Ishihara, *Tetrahedron Letters*, **28**, 6481 (1987).

Zinc–Nickel(II) chloride.

Conjugate hydrogenation.[1]   The combination of zinc and $NiCl_2$ (9:1) effects conjugate reduction of α,β-enones in an aqueous alcohol in which both the enone and product are completely soluble. Ultrasound increases the rate and the yields. Presumably the salt is reduced to a low-valent form that is absorbed on the zinc. No reduction takes place with a 1:1 $Zn-NiCl_2$ couple. The method is not applicable to α,β-unsaturated enals. Isolated double bonds are also reduced by this method, but this hydrogenation can be inhibited by addition of ammonia or triethylamine.

[1] C. Petrier and J.-L. Luche, *Tetrahedron Letters*, **28**, 2347, 2351 (1987).

Zinc borohydride.

Reduction of acid chlorides.[1]   In combination with TMEDA, $Zn(BH_4)_2$ reduces acid chlorides, RCOCl, to primary alcohols, RCH_2OH, in 80–95% yield.

[1] H. Kotsuki, Y. Ushio, N. Yoshimura, and M. Ochi, *Tetrahedron Letters*, **27**, 4213 (1986).

AUTHOR INDEX

SUBJECT INDEX